Student's Solutions Manual
to Accompany

Calculus
with Analytic Geometry
Brief Edition

FIFTH EDITION

HOWARD ANTON
DREXEL UNIVERSITY

Prepared by
Albert Herr
Drexel University

JOHN WILEY & SONS, INC.
New York   Chichester   Brisbane   Toronto   Singapore

Preface

This supplement to the *fifth edition* of **Calculus with Analytic Geometry, Brief Edition**, by Howard Anton is designed to serve as a handy reference for students. It contains answers and, where appropriate, detailed solutions to the more than 2500 odd-numbered exercises in the text. The solutions show intermediate algebraic steps and manipulations that are a frequent cause of trouble for students, and also show details in the use of the various formulas that are encountered in calculus. Diagrams are used throughout the manual to assist in the visualization of problems and to suggest strategies for their solutions. High quality computer-generated figures ensure accuracy where a graph is needed in a solution or answer.

Special thanks are due to everyone at **Techsetters Incorporated** for the outstanding technical assistance that was provided for the production of this manual – typing, generation of art, and many other services that facilitated its preparation.

Contents

CHAPTER 1
Coordinates, Graphs, Lines

EXERCISE SET 1.1

1. **(a)** rational   **(b)** integer, rational   **(c)** integer, rational
   (d) rational   **(e)** integer, rational   **(f)** irrational
   (g) rational   **(h)** integer, rational

3. **(a)**
$$x = 0.123123123\cdots$$
$$1000x = 123.123123123\cdots$$
$$999x = 123$$
$$x = \frac{123}{999} = \frac{41}{333}$$

   (b)
$$x = 12.7777\cdots$$
$$10x = 127.7777\cdots$$
$$9x = 115$$
$$x = \frac{115}{9}$$

   (c)
$$x = 38.07818181\cdots$$
$$100x = 3807.81818181\cdots$$
$$99x = 3769.74$$
$$x = \frac{3769.74}{99} = \frac{376974}{9900} = \frac{20943}{550}$$

   (d)
$$0.4296000\cdots = 0.4296$$
$$= \frac{4296}{10000}$$
$$= \frac{537}{1250}$$

5. **(a)** If r is the radius, then $D = 2r$ so $\left(\frac{8}{9}D\right)^2 = \left(\frac{16}{9}r\right)^2 = \frac{256}{81}r^2$. The area of a circle of radius r is πr^2 so $256/81$ was the approximation used for π.
   (b) $256/81 \approx 3.16049$, $22/7 \approx 3.14268$, and $\pi \approx 3.14159$ so $256/81$ is worse than $22/7$.

7. line 2: blocks 3, 4; line 3: blocks 1, 2
   line 4: blocks 3, 4; line 5: blocks 2, 4, 5
   line 6: blocks 1, 2; line 7: blocks 3, 4

9. **(a)** always correct (add -3 to both sides of $a \leq b$)
   (b) not always correct (correct only if $a = b$)
   (c) not always correct (correct only if $a = b$)
   (d) always correct (multiply both sides of $a \leq b$ by 6)
   (e) not always correct (correct only if $a \geq 0$)
   (f) always correct (multiply both sides of $a \leq b$ by the nonnegative quantity a^2)

11. **(a)** all values because $a = a$ is always valid   **(b)** none

13. **(a)** yes, because $a \leq b$ is true if $a < b$ **(b)** no, because $a < b$ is false if $a = b$ is true

15. (a) $\{x : x$ is a positive odd integer$\}$ (b) $\{x : x$ is an even integer$\}$
    (c) $\{x : x$ is irrational$\}$ (d) $\{x : x$ is an integer and $7 \leq x \leq 10\}$

17. (a) false, there are points inside the triangle that are not inside the circle
    (b) true, all points inside the triangle are also inside the square
    (c) true   (d) false   (e) true
    (f) true, a is inside the circle   (g) true

19. (a) [number line: open ray starting at 4]   (b) [number line: ray starting at -3]

    (c) [number line: segment from -1 to 7]   (d) [number line: segment from -3 through 0 to 3]

    (e) [number line: segment from -3 to 3]   (f) [number line: segment from -3 to 3]

21. (a) $[-2, 2]$ (b) $(-\infty, -2) \cup (2, +\infty)$

23. $3x - 2 < 8$
    $3x < 10$
    $x < \dfrac{10}{3}$
    $S = (-\infty, 10/3)$

25. $4 + 5x \leq 3x - 7$
    $2x \leq -11$
    $x \leq -\dfrac{11}{2}$
    $S = (-\infty, -11/2]$

    [number line: closed point at $-\frac{11}{2}$]

27. $3 \leq 4 - 2x < 7$
    $-1 \leq -2x < 3$
    $\dfrac{1}{2} \geq x > -\dfrac{3}{2}$
    $S = (-3/2, 1/2]$

    [number line: from $-\frac{3}{2}$ open to $\frac{1}{2}$ closed]

29.
$$\frac{x}{x-3} < 4$$

$$\frac{x}{x-3} - 4 < 0$$

$$\frac{x - 4(x-3)}{x-3} < 0$$

$$\frac{12 - 3x}{x-3} < 0$$

$$\frac{4-x}{x-3} < 0$$

$$S = (-\infty, 3) \cup (4, +\infty)$$

31.
$$\frac{3x+1}{x-2} < 1$$

$$\frac{3x+1}{x-2} - 1 < 0$$

$$\frac{3x + 1 - (x-2)}{x-2} < 0$$

$$\frac{2x+3}{x-2} < 0$$

$$\frac{x + 3/2}{x-2} < 0$$

$$S = (-3/2, 2)$$

33.
$$\frac{4}{2-x} \leq 1$$

$$\frac{4 - (2-x)}{2-x} \leq 0$$

$$\frac{x+2}{2-x} \leq 0$$

$$S = (-\infty, -2] \cup (2, +\infty)$$

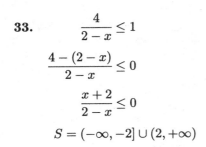

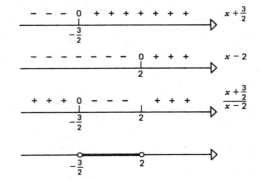

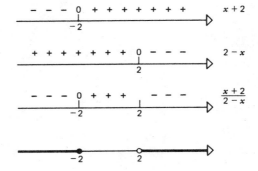

35.
$$x^2 > 9$$
$$x^2 - 9 > 0$$
$$(x+3)(x-3) > 0$$
$$S = (-\infty, -3) \cup (3, +\infty)$$

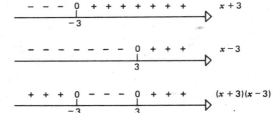

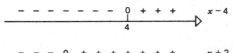

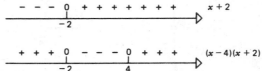

37. $(x-4)(x+2) > 0$
$$S = (-\infty, -2) \cup (4, +\infty)$$

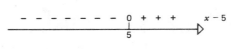

39. $x^2 - 9x + 20 \le 0$
$$(x-4)(x-5) \le 0$$
$$S = [4, 5]$$

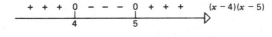

41.

$$\frac{2}{x} < \frac{3}{x-4}$$

$$\frac{2}{x} - \frac{3}{x-4} < 0$$

$$\frac{2(x-4) - 3x}{x(x-4)} < 0$$

$$\frac{-x-8}{x(x-4)} < 0$$

$$\frac{x+8}{x(x-4)} > 0$$

$$S = (-8, 0) \cup (4, +\infty)$$

43. By trial-and-error we find that $x = 2$ is a root of the equation $x^3 - x^2 - x - 2 = 0$ so $x - 2$ is a factor of $x^3 - x^2 - x - 2$. By long division we find that $x^2 + x + 1$ is another factor so $x^3 - x^2 - x - 2 = (x-2)(x^2 + x + 1)$. The linear factors of $x^2 + x + 1$ can be determined by first finding the roots of $x^2 + x + 1 = 0$ by the quadratic formula. These roots are complex numbers so $x^2 + x + 1 \neq 0$ for all real x; thus $x^2 + x + 1$ must be always positive or always negative. Since $x^2 + x + 1$ is positive when $x = 0$, it follows that $x^2 + x + 1 > 0$ for all real x. Hence $x^3 - x^2 - x - 2 > 0$, $(x-2)(x^2 + x + 1) > 0$, $x - 2 > 0$, $x > 2$, so $S = (2, +\infty)$.

45. $\sqrt{x^2 + x - 6}$ is real if $x^2 + x - 6 \geq 0$. Factor to get $(x+3)(x-2) \geq 0$ which has as its solution $x \leq -3$ or $x \geq 2$.

47. $25 \leq \frac{5}{9}(F - 32) \leq 40$, $45 \leq F - 32 \leq 72$, $77 \leq F \leq 104$.

49. (a) Assume m and n are rational, then $m = \frac{p}{q}$ and $n = \frac{r}{s}$ where p, q, r, and s are integers so $m + n = \frac{p}{q} + \frac{r}{s} = \frac{ps + rq}{qs}$ which is rational because $ps + rq$ and qs are integers.

(b) (proof by contradiction)   Assume m is rational and n is irrational, then $m = \frac{p}{q}$ where p and q are integers. Suppose that $m + n$ is rational, then $m + n = \frac{r}{s}$ where r and s are integers so $n = \frac{r}{s} - m = \frac{r}{s} - \frac{p}{q} = \frac{rq - ps}{sq}$. But $rq - ps$ and sq are integers, so n is rational which contradicts the assumption that n is irrational.

51. (by example)   Given that $\sqrt{2}$ is irrational, then
$$(-1)\sqrt{2} = -\sqrt{2} \text{ is irrational (Exercise 50, part (b))}$$
but $\sqrt{2} + (-\sqrt{2}) = 0$, which is rational;

$\sqrt{2} + \sqrt{2} = 2\sqrt{2}$, which is irrational (Exercise 50, part (b));

$(\sqrt{2})(\sqrt{2}) = 2$, which is rational;

$1 + \sqrt{2}$ is irrational (Exercise 49, part (b))

and so $\sqrt{2}(1 + \sqrt{2}) = \sqrt{2} + 2$, which is irrational (Exercise 49, part (b)).

53. **(I)** The average of two rational numbers is rational:

Assume m and n are rational; then $m = \dfrac{p}{q}$ and $n = \dfrac{r}{s}$ where p, q, r, and s are integers. The average of m and n is

$$\frac{1}{2}(m + n) = \frac{1}{2}(p/q + r/s) = \frac{ps + rq}{2qs}$$

which is rational because $ps + rq$ and $2qs$ are integers.

(II) The average of two irrational numbers can be rational or irrational:

By example,

$\sqrt{2}$, and $-\sqrt{2}$ are irrational and their average is $\dfrac{1}{2}(\sqrt{2} + (-\sqrt{2})) = \dfrac{1}{2}(0) = 0$, which is rational;

$\sqrt{2}$ and $\sqrt{2}$ are irrational and their average is $\dfrac{1}{2}(\sqrt{2} + \sqrt{2}) = \sqrt{2}$, which is irrational.

55. $8x^3 - 4x^2 - 2x + 1$ can be factored by grouping terms:

$(8x^3 - 4x^2) - (2x - 1) = 4x^2(2x - 1) - (2x - 1) = (2x - 1)(4x^2 - 1) = (2x - 1)^2(2x + 1)$. The problem, then, is to solve $(2x - 1)^2(2x + 1) < 0$. By inspection, $x = 1/2$ is not a solution. If $x \neq 1/2$, then $(2x - 1)^2 > 0$ and it follows that $2x + 1 < 0$, $2x < -1$, $x < -1/2$, so $S = (-\infty, -1/2)$.

57. If $a < b$, then $ac < bc$ because c is positive; if $c < d$, then $bc < bd$ because b is positive, so $ac < bd$ (part (a), Theorem 1.1.1.)

EXERCISE SET 1.2

1. **(a)** 7   **(b)** $\sqrt{2}$ **(c)** k^2 **(d)** k^2

3. $|x - 3| = |3 - x| = 3 - x$ if $3 - x \geq 0$, which is true if $x \leq 3$.

5. All real values of x because $x^2 + 9 > 0$.

7. $|3x^2 + 2x| = |x(3x + 2)| = |x||3x + 2|$. If $|x||3x + 2| = x|3x + 2|$, then $|x||3x + 2| - x|3x + 2| = 0$, $(|x| - x)|3x + 2| = 0$, so either $|x| - x = 0$ or $|3x + 2| = 0$. If $|x| - x = 0$, then $|x| = x$, which is true for $x \geq 0$. If $|3x + 2| = 0$, then $x = -2/3$. The statement is true for $x \geq 0$ or $x = -2/3$.

9. $\sqrt{(x+5)^2} = |x+5| = x+5$ if $x+5 \geq 0$, which is true if $x \geq -5$.

13. **(a)** $|7-9| = |-2| = 2$ **(b)** $|3-2| = |1| = 1$
   (c) $|6-(-8)| = |14| = 14$ **(d)** $|-3-\sqrt{2}| = |-(3+\sqrt{2})| = 3+\sqrt{2}$
   (e) $|-4-(-11)| = |7| = 7$ **(f)** $|-5-0| = |-5| = 5$

15. **(a)** B is 6 units to the left of A; $b = a - 6 = -3 - 6 = -9$.
   (b) B is 9 units to the right of A; $b = a + 9 = -2 + 9 = 7$.
   (c) B is 7 units from A; either $b = a + 7 = 5 + 7 = 12$ or $b = a - 7 = 5 - 7 = -2$. Since it is given that $b > 0$, it follows that $b = 12$.

17. $|6x - 2| = 7$

Case 1:   Case 2:
$6x - 2 = 7$ $6x - 2 = -7$
$6x = 9$ $6x = -5$
$x = 3/2$ $x = -5/6$

19. $|6x - 7| = |3 + 2x|$

Case 1:   Case 2:
$6x - 7 = 3 + 2x$ $6x - 7 = -(3 + 2x)$
$4x = 10$ $8x = 4$
$x = 5/2$ $x = 1/2$

21. $|9x| - 11 = x$

Case 1:   Case 2:
$9x - 11 = x$ $-9x - 11 = x$
$8x = 11$ $-10x = 11$
$x = 11/8$ $x = -11/10$

23. $\left| \dfrac{x+5}{2-x} \right| = 6$

Case 1:   Case 2:
$\dfrac{x+5}{2-x} = 6$ $\dfrac{x+5}{2-x} = -6$
$x + 5 = 12 - 6x$ $x + 5 = -12 + 6x$
$7x = 7$ $-5x = -17$
$x = 1$ $x = 17/5$

25. $|x + 6| < 3$
$-3 < x + 6 < 3$
$-9 < x < -3$
$S = (-9, -3)$

27. $|2x - 3| \leq 6$
$-6 \leq 2x - 3 \leq 6$
$-3 \leq 2x \leq 9$
$-3/2 \leq x \leq 9/2$
$S = [-3/2, 9/2]$

29. $|x + 2| > 1$

Case 1:   Case 2:
$x + 2 > 1$ $x + 2 < -1$
$x > -1$ $x < -3$
$S = (-\infty, -3) \cup (-1, +\infty)$

31. $|5 - 2x| \geq 4$

Case 1:   Case 2:
$5 - 2x \geq 4$ $5 - 2x \leq -4$
$-2x \geq -1$ $-2x \leq -9$
$x \leq 1/2$ $x \geq 9/2$
$S = (-\infty, 1/2] \cup [9/2, +\infty)$

33. $\dfrac{1}{|x-1|} < 2, x \neq 1$

$|x-1| > 1/2$

Case 1:   Case 2:
$x-1 > 1/2 \quad x-1 < -1/2$
$\quad x > 3/2 \qquad x < 1/2$
$S = (-\infty, 1/2) \cup (3/2, +\infty)$

35. $\dfrac{3}{|2x-1|} \geq 4, x \neq 1/2$

$\dfrac{|2x-1|}{3} \leq \dfrac{1}{4}$

$|2x-1| \leq 3/4$
$-3/4 \leq 2x-1 \leq 3/4$
$1/4 \leq 2x \leq 7/4$
$1/8 \leq x \leq 7/8$
$S = [1/8, 1/2) \cup (1/2, 7/8]$

37. $\sqrt{(x^2-5x+6)^2} = x^2 - 5x + 6$ if $x^2 - 5x + 6 \geq 0$ or, equivalently, if $(x-2)(x-3) \geq 0$; $x \in (-\infty, 2] \cup [3, +\infty)$.

39. If $u = |x-3|$ then $u^2 - 4u = 12$, $u^2 - 4u - 12 = 0$, $(u-6)(u+2) = 0$, so $u = 6$ or $u = -2$. If $u = 6$ then $|x-3| = 6$, so $x = 9$ or $x = -3$. If $u = -2$ then $|x-3| = -2$ which is impossible. The solutions are -3 and 9.

41. $|a-b| = |a+(-b)|$
$\qquad \leq |a| + |-b|$ (triangle inequality)
$\qquad = |a| + |b|$.

43. From Exercise 42
(i) $|a| - |b| \leq |a-b|$; but $|b| - |a| \leq |b-a| = |a-b|$, so (ii) $|a| - |b| \geq -|a-b|$.
Combining (i) and (ii): $-|a-b| \leq |a| - |b| \leq |a-b|$, so $||a| - |b|| \leq |a-b|$.

EXERCISE SET 1.3

1.

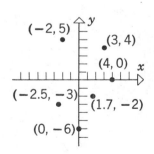

3. **(a)** $x = 2$

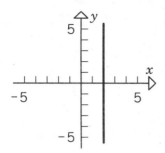

(b) $y = -3$

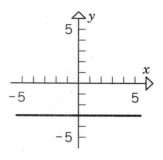

(c) $x \geq 0$

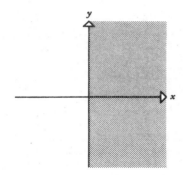

(d) $y = x$

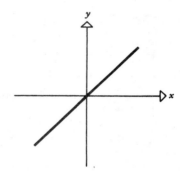

(e) $y \geq x$

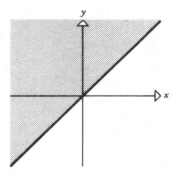

(f) $|x| \geq 1$

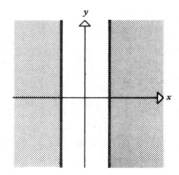

5. **(a)** vertical **(b)** horizontal **(c)** vertical

7. **(a)** $0^2 - 2(0) + 4 = 4$, yes **(b)** $(-3)^2 - 2(-3) + 7 = 22$, no

(c) $(1/2)^2 - 2(1/2) + 19/4 = 1/4 - 1 + 19/4$, yes

(d) $(1 + \sqrt{5-t})^2 - 2(1 + \sqrt{5-t}) + t = 1 + 2\sqrt{5-t} + (5-t) - 2 - 2\sqrt{5-t} + t = 4$, yes

9. (a) origin   **(b)** x-axis   **(c)** y-axis   **(d)** none

11. (a) $x = y^2$, $x = -y^2$ **(b)** $y = x^2$, $y = -x^2$, $y = 1/x^2$, $y = -1/x^2$
     (c) $y = x^3$, $y = \sqrt[3]{x}$, $y = 1/x$, $y = -1/x$

13. (a) y-axis, because $(-x)^4 = 2y^3 + y$ gives $x^4 = 2y^3 + y$.

     (b) origin, because $(-y) = \dfrac{(-x)}{3 + (-x)^2}$ gives $y = \dfrac{x}{3 + x^2}$.

     (c) x-axis, y-axis, and origin because $(-y)^2 = |x| - 5$, $y^2 = |-x| - 5$, and $(-y)^2 = |-x| - 5$
     all give $y^2 = |x| - 5$.

15. $y = 6 - x$

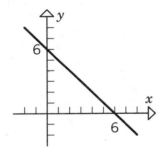

17. $y = 4 - x^2$

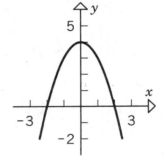

19. $y = \sqrt{x - 4}$

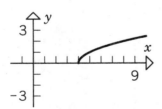

21. $y = |x - 3|$

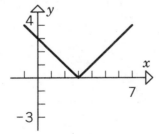

23. $x^2 y = 2$

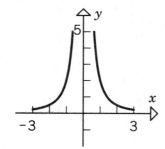

25. $4x^2 + 16y^2 = 16$

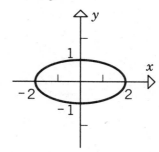

27. $(x - y)(x + y) = 0$
The union of the graphs of
$x - y = 0$ and $x + y = 0$.

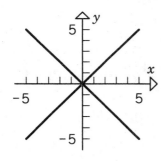

29. $u = 3v^2$

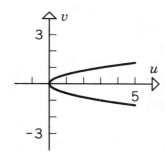

EXERCISE SET 1.4

1. **(a)** $m = \dfrac{4 - 2}{3 - (-1)} = \dfrac{1}{2}$

(b) $m = \dfrac{1 - 3}{7 - 5} = -1$

(c) $m = \dfrac{\sqrt{2} - \sqrt{2}}{-3 - 4} = 0$

(d) $m = \dfrac{12 - (-6)}{-2 - (-2)} = \dfrac{18}{0}$, not defined

3. **(a)** The line through $(1, 1)$ and $(-2, -5)$ has slope $m_1 = \dfrac{-5 - 1}{-2 - 1} = 2$, the line through $(1, 1)$
and $(0, -1)$ has slope $m_2 = \dfrac{-1 - 1}{0 - 1} = 2$. The given points lie on a line because $m_1 = m_2$.

(b) The line through $(-2, 4)$ and $(0, 2)$ has slope $m_1 = \dfrac{2 - 4}{0 + 2} = -1$, the line through $(-2, 4)$
and $(1, 5)$ has slope $m_2 = \dfrac{5 - 4}{1 + 2} = \dfrac{1}{3}$. The given points do not lie on a line because
$m_1 \neq m_2$.

5.

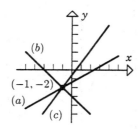

7. (c), (b), (d), (a)

9. **(a)** $m = [0 - (-3)]/(2 - 0) = 3/2$ **(b)** $m = (-2 - 1)/[3 - (-1)] = -3/4$

11. Use the points $(1, 2)$ and (x, y) to calculate the slope: $(y - 2)/(x - 1) = 3$
  (a) if $x = 5$, then $(y - 2)/(5 - 1) = 3$, $y - 2 = 12$, $y = 14$
  (b) if $y = -2$, then $(-2 - 2)/(x - 1) = 3$, $x - 1 = -4/3$, $x = -1/3$

13. Using $(3, k)$ and $(-2, 4)$ to calculate the slope, we find $\dfrac{k - 4}{3 - (-2)} = 5$, $k - 4 = 25$, $k = 29$.

15. $(0 - 2)/(x - 1) = -(0 - 5)/(x - 4)$, $-2x + 8 = 5x - 5$, $7x = 13$, $x = 13/7$.

17. **(a)** $\tan \dfrac{\pi}{6} = 1/\sqrt{3}$ **(b)** $\tan 135° = -1$ **(c)** $\tan 60° = \sqrt{3}$

19. **(a)** $153°$ **(b)** $45°$ **(c)** $117°$ **(d)** $89°$

21. Let $P(x, 0)$ be a point on the x-axis. If AP is perpendicular to BP then $m_{AP}m_{BP} = -1$, so $\dfrac{0 - 2}{x - 1} \cdot \dfrac{0 - 3}{x - 8} = -1$, thus $x^2 - 9x + 14 = 0$, $(x - 7)(x - 2) = 0$, thus $x = 2$ or $x = 7$.

23. Show that opposite sides are parallel by showing that they have the same slope:
  using $(3, -1)$ and $(6, 4)$, $m_1 = 5/3$; using $(6, 4)$ and $(-3, 2)$, $m_2 = 2/9$;
  using $(-3, 2)$ and $(-6, -3)$, $m_3 = 5/3$; using $(-6, -3)$ and $(3, -1)$, $m_4 = 2/9$.
  Opposite sides are parallel because $m_1 = m_3$ and $m_2 = m_4$.

25. **(a)**

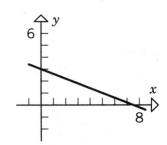

(b)

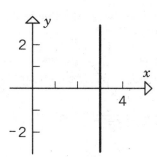

(c)

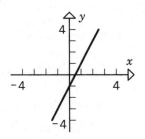

(d)

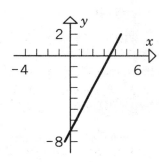

27. **(a)**

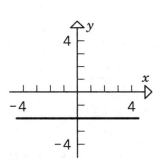

(b)

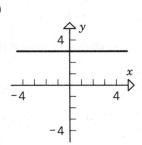

(c)

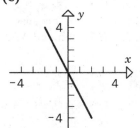

29. **(a)** $m = 3, b = 2$ **(b)** $m = -\dfrac{1}{4}, b = 3$

     (c) $y = -\dfrac{3}{5}x + \dfrac{8}{5}$ so $m = -\dfrac{3}{5}, b = \dfrac{8}{5}$ **(d)** $m = 0, b = 1$

     (e) $y = -\dfrac{b}{a}x + b$ so $m = -\dfrac{b}{a}$, y-intercept b

31. **(a)** $m = \tan\phi = \sqrt{3}, \phi = 60°$ **(b)** $m = \tan\phi = -2, \phi = 117°$

33. $y = -2x + 4$

35. The slope m of the line must equal the slope of $y = 4x - 2$, thus $m = 4$ so the equation is $y = 4x + 7$.

37. The slope m of the line must be the negative reciprocal of the slope of $y = 5x + 9$, thus $m = -1/5$ and the equation is $y = -x/5 + 6$.

39. $y - 4 = \dfrac{-7 - 4}{1 - 2}(x - 2) = 11(x - 2)$, $y = 11x - 18$.

41. $m = \tan\dfrac{\pi}{6} = \dfrac{1}{\sqrt{3}}$ so $y = \dfrac{1}{\sqrt{3}}x - 3$.

43. The line passes through $(0, 2)$ and $(-4, 0)$, thus $m = \dfrac{0 - 2}{-4 - 0} = \dfrac{1}{2}$ so $y = \dfrac{1}{2}x + 2$.

45. $y = 1$

47. The line is vertical because $\phi = \dfrac{\pi}{2}$, so $x = 5$.

49. (a) $m_1 = 4$, $m_2 = 4$; parallel because $m_1 = m_2$.
(b) $m_1 = 2$, $m_2 = -1/2$; perpendicular because $m_1 m_2 = -1$.
(c) $m_1 = 5/3$, $m_2 = 5/3$; parallel because $m_1 = m_2$.
(d) If $A \neq 0$ and $B \neq 0$, then $m_1 = -A/B$, $m_2 = B/A$ and the lines are perpendicular because $m_1 m_2 = -1$. If either A or B (but not both) is zero, then the lines are perpendicular because one is horizontal and the other is vertical.
(e) $m_1 = 4$, $m_2 = 1/4$; neither.

51. $y = (-3/k)x + 4/k$, $k \neq 0$
(a) $-3/k = 2$, $k = -3/2$.   (b) $4/k = 5$, $k = 4/5$.
(c) $3(-2) + k(4) = 4$, $k = 5/2$.
(d) The slope of $2x - 5y = 1$ is $2/5$ so $-3/k = 2/5$, $k = -15/2$.
(e) The slope of $4x + 3y = 2$ is $-4/3$ so the slope of the line perpendicular to it is $3/4$; $-3/k = 3/4$, $k = -4$.

53. slope $\approx 2.4/10 = 0.24$ N/mm $= 240$ N/m.

55. slope $\approx (354 - 330)/(40 - 0) = 0.6$ m/sec/°C; the speed of sound increases at the rate of 0.6 m/sec per °C.

57. (a) If we plot T_C along the horizontal axis and T_F along the vertical axis, then any point on the line relating T_C and T_F can be denoted by (T_C, T_F). From the fact that $(0, 32)$ and $(100, 212)$ are on the line, it follows that the slope is $\dfrac{212 - 32}{100 - 0} = \dfrac{180}{100} = \dfrac{9}{5}$ and that the T_F-intercept is 32. An equation of the line is $T_F = \dfrac{9}{5}T_C + 32$.

   (b) Solving $T_F = \dfrac{9}{5}T_C + 32$ for T_C we get $T_C = \dfrac{5}{9}T_F - \dfrac{160}{9}$, which identifies the slope as $\dfrac{5}{9}$.

   (c) If $T_C = T_F = T$, then $T = \dfrac{9}{5}T + 32$, $T = -40°F = -40°C$.

   (d) $T_C = \dfrac{5}{9}(98.6) - \dfrac{160}{9} = 37°C$

59. (a) slope $= (5.9 - 1)/(50 - 0) = 0.098$ atm/m, $p = 0.098h + 1$.
    (b) If $p = 2$, then $h = 1/0.098 \approx 10.2$ m.

61. (a) $m = (0.75 - 0.80)/4 = -0.0125$ mm/day, $r = -0.0125t + 0.8$.
    (b) If $r = 0$, then $t = 0.8/0.0125 = 64$ days

63. Solve $x = 5t + 2$ for t to get $t = \dfrac{1}{5}x - \dfrac{2}{5}$, so $y = \left(\dfrac{1}{5}x - \dfrac{2}{5}\right) - 3 = \dfrac{1}{5}x - \dfrac{17}{5}$, which is a line.

65. An equation of the line through $(1, 4)$ and $(2, 1)$ is $y = -3x + 7$. It crosses the y-axis at $y = 7$, and the x-axis at $x = 7/3$, so the area of the triangle is $\dfrac{1}{2}(7)(7/3) = 49/6$.

67. Let L_3 be a line that is perpendicular to L_1, then $m_3 = -1/m_1$. But $m_2 = -1/m_1$, so L_2 is parallel to L_3 because they have the same slope. If L_3 is perpendicular to L_1, then so is L_2.

69. To prove that the graph of $Ax + By + C = 0$ is a straight line consider two cases, $B = 0$ and $B \neq 0$. If $B \neq 0$, solve for y to get $y = -\dfrac{A}{B}x - \dfrac{C}{B}$ which is the slope-intercept form of a line with $m = -A/B$ and $b = -C/B$. If $B = 0$, then $A \neq 0$ because A and B cannot both be zero. Thus we get $Ax + C = 0$, or $x = -C/A$ which is an equation of a line parallel to the y-axis. In either case the graph of $Ax + By + C = 0$ is a straight line.

   Conversely, consider any straight line in the xy-plane. If it is vertical, then it has an equation of the form $x = a$, or $x - a = 0$ which is like $Ax + By + C = 0$ with $A = 1$, $B = 0$, and $C = -a$. If the line is not vertical, then it can be written in slope-intercept form as $y = mx + b$, or $mx - y + b = 0$ which is like $Ax + By + C = 0$ with $A = m$, $B = -1$, and $C = b$.

EXERCISE SET 1.5

1. In the proof of Theorem 1.5.1.

3. (a) $d = \sqrt{(1-7)^2 + (9-1)^2} = \sqrt{36 + 64} = \sqrt{100} = 10$

   (b) $\left(\dfrac{7+1}{2}, \dfrac{1+9}{2} \right) = (4, 5)$

5. (a) $d = \sqrt{[-7 - (-2)]^2 + [-4 - (-6)]^2} = \sqrt{25 + 4} = \sqrt{29}$

   (b) $\left(\dfrac{-2 + (-7)}{2}, \dfrac{-6 + (-4)}{2} \right) = (-9/2, -5)$

7. Let $A(5, -2)$, $B(6, 5)$, and $C(2, 2)$ be the given vertices and a, b, and c the lengths of the sides opposite these vertices; then

   $a = \sqrt{(2-6)^2 + (2-5)^2} = \sqrt{25} = 5$ and $b = \sqrt{(2-5)^2 + (2+2)^2} = \sqrt{25} = 5.$

   Triangle ABC is isosceles because it has two equal sides $(a = b)$.

9. $P_1(0, -2)$, $P_2(-4, 8)$, and $P_3(3, 1)$ all lie on a circle whose center is $C(-2, 3)$ if the points P_1, P_2 and P_3 are equidistant from C. Denoting the distances between P_1, P_2, P_3 and C by d_1, d_2 and d_3 we find that $d_1 = \sqrt{(0+2)^2 + (-2-3)^2} = \sqrt{29}$,

   $d_2 = \sqrt{(-4+2)^2 + (8-3)^2} = \sqrt{29}$, and $d_3 = \sqrt{(3+2)^2 + (1-3)^2} = \sqrt{29}.$

   So P_1, P_2 and P_3 lie on a circle whose center is $C(-2, 3)$ because $d_1 = d_2 = d_3$.

11. If $(2, k)$ is equidistant from $(3,7)$ and $(9,1)$, then

    $\sqrt{(2-3)^2 + (k-7)^2} = \sqrt{(2-9)^2 + (k-1)^2}$, $1 + (k-7)^2 = 49 + (k-1)^2$,

    $1 + k^2 - 14k + 49 = 49 + k^2 - 2k + 1$, $-12k = 0$, $k = 0$.

13. The slope of the line segment joining $(2, 8)$ and $(-4, 6)$ is $\dfrac{6-8}{-4-2} = \dfrac{1}{3}$ so the slope of the perpendicular bisector is -3. The midpoint of the line segment is $(-1, 7)$ so an equation of the bisector is $y - 7 = -3(x + 1)$; $y = -3x + 4$.

15. Method (see figure): Find an equation of the perpendicular bisector of the line segment joining $A(3, 3)$ and $B(7, -3)$. All points on this perpendicular bisector are equidistant from A and B, thus find where it intersects the given line.

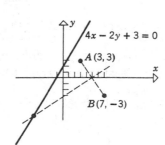

The midpoint of AB is $(5,0)$, the slope of AB is $-3/2$ thus the slope of the perpendicular bisector is $2/3$ so an equation is

$$y - 0 = \frac{2}{3}(x - 5)$$
$$3y = 2x - 10$$
$$2x - 3y - 10 = 0.$$

The solution of the system

$$\begin{cases} 4x - 2y + 3 = 0 \\ 2x - 3y - 10 = 0 \end{cases}$$

gives the point $(-29/8, -23/4)$.

17. Method (see figure): write an equation of the line that goes through the given point and that is perpendicular to the given line; find the point P where this line intersects the given line; find the distance between P and the given point.

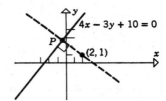

The slope of the given line is $4/3$, so the slope of a line perpendicular to it is $-3/4$.

The line through $(2,1)$ having a slope of $-3/4$ is $y - 1 = -\dfrac{3}{4}(x - 2)$ or, after simplification,

$3x + 4y = 10$ which when solved simultaneously with $4x - 3y + 10 = 0$ yields $(-2/5, 14/5)$ as the point of intersection. The distance d between $(-2/5, 14/5)$ and $(2, 1)$ is

$d = \sqrt{(2 + 2/5)^2 + (1 - 14/5)^2} = 3.$

19. If $B = 0$, then the line $Ax + C = 0$ is vertical and $x = -C/A$ for each point on the line. The line through (x_0, y_0) and perpendicular to the given line is horizontal and intersects the given line at the point $(-C/A, y_0)$. The distance d between $(-C/A, y_0)$ and (x_0, y_0) is

$$d = \sqrt{(x_0 + C/A)^2 + (y_0 - y_0)^2} = \sqrt{\frac{(Ax_0 + C)^2}{A^2}} = \frac{|Ax_0 + C|}{\sqrt{A^2}}$$

which is the value of $\dfrac{|Ax_0 + By_0 + C|}{\sqrt{A^2 + B^2}}$ for $B = 0$.

If $B \neq 0$, then the slope of the given line is $-A/B$ and the line through (x_0, y_0) and perpendicular to the given line is

$$y - y_0 = \frac{B}{A}(x - x_0), \quad Ay - Ay_0 = Bx - Bx_0, \quad Bx - Ay = Bx_0 - Ay_0.$$

The point of intersection of this line and the given line is obtained by solving
$Ax + By = -C$ and $Bx - Ay = Bx_0 - Ay_0$.

Multiply the first equation through by A and the second by B and add the results to get

$$(A^2 + B^2)x = B^2 x_0 - ABy_0 - AC \text{ so } x = \frac{B^2 x_0 - ABy_0 - AC}{A^2 + B^2}$$

Similarly, by multiplying by B and $-A$, we get $y = \dfrac{-ABx_0 + A^2 y_0 - BC}{A^2 + B^2}$.

The square of the distance d between (x, y) and (x_0, y_0) is

$$d^2 = \left[x_0 - \frac{B^2 x_0 - ABy_0 - AC}{A^2 + B^2} \right]^2 + \left[y_0 - \frac{-ABx_0 + A^2 y_0 - BC}{A^2 + B^2} \right]^2$$

$$= \frac{(A^2 x_0 + ABy_0 + AC)^2}{(A^2 + B^2)^2} + \frac{(ABx_0 + B^2 y_0 + BC)^2}{(A^2 + B^2)^2}$$

$$= \frac{A^2 (Ax_0 + By_0 + C)^2 + B^2 (Ax_0 + By_0 + C)^2}{(A^2 + B^2)^2}$$

$$= \frac{(Ax_0 + By_0 + C)^2 (A^2 + B^2)}{(A^2 + B^2)^2} = \frac{(Ax_0 + By_0 + C)^2}{A^2 + B^2}$$

so $d = \dfrac{|Ax_0 + By_0 + C|}{\sqrt{A^2 + B^2}}$.

21. $d = \dfrac{|5(8) + 12(4) - 36|}{\sqrt{5^2 + 12^2}} = \dfrac{|52|}{\sqrt{169}} = \dfrac{52}{13} = 4.$

23. **(a)** center $(0,0)$, radius 5   **(b)** center $(1,4)$, radius 4
     (c) center $(-1, -3)$, radius $\sqrt{5}$ **(d)** center $(0, -2)$, radius 1

25. $(x - 3)^2 + (y - (-2))^2 = 4^2$, $(x - 3)^2 + (y + 2)^2 = 16$

27. $r = 8$ because the circle is tangent to the x-axis, so $(x + 4)^2 + (y - 8)^2 = 64$.

29. $(0,0)$ is on the circle, so $r = \sqrt{(-3 - 0)^2 + (-4 - 0)^2} = 5$; $(x + 3)^2 + (y + 4)^2 = 25$.

31. The center is the midpoint of the line segment joining $(2, 0)$ and $(0, 2)$ so the center is at $(1, 1)$. The radius is $r = \sqrt{(2 - 1)^2 + (0 - 1)^2} = \sqrt{2}$, so $(x - 1)^2 + (y - 1)^2 = 2$.

33. $(x^2 - 2x) + (y^2 - 4y) = 11$, $(x^2 - 2x + 1) + (y^2 - 4y + 4) = 11 + 1 + 4$, $(x - 1)^2 + (y - 2)^2 = 16$; center $(1,2)$ and radius 4.

35. $2(x^2 + 2x) + 2(y^2 - 2y) = 0$, $2(x^2 + 2x + 1) + 2(y^2 - 2y + 1) = 2 + 2$, $(x + 1)^2 + (y - 1)^2 = 2$; center $(-1, 1)$ and radius $\sqrt{2}$.

37. $(x^2 + 2x) + (y^2 + 2y) = -2$, $(x^2 + 2x + 1) + (y^2 + 2y + 1) = -2 + 1 + 1$, $(x+1)^2 + (y+1)^2 = 0$; the point $(-1, -1)$.

39. $x^2 + y^2 = 1/9$; center $(0,0)$ and radius $1/3$.

41. $x^2 + (y^2 + 10y) = -26$, $x^2 + (y^2 + 10y + 25) = -26 + 25$, $x^2 + (y+5)^2 = -1$; no graph

43. $16\left(x^2 + \dfrac{5}{2}x\right) + 16(y^2 + y) = 7$, $16\left(x^2 + \dfrac{5}{2}x + \dfrac{25}{16}\right) + 16\left(y^2 + y + \dfrac{1}{4}\right) = 7 + 25 + 4$,

$(x + 5/4)^2 + (y + 1/2)^2 = 9/4$; center $(-5/4, -1/2)$ and radius $3/2$.

45. (a) $y^2 = 16 - x^2$, so $y = \pm\sqrt{16 - x^2}$. The bottom half is $y = -\sqrt{16 - x^2}$.
   (b) Complete the square in y to get $(y - 2)^2 = 3 - 2x - x^2$, so $y - 2 = \pm\sqrt{3 - 2x - x^2}$, or $y = 2 \pm \sqrt{3 - 2x - x^2}$. The top half is $y = 2 + \sqrt{3 - 2x - x^2}$.

47. (a)

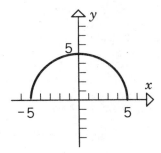

   (b) $y = \sqrt{5 + 4x - x^2}$

$= \sqrt{5 - (x^2 - 4x)}$

$= \sqrt{5 + 4 - (x^2 - 4x + 4)}$

$= \sqrt{9 - (x - 2)^2}$

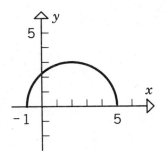

49. The tangent line is perpendicular to the radius at the point. The slope of the radius is $4/3$, so the slope of the perpendicular $is - 3/4$. An equation of the tangent line is $y - 4 = -\dfrac{3}{4}(x - 3)$, or $y = -\dfrac{3}{4}x + \dfrac{25}{4}$.

51. **(a)** The center of the circle is at $(0,0)$ and its radius is $\sqrt{20} = 2\sqrt{5}$. The distance between P and the center is $\sqrt{(-1)^2 + (2)^2} = \sqrt{5}$ which is less than $2\sqrt{5}$, so P is inside the circle.

(b) Draw the diameter of the circle that passes through P, then the shorter segment of the diameter is the shortest line that can be drawn from P to the circle, and the longer segment is the longest line that can be drawn from P to the circle (can you prove it?). Thus, the smallest distance is $2\sqrt{5} - \sqrt{5} = \sqrt{5}$, and the largest is $2\sqrt{5} + \sqrt{5} = 3\sqrt{5}$.

53. Let (a, b) be the coordinates of T (or T'). The radius from $(0,0)$ to T (or T') will be perpendicular to L (or L') so, using slopes, $b/a = -(a-3)/b$, $a^2 + b^2 = 3a$. But (a, b) is on the circle so $a^2 + b^2 = 1$, thus $3a = 1$, $a = 1/3$. Let $a = 1/3$ in $a^2 + b^2 = 1$ to get $b^2 = 8/9$, $b = \pm\sqrt{8}/3$. The coordinates of T and T' are $(1/3, \sqrt{8}/3)$ and $(1/3, -\sqrt{8}/3)$.

55. **(a)** $[(x-4)^2 + (y-1)^2] + [(x-2)^2 + (y+5)^2] = 45$
$x^2 - 8x + 16 + y^2 - 2y + 1 + x^2 - 4x + 4 + y^2 + 10y + 25 = 45$
$2x^2 + 2y^2 - 12x + 8y + 1 = 0$, which is a circle.

(b) $2(x^2 - 6x) + 2(y^2 + 4y) = -1$, $2(x^2 - 6x + 9) + 2(y^2 + 4y + 4) = -1 + 18 + 8$,
$(x-3)^2 + (y+2)^2 = 25/2$; center $(3, -2)$, radius $5/\sqrt{2}$.

57. $y = x^2 + 2$

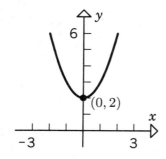

59. $y = x^2 + 2x - 3$

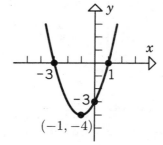

61. $y = -x^2 + 4x + 5$

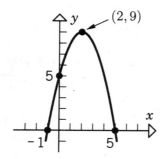

63. $y = (x - 2)^2$

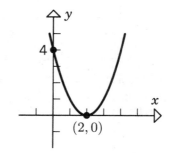

65. $x^2 - 2x + y = 0$ **67.** $y = 3x^2 - 2x + 1$

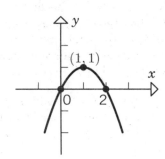   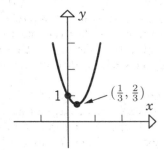

69. $x = -y^2 + 2y + 2$

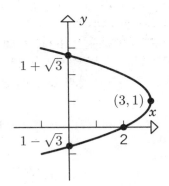

71.   **(a)** $x^2 = 3 - y$, $x = \pm\sqrt{3-y}$. The right half is $x = \sqrt{3-y}$.

        (b)   Complete the square in x to get $(x-1)^2 = y + 1$, $x = 1 \pm \sqrt{y+1}$. The left half is $x = 1 - \sqrt{y+1}$.

73.   **(a)**   **(b)**

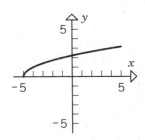   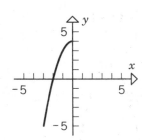

75. **(a)** $s = 32t - 16t^2$

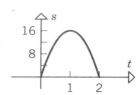

(b) The ball will be at its highest point when $t = 1$ sec; it will rise 16 ft.

77. **(a)** $(3)(2x) + (2)(2y) = 600$, $6x + 4y = 600$, $y = 150 - 3x/2$.

(b) $A = xy = x(150 - 3x/2) = 150x - 3x^2/2$.

(c) The graph of A versus x is a parabola with its vertex (high point) at $x = -b/(2a) = -150/(-3) = 50$, so the maximum value of A is $A = 150(50) - 3(50)^2/2 = 3,750$ ft^2.

79. **(a)** The parabola $y = 2x^2 + 5x - 1$ opens upward and has x-intercepts of $x = (-5 \pm \sqrt{33})/4$, so $2x^2 + 5x - 1 < 0$ if $(-5 - \sqrt{33})/4 < x < (-5 + \sqrt{33})/4$.

(b) The parabola $y = x^2 - 2x + 3$ opens upward and has no x-intercepts, so $x^2 - 2x + 3 > 0$ if $-\infty < x < +\infty$.

81. **(a)** The t-coordinate of the vertex is $t = -40/[(2)(-16)] = 5/4$, so the maximum height is $s = 5 + 40(5/4) - 16(5/4)^2 = 30$ ft.

(b) $s = 5 + 40t - 16t^2 = 0$ if $t \approx 2.6$ sec.

(c) $s = 5 + 40t - 16t^2 > 12$ if $16t^2 - 40t + 7 < 0$, which is true if $(5 - 3\sqrt{2})/4 < t < (5 + 3\sqrt{2})/4$. The length of this interval is $(5 + 3\sqrt{2})/4 - (5 - 3\sqrt{2})/4 = 3\sqrt{2}/2 \approx 2.1$ sec.

TECHNOLOGY EXERCISES 1

1. $x = 1.115088$, 3.934317

3. $x = -0.845838$, 0, 2, 3.220645

5. $x < -0.871925$, $x > 4.205258$

7. $x < 2.478052$, $x > 3$

9. $-0.770917 < x < 1.670244$

11. $\sqrt{(x-1)^2 + (\sqrt{x} - 2)^2}$; $x = 1.358094$

13. $63.9\pi h^2(2.5 - h/3) = 108$, $h = 0.48$ ft.

15. Let m_E and m_M be the mass of the earth and the moon, respectively, D the distance between the earth and moon, and x the distance of the body from the earth, then

$$Gm_E m/x^2 = Gm_M m/(D-x)^2, \ (D-x)^2 = (m_M/m_E)x^2, \ D - x = \sqrt{m_M/m_E}\,x,$$

$$x = D/(1 + \sqrt{m_M/m_E}) = 3.84 \times 10^8/(1 + \sqrt{7.35 \times 10^{22}/5.98 \times 10^{24}}) = 3.46 \times 10^8 \text{ meters.}$$

17. **(a)** $N = 80$ when $t = 9.35$ years.

    (b) 220 sheep

CHAPTER 2
Functions and Limits

EXERCISE SET 2.1

1. (a) 14   (b) 50   (c) 2
   (d) 11   (e) $3a^2 + 6a + 5$ (f) $27t^2 + 2$

3. (a) $2(-4) = -8$ (b) $1/4$ (c) $2(0) = 0$
   (d) $2(3) = 6$ (e) $2(2.9) = 5.8$ (f) $1/(t^2 + 5)$

5. $(-\infty, 3) \cup (3, +\infty)$ 7. $(-\infty, -\sqrt{3}] \cup [\sqrt{3}, +\infty)$

9. $\dfrac{x-1}{x+2} \geq 0$ if $x < -2$ or $x \geq 1$; domain: $(-\infty, -2) \cup [1, +\infty)$

11. $(-\infty, +\infty)$ 13. $[5, +\infty) \cap (-\infty, 8] = [5, 8]$

15. $x^2 - 2x + 5 = 0$ has no real solutions so $x^2 - 2x + 5$ is always positive or always negative. If $x = 0$, then $x^2 - 2x + 5 = 5 > 0$; domain: $(-\infty, +\infty)$.

17. $(-\infty, 0) \cup (0, +\infty)$ 19. $[0, +\infty)$

21. all real values of x except those for which $\sin x = 1$; domain: all x except $x = \pi/2 + 2k\pi$, $k = 0, \pm 1, \pm 2, \cdots$

23. domain: $(-\infty, 3]$; range: $[0, +\infty)$ 25. domain: $[-2, 2]$; range: $[0, 2]$

27. domain: $[0, +\infty)$; range: $[3, +\infty)$ 29. domain: $(-\infty, +\infty)$; range: $[3, +\infty)$

31. domain: $(-\infty, +\infty)$; range: $(-\infty, +\infty)$ 33. domain: $(-\infty, +\infty)$; range: $[-3, 3]$

35. domain: $(-\infty, +\infty)$; range: $[1, 3]$

37. If $x < 0$, then $|x| = -x$ so $f(x) = -x + 3x + 1 = 2x + 1$. If $x \geq 0$, then $|x| = x$ so $f(x) = x + 3x + 1 = 4x + 1$;

$$f(x) = \begin{cases} 2x + 1, & x < 0 \\ 4x + 1, & x \geq 0 \end{cases}$$

39. If $x < 0$, then $|x| = -x$ and $|x - 1| = 1 - x$ so $g(x) = -x + 1 - x = 1 - 2x$. If $0 \le x < 1$, then $|x| = x$ and $|x - 1| = 1 - x$ so $g(x) = x + 1 - x = 1$. If $x \ge 1$, then $|x| = x$ and $|x - 1| = x - 1$ so $g(x) = x + x - 1 = 2x - 1$;

$$g(x) = \begin{cases} 1 - 2x, & x < 0 \\ 1, & 0 \le x < 1 \\ 2x - 1, & x \ge 1 \end{cases}.$$

41. $\sqrt{3x - 2} = 6$, $3x - 2 = 36$, $3x = 38$, $x = 38/3$.

43. $x^2 + 5 = 7$, $x^2 = 2$, $x = \pm\sqrt{2}$.   **45.** $\cos x = 1$, $x = 2k\pi$, $k = 0, \pm1, \pm2, \cdots$

47. $\sin\sqrt{x} = 1/2$, $\sqrt{x} = \pi/6 + 2k\pi$ or $5\pi/6 + 2k\pi$ for $k = 0, 1, 2, \cdots$ so $x = (1/6 + 2k)^2\pi^2$ or $(5/6 + 2k)^2\pi^2$ for $k = 0, 1, 2, \cdots$

49. $A = \pi r^2$, but $C = 2\pi r$ so $r = \dfrac{C}{2\pi}$; $A = \dfrac{C^2}{4\pi}$.

51. (a) $S = 6x^2$
   (b) $V = x^3$ so $x = V^{1/3}$; substitute into (a) to get $S = 6V^{2/3}$.

53. $V = x(8 - 2x)(15 - 2x) = x(120 - 46x + 4x^2) = 4x^3 - 46x^2 + 120x$.

55. $h = L - L\cos\theta = L(1 - \cos\theta)$.

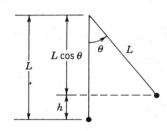

57. (a) $25°$F   (b) $2°$F   (c) $-15°$F.

59. $91.4 + (91.4 - T)[0.0203(8) - 0.304\sqrt{8} - 0.474] = -10$, solve for T to get $T \approx 5°$F.

61. Multiplication of the numerator and denominator of a fraction by the same number is valid only if the number is not zero, so $(1 - 1/x)/(1 + 1/x) = (x - 1)/(x + 1)$ if $x \ne 0$.

63. $\dfrac{(x + 2)(x^2 - 1)}{(x + 2)(x + 1)} = \dfrac{(x + 2)(x + 1)(x - 1)}{(x + 2)(x + 1)} = x - 1$, $x \ne -1$ or -2

65. $\dfrac{x+1+\sqrt{x+1}}{\sqrt{x+1}} = \dfrac{\sqrt{x+1}(\sqrt{x+1}+1)}{\sqrt{x+1}} = \sqrt{x+1}+1,\ x \neq -1$

67. $\dfrac{x^3+2x^2-3x}{(x-1)(x+3)} = \dfrac{x(x^2+2x-3)}{(x-1)(x+3)} = \dfrac{x(x-1)(x+3)}{(x-1)(x+3)} = x,\ x \neq -3\ \text{or}\ 1$

EXERCISE SET 2.2

1. (a) $f(t) = t^2 + 1$
(b) $f(t+2) = (t+2)^2 + 1 = t^2 + 4t + 5$
(c) $f(x+2) = (x+2)^2 + 1 = x^2 + 4x + 5$
(d) $f(1/x) = (1/x)^2 + 1 = 1/x^2 + 1$
(e) $f(x+h) = (x+h)^2 + 1 = x^2 + 2hx + h^2 + 1$
(f) $f(-x) = (-x)^2 + 1 = x^2 + 1$
(g) $f(\sqrt{x}) = (\sqrt{x})^2 + 1 = x + 1,\ x \geq 0$
(h) $f(3x) = (3x)^2 + 1 = 9x^2 + 1$

3. (a) $f(-1) - g(-1) = 4 - 3 = 1$ (b) $f(-1) \cdot g(-1) = (4)(3) = 12$
(c) $f(2)/g(2) = 5/(-1) = -5$ (d) $f(g(2)) = f(-1) = 4$

5. (a) $x^2 + 2x + 1$ (b) $-x^2 + 2x - 1$ (c) $2x(x^2+1)$
(d) $\dfrac{2x}{x^2+1}$ (e) $2(x^2+1)$ (f) $4x^2 + 1$

7. (a) $\sqrt{x+1} + x - 2$ (b) $\sqrt{x+1} - x + 2$ (c) $(x-2)\sqrt{x+1}$
(d) $\dfrac{\sqrt{x+1}}{x-2}$ (e) $\sqrt{x-1}$ (f) $\sqrt{x+1} - 2$

9. (a) $\sqrt{x-2} + \sqrt{x-3}$ (b) $\sqrt{x-2} - \sqrt{x-3}$ (c) $\sqrt{x-2}\sqrt{x-3}$
(d) $\dfrac{\sqrt{x-2}}{\sqrt{x-3}}$ (e) $\sqrt{\sqrt{x-3}-2}$ (f) $\sqrt{\sqrt{x-2}-3}$

11. (a) $\sqrt{1-x^2} + \sin 3x$ (b) $\sqrt{1-x^2} - \sin 3x$ (c) $\sqrt{1-x^2}\sin 3x$
(d) $\sqrt{1-x^2}/\sin 3x$ (e) $\sqrt{1-\sin^2 3x} = \sqrt{\cos^2 3x}$ (f) $\sin 3\sqrt{1-x^2}$
                                                          $= |\cos 3x|$

13. $\dfrac{f(x+h) - f(x)}{h} = \dfrac{3(x+h)^2 - 5 - (3x^2 - 5)}{h} = \dfrac{6hx + 3h^2}{h} = 6x + 3h,\quad h \neq 0$

15. $\dfrac{f(x+h)-f(x)}{h} = \dfrac{1/(x+h)-1/x}{h} = \dfrac{x-(x+h)}{hx(x+h)} = -\dfrac{h}{hx(x+h)} = -\dfrac{1}{x(x+h)}$, $\quad h \neq 0$

17. $(f \circ g)(x) = \dfrac{4}{x+5}, x \geq 0; (g \circ f)(x) = \dfrac{2}{\sqrt{x^2+5}}$

19. **(a)** $4x - 15$ **(b)** $4x^2 - 20x + 25$

21. **(a)** $f(g(x)) = 1/(x^2+1)$, which is defined for all x.
    (b) $g(x) = x^2 + 2$, for example
    (c) g must be defined for all x and $g(x)$ must never equal zero.

23. Show that $(f \circ (g \circ h))(x) = ((f \circ g) \circ h)(x)$: $(f \circ (g \circ h))(x) = f((g \circ h)(x)) = f(g(h(x)))$, $((f \circ g) \circ h)(x) = (f \circ g)(h(x)) = f(g(h(x)))$. Thus $(f \circ (g \circ h))(x) = ((f \circ g) \circ h)(x)$ so $f \circ (g \circ h) = (f \circ g) \circ h$.

25. $g(x) = \sqrt{x}, h(x) = x + 2$ **27.** $g(x) = x^7, h(x) = x - 5$

29. $g(x) = |x|, h(x) = x^2 - 3x + 5$ **31.** $g(x) = x^2, h(x) = \sin x$

33. $g(x) = \dfrac{3}{5+x}, h(x) = \cos x$ **35.** $g(x) = \dfrac{x}{3+x}, h(x) = \tan x$

37. $f(x) = \sqrt{x}, g(x) = 3 - x^2, h(x) = \sin x$

39. $u = x + 1$ so $x = u - 1$; $f(u) = (u-1)^2 + 3(u-1) + 5 = u^2 + u + 3$; $f(x) = x^2 + x + 3$.

41. $f(g(x)) = f(2x - 1) = 0$ only if $2x - 1 = -1$ or $2x - 1 = 2$, so $x = 0$ or $x = 3/2$.

43. $f(g(x)) = \sqrt{g(x)+5} = 3|x|$, $g(x) + 5 = 9x^2$, $g(x) = 9x^2 - 5$.

45. $f(x) = f(x/2 + x/2) = f(x/2) - f(x/2) = 0$

47. **(a)** monomial, polynomial, rational, explicit algebraic
    (b) explicit algebraic
    (c) rational, explicit algebraic
    (d) polynomial, rational, explicit algebraic

49. **(a)** explicit algebraic
    (b) rational, explicit algebraic
    (c) monomial, polynomial, rational, explicit algebraic
    (d) explicit algebraic $\left(|x| = \sqrt{x^2}\right)$

EXERCISE SET 2.3

1. (a) $-4, -3, -2, 2, 3$ (b) $0, 4$
   (c) $-4 \leq x \leq -3,\ -2 \leq x \leq 2,\ x \geq 3$ (d) $x \leq -4,\ -3 \leq x \leq -2,\ 2 \leq x \leq 3$

3.

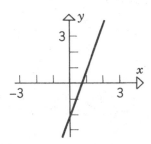

5.

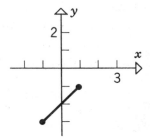

7.

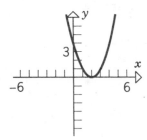

9.

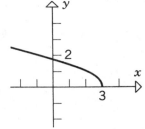

11.

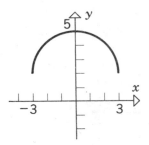

13.

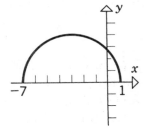

15.

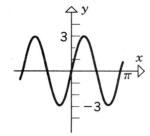

17.

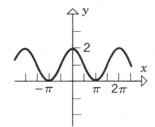

19.

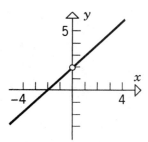

21.

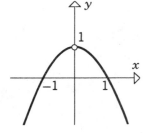

23.

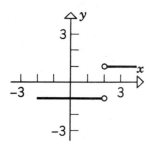

25.

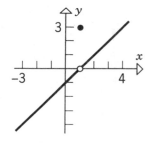

27.

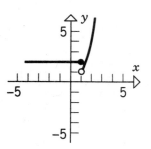

29.

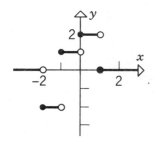

31. $x \leq 2 : f(x) = 2x + (2 - x) = x + 2,$
$x > 2 : f(x) = 2x + (x - 2) = 3x - 2;$
$$f(x) = \begin{cases} x + 2, & x \leq 2 \\ 3x - 2, & x > 2 \end{cases}$$

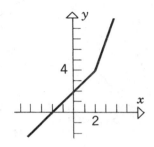

33. $x < 3 : g(x) = (5 - x) - (3 - x) = 2,$
$3 \leq x < 5 : g(x) = (5 - x) - (x - 3) = 8 - 2x,$
$x \geq 5 : g(x) = (x - 5) - (x - 3) = -2;$
$$g(x) = \begin{cases} 2, & x < 3 \\ 8 - 2x, & 3 \leq x < 5 \\ -2, & x \geq 5 \end{cases}$$

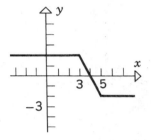

35. **(a)** $(f + g)(x) = \begin{cases} 0, & x < 0 \\ 2x, & x \geq 0 \end{cases}$ **(b)** $(f - g)(x) = \begin{cases} -2x, & x < 0 \\ 0, & x \geq 0 \end{cases}$

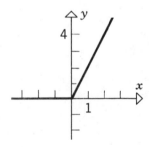

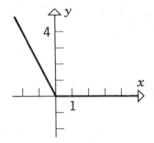

(c) $(f \cdot g)(x) = \begin{cases} -x^2, & x < 0 \\ x^2, & x \geq 0 \end{cases}$ **(d)** $(f/g)(x) = \begin{cases} -1, & x < 0 \\ 1, & x > 0 \end{cases}$

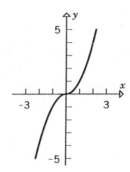   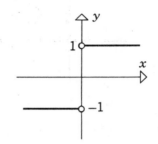

37. If $0 \leq x \leq 1$, $A = \dfrac{1}{2}(2x)(x) = x^2$;

if $x > 1$, $A = 1 + 2(x - 1) = 2x - 1$;

$A = \begin{cases} x^2, & 0 \leq x \leq 1 \\ 2x - 1, & x > 1 \end{cases}$

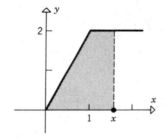

39. $g(x) = \begin{cases} 2, & x < -1 \\ 1 - x, & -1 \leq x < 1 \\ \dfrac{1}{2}(x - 1), & x \geq 1 \end{cases}$

41. (a)

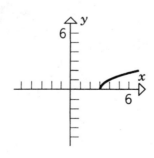

(b)

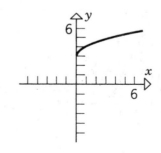

(c)

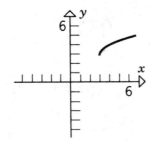

(d)

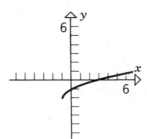

43. (a)

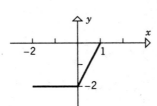

(b)

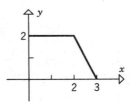

(c)

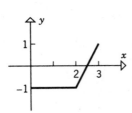

(d)

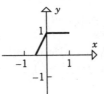

45.

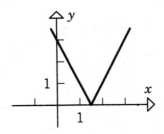

47. (a)

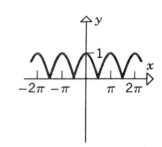

(b)

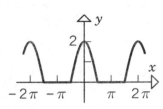

49. (a) $f(-x) = (-x)^2 = x^2 = f(x)$, even   **(b)** $f(-x) = (-x)^3 = -x^3 = -f(x)$, odd
(c) $f(-x) = |-x| = |x| = f(x)$, even   **(d)** $f(-x) = -x + 1$, neither

(e) $f(-x) = \dfrac{(-x)^5 - (-x)}{1 + (-x)^2} = \dfrac{-x^5 + x}{1 + x^2} = -\dfrac{x^5 - x}{1 + x^2} = -f(x)$, odd

(f) $f(-x) = 2 = f(x)$, even

51. (a) even   **(b)** odd   **(c)** odd   **(d)** neither

53. (a) If f and g are even, then $(f \cdot g)(-x) = f(-x)g(-x) = f(x)g(x) = (f \cdot g)(x)$ so $f \cdot g$ is even.

(b) If f and g are odd, then
$(f \cdot g)(-x) = f(-x)g(-x) = [-f(x)][-g(x)] = f(x)g(x) = (f \cdot g)(x)$ so $f \cdot g$ is even.

(c) If f is even and g is odd, then
$(f \cdot g)(-x) = f(-x)g(-x) = f(x)[-g(x)] = -f(x)g(x) = -(f \cdot g)(x)$ so $f \cdot g$ is odd.

55. (a) both; $y = -2x - 4$, $x = -\dfrac{1}{2}y - 2$ **(b)** y is a function of x; $y = \dfrac{1}{x^{2/3}}$

(c) neither   **(d)** both; $y = \dfrac{1}{2x}$, $x = \dfrac{1}{2y}$

57. (a) $y = 1/x^2$ **(b)**

$$x = \dfrac{1-y}{1+y}$$
$$x(1+y) = 1 - y$$
$$x + xy = 1 - y$$
$$xy + y = 1 - x$$
$$y(x+1) = 1 - x$$
$$y = \dfrac{1-x}{1+x}$$

(c) $y^2 + 2xy + x^2 = 0$
$(y + x)^2 = 0$
$y + x = 0$
$y = -x$

59. Treat $y^2 + 4xy + 1 = 0$ as a quadratic equation in y and use the quadratic formula to solve for y :

$$y = \frac{-4x \pm \sqrt{16x^2 - 4}}{2} = -2x \pm \sqrt{4x^2 - 1}.$$

There are two real values of y for each x for which $|x| > 1/2$.

61. **(a)** both   **(b)** a function of x
    (c) a function of y **(d)** neither

EXERCISE SET 2.4

1. **(a)** -1 **(b)** 3 **(c)** does not exist
    (d) 1 **(e)** -1 **(f)** 3

3. **(a)** 1 **(b)** 1 **(c)** 1
    (d) 1 **(e)** $-\infty$ **(f)** $+\infty$

5. **(a)** 0 **(b)** 0 **(c)** 0
    (d) 3 **(e)** $+\infty$ **(f)** $+\infty$

7. **(a)** $-\infty$ **(b)** $+\infty$ **(c)** does not exist
    (d) not defined   **(e)** 2 **(f)** 0

9. **(a)** $-\infty$ **(b)** $-\infty$ **(c)** $-\infty$
    (d) 1 **(e)** 2 **(f)** 2

11. **(a)** 0 **(b)** 0 **(c)** 0
    (d) 0 **(e)** does not exist   **(f)** does not exist

13. all values except -4

EXERCISE SET 2.5

1. 7 3. π 5. 36

7. $\sqrt{109}$ 9. 14 11. 0

13. $\lim_{x \to 4} \dfrac{x^2 - 16}{x - 4} = \lim_{x \to 4} \dfrac{(x + 4)(x - 4)}{x - 4} = \lim_{x \to 4} (x + 4) = 8$

15. $\lim_{x \to 1^+} \dfrac{x^4 - 1}{x - 1} = \lim_{x \to 1^+} \dfrac{(x^2 - 1)(x^2 + 1)}{x - 1}$

$$= \lim_{x \to 1^+} \dfrac{(x - 1)(x + 1)(x^2 + 1)}{x - 1} = \lim_{x \to 1^+} (x + 1)(x^2 + 1) = 4$$

17. $\lim_{x \to -1} \dfrac{x^2 + 6x + 5}{x^2 - 3x - 4} = \lim_{x \to -1} \dfrac{(x + 1)(x + 5)}{(x + 1)(x - 4)} = \lim_{x \to -1} \dfrac{x + 5}{x - 4} = -\dfrac{4}{5}$

19. $\lim_{x \to +\infty} \dfrac{3x + 1}{2x - 5} = \lim_{x \to +\infty} \dfrac{3 + 1/x}{2 - 5/x} = \dfrac{3}{2}$ **21.**   0

23. $\lim_{x \to -\infty} \dfrac{x - 2}{x^2 + 2x + 1} = \lim_{x \to -\infty} \dfrac{1/x - 2/x^2}{1 + 2/x + 1/x^2} = 0$

25. $\lim_{x \to -\infty} \dfrac{\sqrt{5x^2 - 2}}{x + 3} = \lim_{x \to -\infty} \dfrac{-\sqrt{5 - 2/x^2}}{1 + 3/x} = -\sqrt{5}$

27. $\lim_{y \to -\infty} \dfrac{2 - y}{\sqrt{7 + 6y^2}} = \lim_{y \to -\infty} \dfrac{2/y - 1}{-\sqrt{7/y^2 + 6}} = 1/\sqrt{6}$

29. $\lim_{x \to -\infty} \dfrac{\sqrt{3x^4 + x}}{x^2 - 8} = \lim_{x \to -\infty} \dfrac{\sqrt{3 + 1/x^3}}{1 - 8/x^2} = \sqrt{3}$

31. $+\infty$ **33.**   does not exist   **35.** $-\infty$

37. $+\infty$ **39.**   does not exist

41. $\lim_{x \to 4^-} \dfrac{3 - x}{x^2 - 2x - 8} = \lim_{x \to 4^-} \dfrac{3 - x}{(x - 4)(x + 2)} = +\infty$

43. $\lim_{x \to +\infty} \dfrac{7 - 6x^5}{x + 3} = \lim_{x \to +\infty} \dfrac{7/x - 6x^4}{1 + 3/x} = -\infty$

45. $\lim_{t \to +\infty} \dfrac{6 - t^3}{7t^3 + 3} = \lim_{t \to +\infty} \dfrac{6/t^3 - 1}{7 + 3/t^3} = -1/7$

47. $\lim_{x \to 0^-} \dfrac{x}{|x|} = \lim_{x \to 0^-} \dfrac{x}{(-x)} = \lim_{x \to 0^-} (-1) = -1$

49. $\lim\limits_{x\to 9} \dfrac{x-9}{\sqrt{x}-3} = \lim\limits_{x\to 9} \dfrac{(\sqrt{x}-3)(\sqrt{x}+3)}{\sqrt{x}-3} = \lim\limits_{x\to 9}(\sqrt{x}+3) = 6$

51. $+\infty$ **53.** $+\infty$ **55.** $-\infty$

57. If $a \neq 0$, then $\lim\limits_{x\to a} \dfrac{x}{x+a} = \lim\limits_{x\to a} \dfrac{a}{2a} = \dfrac{1}{2}$; if $a = 0$, then $\lim\limits_{x\to 0} \dfrac{x}{x} = \lim\limits_{x\to 0}(1) = 1$.

59. (a) $\lim\limits_{x\to 3^-} f(x) = \lim\limits_{x\to 3^-}(x-1) = 2$ (b) $\lim\limits_{x\to 3^+} f(x) = \lim\limits_{x\to 3^+}(3x-7) = 2$
     (c) 2

61. $\lim\limits_{x\to 3} h(x) = \lim\limits_{x\to 3}(x^2 - 2x + 1) = 4$

63. (a) The limit of a difference is the difference of limits if the latter limits exist, in this problem these limits do not exist.

     (b) $\lim\limits_{x\to 0^+}\left(\dfrac{1}{x} - \dfrac{1}{x^2}\right) = \lim\limits_{x\to 0^+} \dfrac{x-1}{x^2} = -\infty$

65. $\lim\limits_{x\to 0} \dfrac{\sqrt{x+4}-2}{x} = \lim\limits_{x\to 0} \dfrac{(x+4)-4}{x(\sqrt{x+4}+2)} = \lim\limits_{x\to 0} \dfrac{1}{\sqrt{x+4}+2} = \dfrac{1}{4}$

67. $\lim\limits_{x\to 0} \dfrac{\sqrt{5x+9}-3}{x} = \lim\limits_{x\to 0} \dfrac{5x}{x(\sqrt{5x+9}+3)} = \lim\limits_{x\to 0} \dfrac{5}{\sqrt{5x+9}+3} = \dfrac{5}{6}$

69. $\lim\limits_{x\to +\infty}(\sqrt{x^2+3}-x) = \lim\limits_{x\to +\infty} \dfrac{(x^2+3)-x^2}{\sqrt{x^2+3}+x} = \lim\limits_{x\to +\infty} \dfrac{3}{\sqrt{x^2+3}+x} = 0$

71. $\lim\limits_{x\to +\infty}(\sqrt{x^2+5x}-x) = \lim\limits_{x\to +\infty} \dfrac{5x}{\sqrt{x^2+5x}+x} = \lim\limits_{x\to +\infty} \dfrac{5}{\sqrt{1+5/x}+1} = \dfrac{5}{2}$

73. $\lim\limits_{x\to +\infty}(\sqrt{x^2+ax}-x) = \lim\limits_{x\to +\infty} \dfrac{(x^2+ax)-x^2}{\sqrt{x^2+ax}+x} = \lim\limits_{x\to +\infty} \dfrac{ax}{\sqrt{x^2+ax}+x}$

$$= \lim\limits_{x\to +\infty} \dfrac{a}{\sqrt{1+a/x}+1} = a/2$$

75. $r(x) = p(x)/q(x)$ if $r(x)$ is a rational function, where $p(x)$ and $q(x)$ are polynomials. We know that $\lim\limits_{x\to a} p(x) = p(a)$ and $\lim\limits_{x\to a} q(x) = q(a)$ so if $q(a) \neq 0$ then

$$\lim\limits_{x\to a} r(x) = \lim\limits_{x\to a} \dfrac{p(x)}{q(x)} = \dfrac{\lim\limits_{x\to a} p(x)}{\lim\limits_{x\to a} q(x)} = \dfrac{p(a)}{q(a)} = r(a). \text{ If } q(a) = 0, \text{ then } r(a) \text{ is not defined and}$$

thus cannot equal $\lim\limits_{x\to a} r(x)$, so $\lim\limits_{x\to a} r(x) = r(a)$ only if $r(a)$ is defined.

EXERCISE SET 2.6

1. $|2x - 8| < 0.1$
    if $2|x - 4| < 0.1$
    or if $|x - 4| < 0.05$,
    so $\delta = 0.05$

3. $|(7x + 5) - (-2)| < 0.01$
    if $|7x + 7| < 0.01$
    or if $7|x + 1| < 0.01$
    or if $|x + 1| < \dfrac{1}{700}$
    so $\delta = \dfrac{1}{700}$.

5. Suppose $x \neq 2$, then $\left| \dfrac{x^2 - 4}{x - 2} - 4 \right| = |(x + 2) - 4| = |x - 2|$. Thus $\left| \dfrac{x^2 - 4}{x - 2} - 4 \right| < 0.05$ if

    $0 < |x - 2| < 0.05$, so $\delta = 0.05$.

7. $|x^2 - 16| = |(x + 4)(x - 4)| = |x + 4|\,|x - 4|$. If we restrict δ so that $\delta \leq 1$, then

    $$|x - 4| < 1$$
    $$3 < x < 5$$
    $$7 < x + 4 < 9$$
    $$|x + 4| < 9$$
    $$|x + 4|\,|x - 4| \leq 9|x - 4|.$$

    Thus $|x^2 - 16| < 0.001$ if $9|x - 4| < 0.001$, or if $|x - 4| < 1/9000$, so $\delta = 1/9000$.

9. $\left| \dfrac{1}{x} - \dfrac{1}{5} \right| = \left| \dfrac{5 - x}{5x} \right| = \dfrac{|x - 5|}{|5x|}$. If we restrict δ so that $\delta \leq 1$, then

    $$|x - 5| < 1$$
    $$4 < x < 6$$
    $$20 < 5x < 30$$
    $$\frac{1}{20} > \frac{1}{5x} > \frac{1}{30}$$
    $$\frac{1}{|5x|} < \frac{1}{20}$$
    $$\frac{|x - 5|}{5|x|} \leq \frac{|x - 5|}{20}.$$

    Thus $\left| \dfrac{1}{x} - \dfrac{1}{5} \right| < 0.05$ if $\dfrac{|x - 5|}{20} < 0.05$, or if $|x - 5| < 1$, so $\delta = 1$.

11. $|3x - 15| = 3|x - 5| < \epsilon$, if $|x - 5| < \epsilon/3$, so $\delta = \epsilon/3$.

13. $|(2x - 7) + 3| = |2x - 4| = 2|x - 2| < \epsilon$ if $|x - 2| < \epsilon/2$, so $\delta = \epsilon/2$.

15. Suppose $x \neq 0$, then $\left| \dfrac{x^2 + x}{x} - 1 \right| = |(x + 1) - 1| = |x|$. Thus $\left| \dfrac{x^2 + 1}{x} - 1 \right| < \epsilon$ if $0 < |x| < \epsilon$, so $\delta = \epsilon$.

17. $|2x^2 - 2| = 2|x + 1||x - 1|$. If we restrict δ so that $\delta \leq 1$, then

$$|x - 1| < 1$$
$$0 < x < 2$$
$$1 < x + 1 < 3$$
$$|x + 1| < 3$$
$$2|x + 1||x - 1| \leq 6|x - 1|.$$

Thus $|2x^2 - 2| < \epsilon$ if $6|x - 1| < \epsilon$, or if $|x - 1| < \epsilon/6$, so $\delta = \min(\epsilon/6, 1)$.

19. $\left| \dfrac{1}{x} - 3 \right| = \left| \dfrac{3}{x} \right| \left| \dfrac{1}{3} - x \right| = \left| \dfrac{3}{x} \right| \left| x - \dfrac{1}{3} \right|$. If we restrict δ so that $\delta \leq \dfrac{1}{4}$, then

$$\left| x - \frac{1}{3} \right| < \frac{1}{4}$$
$$\frac{1}{12} < x < \frac{7}{12}$$
$$12 > \frac{1}{x} > \frac{12}{7}$$
$$36 > \frac{3}{x} > \frac{36}{7}$$
$$\left| \frac{3}{x} \right| < 36$$
$$\left| \frac{3}{x} \right| \left| x - \frac{1}{3} \right| \leq 36 \left| x - \frac{1}{3} \right|.$$

Thus $\left| \dfrac{1}{x} - 3 \right| < \epsilon$ if $36 \left| x - \dfrac{1}{3} \right| < \epsilon$, or if $\left| x - \dfrac{1}{3} \right| < \dfrac{\epsilon}{36}$, so $\delta = \min(\epsilon/36, 1/4)$.

21. $|\sqrt{x} - 2| = \left| \dfrac{\sqrt{x} - 2}{1} \dfrac{\sqrt{x} + 2}{\sqrt{x} + 2} \right| = \dfrac{|x - 4|}{\sqrt{x} + 2}$. If we restrict δ so that $\delta \leq 4$, then

$$|x - 4| < 4$$
$$0 < x < 8$$
$$0 < \sqrt{x} < \sqrt{8}$$
$$2 < \sqrt{x} + 2 < \sqrt{8} + 2$$
$$\frac{1}{2} > \frac{1}{\sqrt{x} + 2} > \frac{1}{\sqrt{8} + 2}$$

$$\frac{1}{\sqrt{x}+2} < \frac{1}{2}$$

$$\frac{|x-4|}{\sqrt{x}+2} \le \frac{1}{2}|x-4|$$

Thus $|\sqrt{x}-2| < \epsilon$ if $\frac{1}{2}|x-4| < \epsilon$, or if $|x-4| < 2\epsilon$, so $\delta = \min(2\epsilon, 4)$.

23. If $x \ne 1$, then $|f(x)-3| = |(x+2)-3| = |x-1|$. Thus $|f(x)-3| < \epsilon$ if $0 < |x-1| < \epsilon$ so $\delta = \epsilon$.

25. Assume there is a number L such that $\lim_{x \to 0} f(x) = L$. Then there exists a number $\delta > 0$ such that $|f(x)-L| < \frac{1}{8}$ whenever $0 < |x-0| < \delta$; in particular, $x = \frac{\delta}{2}$ and $x = -\frac{\delta}{2}$ are two such values of x. But $f(\delta/2) = \frac{1}{8}$ and $f(-\delta/2) = -\frac{1}{8}$ so $\left|\frac{1}{8} - L\right| < \frac{1}{8}$ and $\left|-\frac{1}{8} - L\right| < \frac{1}{8}$ or, equivalently, $0 < L < \frac{1}{4}$ and $-\frac{1}{4} < L < 0$, which is impossible.

27. Assume there is a number L such that $\lim_{x \to 1} \frac{1}{x-1} = L$. Then there exists a number $\delta > 0$ such that $\left|\frac{1}{x-1} - L\right| < 1$ whenever $0 < |x-1| < \delta$; in particular, $x = 1 + \frac{\delta}{\delta+1}$ and $x = 1 - \frac{\delta}{\delta+1}$ are two such values of x. So $\left|\frac{\delta+1}{\delta} - L\right| < 1$ and $\left|-\frac{\delta+1}{\delta} - L\right| < 1$ or equivalently, $\frac{1}{\delta} < L < \frac{1}{\delta} + 2$ and $-2 - \frac{1}{\delta} < L < -\frac{1}{\delta}$, which is impossible.

29. $|x^2 - 9| = |x+3||x-3|$. If $\delta \le 2$, then

$$|x-3| < 2$$
$$1 < x < 5$$
$$4 < x+3 < 8$$
$$|x+3| < 8$$
$$|x+3|\,|x-3| \le 8|x-3|.$$

Thus $|x^2 - 9| < \epsilon$ if $8|x-3| < \epsilon$, or if $|x-3| < \epsilon/8$, so $\delta = \min(\epsilon/8, 2)$

EXERCISE SET 2.7

1. continuous on (d), (e), (f); discontinuous at $x = 2$ on (a), (b), (c)

3. continuous on (b), (d), (f); discontinuous at $x = 1, 3$ on (a), $x = 1$ on (c), $x = 3$ on (e)

5. none   7. none   9. $x = \pm 4$

11. $x = \pm 3$ 13. none   15. none

17. **(a)** f is continuous everywhere for any k, except perhaps at $x = 1$;
$\lim\limits_{x \to 1^-} f(x) = \lim\limits_{x \to 1^-} (7x - 2) = 5$, $\lim\limits_{x \to 1^+} f(x) = \lim\limits_{x \to 1^+} kx^2 = k$, and $f(1) = 5$ thus
$\lim\limits_{x \to 1} f(x) = f(1)$ if $k = 5$, so f is continuous everywhere if $k = 5$.

    (b) $\lim\limits_{x \to 2^-} f(x) = \lim\limits_{x \to 2^-} kx^2 = 4k$, $\lim\limits_{x \to 2^+} f(x) = \lim\limits_{x \to 2^+} (2x + k) = 4 + k$, and $f(2) = 4k$,
so $\lim\limits_{x \to 2} f(x) = f(2)$ if $4k = 4 + k$, $k = 4/3$.

19. **(a)** If $c > 0$, then $\lim\limits_{x \to c} f(x) = \lim\limits_{x \to c} \sqrt{x} = \sqrt{c} = f(c)$; also $\lim\limits_{x \to 0^+} f(x) = \lim\limits_{x \to 0^+} \sqrt{x} = 0 = f(0)$ so
$f(x)$ is continuous on $[0, +\infty)$.

    (b) $\sqrt{g(x)}$ is continuous by Theorem 2.7.6 because it is the composition of $\sqrt{x}$ with $g(x)$
where $\sqrt{x}$ is continuous on $[0, +\infty)$ and $g(x)$ is continuous and nonnegative.

21. $x = 0, \pm 1, \pm 2, \cdots$

23. **(a)** $x = 0$; not removable because $\lim\limits_{x \to 0} \dfrac{|x|}{x}$ does not exist.

    (b) $x = -3$; removable because $\lim\limits_{x \to -3} \dfrac{x^2 + 3x}{x + 3} = -3$.

    (c) $x = \pm 2$; removable at $x = 2$ because $\lim\limits_{x \to 2} \dfrac{x - 2}{|x| - 2} = 1$, not removable at $x = -2$ because
$\lim\limits_{x \to -2} \dfrac{x - 2}{|x| - 2}$ does not exist.

25. If f and g are continuous at c, then $\lim\limits_{x \to c} f(x) = f(c)$ and $\lim\limits_{x \to c} g(x) = g(c)$, so

(a) $\lim\limits_{x \to c}(f + g)(x) = \lim\limits_{x \to c}[f(x) + g(x)] = \lim\limits_{x \to c} f(x) + \lim\limits_{x \to c} g(x) = f(c) + g(c) = (f + g)(c)$
therefore $f + g$ is continuous at c.

(b) Similar to part (a) with $+$ replaced by $-$.

(c) $\lim\limits_{x \to c}(f \cdot g)(x) = \lim\limits_{x \to c}[f(x)g(x)] = \left[\lim\limits_{x \to c} f(x)\right]\left[\lim\limits_{x \to c} g(x)\right] = f(c)g(c) = (f \cdot g)(c)$
therefore $f \cdot g$ is continuous at c.

27. (a) Let $f(x) = \begin{cases} 0, & x < 2 \\ 1, & x \geq 2 \end{cases}$ and $g(x) = \begin{cases} 1, & x < 2 \\ 0, & x \geq 2 \end{cases}$; f and g are discontinuous at $x = 2$,

but $f + g$ is continuous at $x = 2$. If $f(x) = \begin{cases} 0, & x < 2 \\ 1, & x \geq 2 \end{cases}$ and $g(x) = \begin{cases} 1, & x < 2 \\ 2, & x \geq 2 \end{cases}$;

then f, g, and $f + g$ are discontinuous at $x = 2$.

(b) Replace $f + g$ by $f \cdot g$ everywhere in part (a).

29. If $f(a)$ and $f(b)$ have opposite signs then 0 is between $f(a)$ and $f(b)$. From Theorem 2.7.9 there is at least one number x in $[a, b]$ such that $f(x) = 0$. But $f(a) \neq 0$ and $f(b) \neq 0$ by assumption, so there is at least one solution of $f(x) = 0$ in the interval (a, b).

31. $f(x) = x^3 + x^2 - 2x - 1$ is continuous on $[-1, 1]$, $f(-1) = 1$ and $f(1) = -1$ have opposite signs so Theorem 2.7.10 applies.

33. From the graph there are two real solutions, one in $(-2, -1.5)$ and the other in $(1, 1.5)$. For the one in $(-2, -1.5)$ we find

x	-2.0	-1.9	-1.8	-1.7	-1.6
y	$-9.$	-6.13	-3.70	-1.65	0.05

so one solution is in $(-1.7, -1.6)$; use -1.65 to approximate it. For the one in $(1, 1.5)$ we find

x	1.0	1.1	1.2	1.3	1.4
y	$3.$	2.44	1.73	0.84	-0.24

so this solution is in $(1.3, 1.4)$; use 1.35 to approximate it.

35. If $f(x) = x^2 - 5$, then $f(2) = -1$ and $f(3) = 4$ have opposite signs so $\sqrt{5}$ is in $(2, 3)$.

(a) $f(2.2) = -0.16$ and $f(2.3) = 0.29$ so $\sqrt{5}$ is in $(2.2, 2.3)$; use 2.25 to approximate it with an error of at most 0.05.

(b) $f(2.23) = -0.03$ and $f(2.24) = 0.02$ so $\sqrt{5}$ is in $(2.23, 2.24)$; use 2.235 to approximate it with an error of at most 0.005.

37. $f(x) = \begin{cases} x+1, & 0 \le x \le 1 \\ x-3, & 1 < x \le 2 \end{cases}$, for example.

39. If $p(x) = a_n x^n + a_{n-1} x^{n-1} + \cdots + a_1 x + a_0$ where $a_n \ne 0$ and n is odd, then either

$\lim\limits_{x \to +\infty} p(x) = +\infty$ and $\lim\limits_{x \to -\infty} p(x) = -\infty$, or $\lim\limits_{x \to +\infty} p(x) = -\infty$ and $\lim\limits_{x \to -\infty} p(x) = +\infty$,

depending on whether $a_n > 0$ or $a_n < 0$, respectively. In either case, there are numbers a and b with $p(x)$ continuous on $[a, b]$ where $p(a)$ and $p(b)$ have opposite signs so that $p(x) = 0$ has at least one real solution in (a, b).

EXERCISE SET 2.8

1. none

3. $x = n\pi$; $n = 0, \pm 1, \pm 2, \cdots$

5. $x = n\pi$; $n = 0, \pm 1, \pm 2, \cdots$

7. none

9. discontinuous if $\sin x = 1/2$, so $x = \pi/6 + 2n\pi$ or $x = 5\pi/6 + 2n\pi$; $n = 0, \pm 1, \pm 2, \cdots$

11. $\sin(g(x))$ is the composition of $\sin x$ with $g(x)$; $\sin x$ is continuous, so by Theorem 2.7.6 $\sin(g(x))$ is continuous at every point where $g(x)$ is continuous.

13. $\lim\limits_{x \to +\infty} \cos(1/x) = \cos(0) = 1$

15. $\lim\limits_{x \to +\infty} \sin\left(\dfrac{\pi x}{2 - 3x}\right) = \sin\left(\lim\limits_{x \to +\infty} \dfrac{\pi x}{2 - 3x}\right) = \sin(-\pi/3) = -\sqrt{3}/2$

17. $\lim\limits_{\theta \to 0} \dfrac{\sin 3\theta}{\theta} = \lim\limits_{\theta \to 0} 3\dfrac{\sin 3\theta}{3\theta} = 3 \lim\limits_{\theta \to 0} \dfrac{\sin 3\theta}{3\theta} = 3(1) = 3.$

19. $\lim\limits_{x \to 0^-} \dfrac{\sin x}{|x|} = \lim\limits_{x \to 0^-} \left(-\dfrac{\sin x}{x}\right) = -1$

21. $\lim\limits_{x \to 0^+} \dfrac{\sin x}{5\sqrt{x}} = \dfrac{1}{5} \lim\limits_{x \to 0^+} \sqrt{x} \left(\dfrac{\sin x}{x}\right) = 0$

23. $\lim\limits_{x \to 0} \dfrac{\tan 7x}{\sin 3x} = \lim\limits_{x \to 0} \dfrac{\frac{\sin 7x}{\cos 7x}}{\sin 3x} = \lim\limits_{x \to 0} \dfrac{1}{\cos 7x} \dfrac{\sin 7x}{\sin 3x} = \lim\limits_{x \to 0} \dfrac{1}{\cos 7x} \dfrac{7\frac{\sin 7x}{7x}}{3\frac{\sin 3x}{3x}}$

$= \dfrac{7}{3} \lim\limits_{x \to 0} \dfrac{1}{\cos 7x} \dfrac{\lim\limits_{x \to 0} \frac{\sin 7x}{7x}}{\lim\limits_{x \to 0} \frac{\sin 3x}{3x}} = \dfrac{7}{3}(1)\dfrac{(1)}{(1)} = \dfrac{7}{3}.$

25. $\lim\limits_{h\to 0}\dfrac{h}{\tan h} = \lim\limits_{h\to 0}\dfrac{h}{\dfrac{\sin h}{\cos h}} = \lim\limits_{h\to 0}\dfrac{h\cos h}{\sin h} = \lim\limits_{h\to 0}\dfrac{\cos h}{\dfrac{\sin h}{h}} = \dfrac{\lim\limits_{h\to 0}\cos h}{\lim\limits_{h\to 0}\dfrac{\sin h}{h}} = \dfrac{1}{1} = 1.$

27. $\lim\limits_{\theta\to 0}\dfrac{\theta^2}{1-\cos\theta} = \lim\limits_{\theta\to 0}\dfrac{\theta^2}{1-\cos\theta}\dfrac{1+\cos\theta}{1+\cos\theta} = \lim\limits_{\theta\to 0}\dfrac{\theta^2(1+\cos\theta)}{1-\cos^2\theta}$

$\qquad = \lim\limits_{\theta\to 0}\dfrac{\theta^2(1+\cos\theta)}{\sin^2\theta} = \lim\limits_{\theta\to 0}\dfrac{1+\cos\theta}{\dfrac{\sin^2\theta}{\theta^2}} = \dfrac{\lim\limits_{\theta\to 0}(1+\cos\theta)}{\lim\limits_{\theta\to 0}\left(\dfrac{\sin\theta}{\theta}\right)^2} = \dfrac{1+1}{1^2} = 2.$

29. $\lim\limits_{\theta\to 0}\dfrac{\theta}{\cos\theta} = \dfrac{\lim\limits_{\theta\to 0}\theta}{\lim\limits_{\theta\to 0}\cos\theta} = \dfrac{0}{1} = 0.$

31. Use the identity $1-\cos\theta = 2\sin^2\dfrac{\theta}{2}$:

$\lim\limits_{h\to 0}\dfrac{1-\cos 5h}{\cos 7h - 1} = \lim\limits_{h\to 0}\dfrac{2\sin^2(5h/2)}{-2\sin^2(7h/2)} = -\lim\limits_{h\to 0}\dfrac{(5/2)^2\dfrac{\sin^2(5h/2)}{(5h/2)^2}}{(7/2)^2\dfrac{\sin^2(7h/2)}{(7h/2)^2}}$

$\qquad = -\dfrac{25}{49}\dfrac{\lim\limits_{h\to 0}\left[\dfrac{\sin(5h/2)}{5h/2}\right]^2}{\lim\limits_{h\to 0}\left[\dfrac{\sin(7h/2)}{7h/2}\right]^2} = -\dfrac{25}{49}\cdot\dfrac{1^2}{1^2} = -\dfrac{25}{49}.$

33. $\lim\limits_{x\to 0^+}\cos(1/x)$ does not exist due to oscillation

35. $\lim\limits_{x\to 0}\dfrac{2x+\sin x}{x} = \lim\limits_{x\to 0}\left(2+\dfrac{\sin x}{x}\right) = 2+1 = 3$

37. $\lim\limits_{x\to 0^-}f(x) = \lim\limits_{x\to 0^-}\dfrac{\tan kx}{x} = \lim\limits_{x\to 0^-}\dfrac{k}{\cos kx}\dfrac{\sin kx}{kx} = k;\ \lim\limits_{x\to 0^+}f(x) = \lim\limits_{x\to 0^+}(3x+2k^2) = 2k^2;$

$\qquad f(0) = 2k^2;$ so $\lim\limits_{x\to 0}f(x) = f(0)$ if $2k^2 = k,\ 2k^2 - k = 0,\ k(2k-1) = 0;\ k = 1/2.$

39. **(a)** If $t = \dfrac{1}{x}$, then $x = \dfrac{1}{t}$ and $x \to +\infty$ as $t \to 0^+$ so $\lim\limits_{x\to +\infty}x\sin\dfrac{1}{x} = \lim\limits_{t\to 0^+}\dfrac{\sin t}{t} = 1.$

$\qquad$ **(b)** If $t = \dfrac{1}{x}$, then $x = \dfrac{1}{t}$ and $x \to -\infty$ as $t \to 0^-$ so $\lim\limits_{x\to -\infty}\left(1-\cos\dfrac{1}{x}\right) = \lim\limits_{t\to 0^-}\dfrac{1-\cos t}{t} = 0.$

(c) If $t = \pi - x$, then $x = \pi - t$ and $x \to \pi$ as $t \to 0$

so $\displaystyle\lim_{x \to \pi} \frac{\pi - x}{\sin x} = \lim_{t \to 0} \frac{t}{\sin(\pi - t)} = \lim_{t \to 0} \frac{t}{\sin t} = 1.$

41. Let $t = x - 1$, then $x = t + 1$ and

$\displaystyle\lim_{x \to 1} \frac{\sin(\pi x)}{x - 1} = \lim_{t \to 0} \frac{\sin(\pi t + \pi)}{t} = \lim_{t \to 0} \frac{-\sin \pi t}{t} = -\pi \lim_{t \to 0} \frac{\sin \pi t}{\pi t} = -\pi$

43. (a) $-1 \leq \sin x \leq 1$ so $-\dfrac{1}{x} \leq \dfrac{\sin x}{x} \leq \dfrac{1}{x}$ if $x > 0$.

But $-1/x$ and $1/x \to 0$ as $x \to +\infty$, thus $\displaystyle\lim_{x \to +\infty} \frac{\sin x}{x} = 0.$

(b) Replace $\sin x$ by $\cos x$ in part (a).

45. $\displaystyle\lim_{x \to 0}(1 - x^2) = 1$ and $\displaystyle\lim_{x \to 0} \cos x = 1$ so $\displaystyle\lim_{x \to 0} f(x) = 1.$

47. If $x > 0$, then $xL \leq xf(x) \leq xM$; if $x < 0$, then $xL \geq xf(x) \geq xM$. So $\displaystyle\lim_{x \to 0} xf(x) = 0$ because $\displaystyle\lim_{x \to 0^+} xL = \lim_{x \to 0^+} xM = 0$ and $\displaystyle\lim_{x \to 0^-} xL = \lim_{x \to 0^-} xM = 0.$

49. If $h < 0$, then $-h > 0$ so from (8) $\cos(-h) < \dfrac{\sin(-h)}{-h} < 1, \cos h < \dfrac{-\sin h}{-h} < 1,$

$\cos h < \dfrac{\sin h}{h} < 1.$

51. $\displaystyle\lim_{h \to 0} \cos(c + h) = \lim_{h \to 0}[\cos c \cos h - \sin c \sin h] = \cos c \lim_{h \to 0} \cos h - \sin c \lim_{h \to 0} \sin h$

$= (\cos c)(1) - (\sin c)(0) = \cos c$

so $\cos x$ is continuous.

53. (a) $\sin 10° \approx 0.17365$ $\qquad\qquad\qquad$ **(b)** $\pi/18 \approx 0.17453$

55. (a) $\tan 5° \approx 0.08749$ $\qquad\qquad\qquad$ **(b)** $\pi/36 \approx 0.08727$

57. $f(x) = x - \cos x$ is continuous on $[0, \pi/2]$; $f(0) = -1$ and $f(\pi/2) = \pi/2$ have opposite signs so Theorem 2.7.10 applies.

TECHNOLOGY EXERCISES 2

1. 2.718282 (exact value: e) $\qquad$ **3.** 0.540302 (exact value: $\cos 1$) $\qquad$ **5.** 1/2

7. (b) 3°F, -11°F, -18°F, -22°F

$\quad$ **(c)** 34 mi/hr, 19 mi/hr, 12 mi/hr, 7 mi/hr

9. domain: $[-0.724492, 1.220744]$
   range: $[-1.055084, 1.490152]$

11. $\delta = 0.077471$

13. **(b)** 24.61 ft/sec
    (c) Solve $v = 0.98(24.61)$ for t to get $t = 3.01$ sec

15. **(d)** 1.324702   17. Use $x_{n+1} = (x_n + 2)^{1/5}$ to get 1.267168

CHAPTER 3
Differentiation

EXERCISE SET 3.1

1. **(a)** $m_{sec} = \dfrac{f(4) - f(3)}{4 - 3} = \dfrac{(4)^2/2 - (3)^2/2}{1} = \dfrac{7}{2}$.

(b) $m_{tan} = \lim\limits_{x_1 \to 3} \dfrac{f(x_1) - f(3)}{x_1 - 3} = \lim\limits_{x_1 \to 3} \dfrac{x_1^2/2 - 9/2}{x_1 - 3}$

$= \lim\limits_{x_1 \to 3} \dfrac{x_1^2 - 9}{2(x_1 - 3)} = \lim\limits_{x_1 \to 3} \dfrac{(x_1 + 3)(x_1 - 3)}{2(x_1 - 3)} = \lim\limits_{x_1 \to 3} \dfrac{x_1 + 3}{2} = 3$.

(c) $m_{tan} = \lim\limits_{x_1 \to x_0} \dfrac{f(x_1) - f(x_0)}{x_1 - x_0}$

$= \lim\limits_{x_1 \to x_0} \dfrac{x_1^2/2 - x_0^2/2}{x_1 - x_0}$

$= \lim\limits_{x_1 \to x_0} \dfrac{x_1^2 - x_0^2}{2(x_1 - x_0)}$

$= \lim\limits_{x_1 \to x_0} \dfrac{x_1 + x_0}{2} = x_0$

(d)

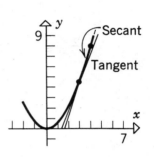

3. **(a)** $m_{sec} = \dfrac{f(3) - f(2)}{3 - 2} = \dfrac{1/3 - 1/2}{1} = -\dfrac{1}{6}$.

(b) $m_{tan} = \lim\limits_{x_1 \to 2} \dfrac{f(x_1) - f(2)}{x_1 - 2} = \lim\limits_{x_1 \to 2} \dfrac{1/x_1 - 1/2}{x_1 - 2}$

$= \lim\limits_{x_1 \to 2} \dfrac{2 - x_1}{2x_1(x_1 - 2)} = \lim\limits_{x_1 \to 2} \dfrac{-1}{2x_1} = -\dfrac{1}{4}$.

(c) $m_{tan} = \lim\limits_{x_1 \to x_0} \dfrac{f(x_1) - f(x_0)}{x_1 - x_0}$

$= \lim\limits_{x_1 \to x_0} \dfrac{1/x_1 - 1/x_0}{x_1 - x_0}$

$= \lim\limits_{x_1 \to x_0} \dfrac{x_0 - x_1}{x_0 x_1(x_1 - x_0)}$

$= \lim\limits_{x_1 \to x_0} \dfrac{-1}{x_0 x_1} = -\dfrac{1}{x_0^2}$.

(d)

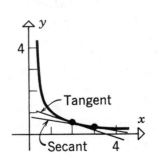

5. (a) $m_{\tan} = \lim\limits_{x_1 \to x_0} \dfrac{f(x_1) - f(x_0)}{x_1 - x_0} = \lim\limits_{x_1 \to x_0} \dfrac{(x_1^2 + 1) - (x_0^2 + 1)}{x_1 - x_0}$

$\qquad = \lim\limits_{x_1 \to x_0} \dfrac{x_1^2 - x_0^2}{x_1 - x_0} = \lim\limits_{x_1 \to x_0} (x_1 + x_0) = 2x_0.$

(b) $m_{\tan} = 2(2) = 4.$

7. (a) $m_{\tan} = \lim\limits_{x_1 \to x_0} \dfrac{f(x_1) - f(x_0)}{x_1 - x_0} = \lim\limits_{x_1 \to x_0} \dfrac{\sqrt{x_1} - \sqrt{x_0}}{x_1 - x_0}$

$\qquad = \lim\limits_{x_1 \to x_0} \dfrac{1}{\sqrt{x_1} + \sqrt{x_0}} = \dfrac{1}{2\sqrt{x_0}}.$

(b) $m_{\tan} = \dfrac{1}{2\sqrt{1}} = \dfrac{1}{2}.$

9. (a) $m_{\tan} = (50 - 10)/(15 - 5)$
$\qquad = 40/10$
$\qquad = 4 \text{ m/sec}$

(b)

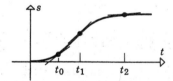

11. From the figure:

(a) The particle is moving faster at time t_0 because the slope of the tangent to the curve at t_0 is greater than that at t_2.

(b) The initial velocity is 0 because the slope of a horizontal line is 0.

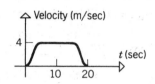

(c) The particle is speeding up because the slope increases as t increases from t_0 to t_1.

(d) The particle is slowing down because the slope decreases as t increases from t_1 to t_2.

13. It is a straight line with slope equal to the velocity.

15. (a) 72°F at about 4:30 P.M.

(b) about $(67 - 43)/6 = 4°\text{F/hr}$

(c) decreasing most rapidly at about 9 P.M.; rate of change of temperature is about $-7°\text{F/hr}$ (slope of estimated tangent line to curve at 9 P.M.)

17. (a) during the first year after birth
    (b) about 6 cm/year (slope of estimated tangent line at age 5)
    (c) the growth rate is greatest at about age 14; about 10 cm/year
    (d)

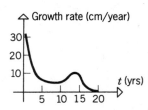

19. (a) $5(40)^3 = 320,000$ ft.
    (b) $v_{ave} = 320,000/40 = 8,000$ ft/sec.
    (c) $5t^3 = 135$ when the rocket has gone 135 ft, so $t^3 = 27$, $t = 3$ sec;
       $v_{ave} = 135/3 = 45$ ft/sec.
    (d) $v_{inst} = \lim\limits_{t_1 \to 40} \dfrac{5t_1^3 - 5(40)^3}{t_1 - 40} = \lim\limits_{t_1 \to 40} \dfrac{5(t_1^3 - 40^3)}{t_1 - 40}$

    $= \lim\limits_{t_1 \to 40} 5(t_1^2 + 40t_1 + 1600) = 24,000$ ft/sec.

21. (a) $v_{ave} = \dfrac{6(4)^4 - 6(2)^4}{4 - 2} = 720$ ft/min.

    (b) $v_{inst} = \lim\limits_{t_1 \to 2} \dfrac{6t_1^4 - 6(2)^4}{t_1 - 2} = \lim\limits_{t_1 \to 2} \dfrac{6(t_1^4 - 16)}{t_1 - 2}$

    $= \lim\limits_{t_1 \to 2} \dfrac{6(t_1^2 + 4)(t_1^2 - 4)}{t_1 - 2} = \lim\limits_{t_1 \to 2} 6(t_1^2 + 4)(t_1 + 2) = 192$ ft/min.

23. (a) $r_0 = 1$, $r_1 = 2$; $A_0 = \pi(1)^2 = \pi$, $A_1 = \pi(2)^2 = 4\pi$, $\dfrac{A_1 - A_0}{r_1 - r_0} = \dfrac{4\pi - \pi}{2 - 1} = 3\pi$.

    (b) $\lim\limits_{r_1 \to 2} \dfrac{\pi r_1^2 - \pi(2)^2}{r_1 - 2} = \lim\limits_{r_1 \to 2} \dfrac{\pi(r_1^2 - 4)}{r_1 - 2} = \lim\limits_{r_1 \to 2} \pi(r_1 + 2) = 4\pi$.

25. $m_{sec} = \dfrac{f(x_1) - f(x_0)}{x_1 - x_0} = \dfrac{x_1^2 - x_0^2}{x_1 - x_0} = \dfrac{(x_1 + x_0)(x_1 - x_0)}{x_1 - x_0} = x_1 + x_0$, $x_1 \neq x_0$
    m_{sec} approaches $2x_0$ as x_1 approaches x_0, thus $m_{tan} = 2x_0$ so
    $|m_{tan} - m_{sec}| = |2x_0 - (x_1 + x_0)| = |x_0 - x_1| = |x_1 - x_0|$.

EXERCISE SET 3.2

1. $f'(x) = \lim_{h \to 0} \dfrac{3(x+h)^2 - 3x^2}{h} = \lim_{h \to 0} \dfrac{3(x^2 + 2xh + h^2) - 3x^2}{h}$

$= \lim_{h \to 0} \dfrac{6xh + 3h^2}{h} = \lim_{h \to 0} (6x + 3h) = 6x.$

3. $f'(x) = \lim_{h \to 0} \dfrac{(x+h)^3 - x^3}{h} = \lim_{h \to 0} \dfrac{x^3 + 3x^2h + 3xh^2 + h^3 - x^3}{h}$

$= \lim_{h \to 0} \dfrac{3x^2h + 3xh^2 + h^3}{h} = \lim_{h \to 0} (3x^2 + 3xh + h^2) = 3x^2.$

5. $f'(x) = \lim_{h \to 0} \dfrac{\sqrt{x+h+1} - \sqrt{x+1}}{h}$

$= \lim_{h \to 0} \dfrac{\sqrt{x+h+1} - \sqrt{x+1}}{h} \cdot \dfrac{\sqrt{x+h+1} + \sqrt{x+1}}{\sqrt{x+h+1} + \sqrt{x+1}}$

$= \lim_{h \to 0} \dfrac{(x+h+1) - (x+1)}{h(\sqrt{x+h+1} + \sqrt{x+1})} = \lim_{h \to 0} \dfrac{h}{h(\sqrt{x+h+1} + \sqrt{x+1})}$

$= \lim_{h \to 0} \dfrac{1}{\sqrt{x+h+1} + \sqrt{x+1}} = \dfrac{1}{2\sqrt{x+1}}.$

7. $f'(x) = \lim_{h \to 0} \dfrac{\dfrac{1}{x+h} - \dfrac{1}{x}}{h} = \lim_{h \to 0} \dfrac{\dfrac{x - (x+h)}{x(x+h)}}{h}$

$= \lim_{h \to 0} \dfrac{-h}{hx(x+h)} = \lim_{h \to 0} -\dfrac{1}{x(x+h)} = -\dfrac{1}{x^2}.$

9. $f'(x) = \lim_{h \to 0} \dfrac{[a(x+h)^2 + b] - [ax^2 + b]}{h} = \lim_{h \to 0} \dfrac{ax^2 + 2axh + ah^2 + b - ax^2 - b}{h}$

$= \lim_{h \to 0} \dfrac{2axh + ah^2}{h} = \lim_{h \to 0} (2ax + ah) = 2ax.$

11. $f'(x) = \lim_{h \to 0} \dfrac{\dfrac{1}{\sqrt{x+h}} - \dfrac{1}{\sqrt{x}}}{h} = \lim_{h \to 0} \dfrac{\sqrt{x} - \sqrt{x+h}}{h\sqrt{x}\sqrt{x+h}}$

$= \lim_{h \to 0} \dfrac{x - (x+h)}{h\sqrt{x}\sqrt{x+h}(\sqrt{x} + \sqrt{x+h})} = \lim_{h \to 0} \dfrac{-1}{\sqrt{x}\sqrt{x+h}(\sqrt{x} + \sqrt{x+h})} = -\dfrac{1}{2x^{3/2}}.$

13. $f'(3) = 6(3) = 18$; $f(3) = 3(3)^2 = 27$ so $y - 27 = 18(x - 3)$, $y = 18x - 27$.

15. $f'(0) = 3(0)^2 = 0$; $f(0) = 0^3 = 0$ so $y - 0 = (0)(x - 0)$, $y = 0$.

17. $f'(8) = \dfrac{1}{2\sqrt{8+1}} = \dfrac{1}{6}$; $f(8) = \sqrt{8+1} = 3$ so $y - 3 = \dfrac{1}{6}(x - 8)$, $y = \dfrac{1}{6}x + \dfrac{5}{3}$.

19. $y - (-1) = 5(x - 3)$, $y = 5x - 16$.

21. **(a)** $\dfrac{dy}{dx} = \lim\limits_{h \to 0} \dfrac{[4(x+h)^2 + 2] - [4x^2 + 2]}{h}$

$= \lim\limits_{h \to 0} \dfrac{4x^2 + 8xh + 4h^2 + 2 - 4x^2 - 2}{h} = \lim\limits_{h \to 0}(8x + 4h) = 8x$.

(b) $\dfrac{dy}{dx}\bigg|_{x=1} = 8(1) = 8$.

23. $f'(t) = \lim\limits_{h \to 0} \dfrac{f(t+h) - f(t)}{h} = \lim\limits_{h \to 0} \dfrac{[4(t+h)^2 + (t+h)] - [4t^2 + t]}{h}$

$= \lim\limits_{h \to 0} \dfrac{4t^2 + 8th + 4h^2 + t + h - 4t^2 - t}{h}$

$= \lim\limits_{h \to 0} \dfrac{8th + 4h^2 + h}{h} = \lim\limits_{h \to 0}(8t + 4h + 1) = 8t + 1$.

25. $\dfrac{dA}{d\lambda} = \lim\limits_{h \to 0} \dfrac{[3(\lambda + h)^2 - (\lambda + h)] - [3\lambda^2 - \lambda]}{h} = \lim\limits_{h \to 0} \dfrac{3\lambda^2 + 6\lambda h + 3h^2 - \lambda - h - 3\lambda^2 + \lambda}{h}$

$= \lim\limits_{h \to 0} \dfrac{6\lambda h + 3h^2 - h}{h} = \lim\limits_{h \to 0}(6\lambda + 3h - 1) = 6\lambda - 1$.

27. **(a)** D $\qquad$ **(b)** F $\qquad$ **(c)** B $\qquad$ **(d)** C $\qquad$ **(e)** A $\qquad$ **(f)** E

29. Estimate the slope of the tangent lines at $t = 0$ and $t = 15$ to get $\dfrac{dC}{dt}\bigg|_{t=0} \approx 0.08$ mol/L

per second and $\dfrac{dC}{dt}\bigg|_{t=15} \approx 0.018$ mol/L per second.

31. **(a)** $F \approx 200$ lb, $dF/d\theta \approx 60$ lb/rad $\qquad$ **(b)** $\mu = (dF/d\theta)/F \approx 60/200 = 0.3$

33.

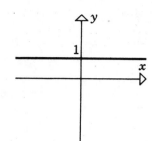

35.

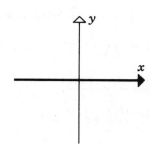

37.

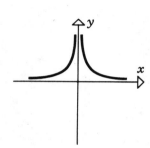

39.

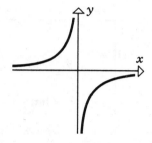

41. $\lim_{x\to 0} f(x) = \lim_{x\to 0} \sqrt[3]{x} = 0 = f(0)$,
so f is continuous at $x = 0$.

$$\lim_{h\to 0} \frac{f(0+h) - f(0)}{h} = \lim_{h\to 0} \frac{\sqrt[3]{h} - 0}{h}$$

$$= \lim_{h\to 0} \frac{1}{h^{2/3}} = +\infty,$$

so $f'(0)$ does not exist.

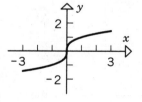

43. $\lim_{x\to 1^-} f(x) = \lim_{x\to 1^+} f(x) = f(1)$, so
f is continuous at $x = 1$.

$$\lim_{h\to 0^-} \frac{f(1+h) - f(1)}{h} = \lim_{h\to 0^-} \frac{[(1+h)^2 + 1] - 2}{h}$$

$$= \lim_{h\to 0^-} (2 + h) = 2;$$

$$\lim_{h\to 0^+} \frac{f(1+h) - f(1)}{h} = \lim_{h\to 0^+} \frac{2(1+h) - 2}{h}$$

$$= \lim_{h\to 0^+} 2 = 2,$$

so $f'(1) = 2$.

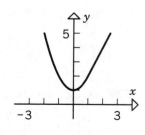

45. f is continuous at $x = 1$ because it is differentiable there, thus $\lim\limits_{h\to 0} f(1+h) = f(1)$ and so

$f(1) = 0$ because $\lim\limits_{h\to 0} \dfrac{f(1+h)}{h}$ exists; $f'(1) = \lim\limits_{h\to 0} \dfrac{f(1+h) - f(1)}{h} = \lim\limits_{h\to 0} \dfrac{f(1+h)}{h} = 5$.

47. $f'(x) = \lim\limits_{h\to 0} \dfrac{f(x+h) - f(x)}{h}$

$\qquad = \lim\limits_{h\to 0} \dfrac{f(x)f(h) - f(x)}{h}$

$\qquad = \lim\limits_{h\to 0} \dfrac{f(x)[f(h) - 1]}{h} = f(x) \lim\limits_{h\to 0} \dfrac{f(h) - f(0)}{h} = f(x)f'(0) = f(x)$

EXERCISE SET 3.3

1. $28x^6$

3. $24x^7 + 2$

5. 0

7. $-\dfrac{1}{3}(7x^6 + 2)$

9. $3ax^2 + 2bx + c$

11. $24x^{-9} + 1/\sqrt{x}$

13. $y = x^{-3} + x^{-7}$ so $\dfrac{dy}{dx} = -3x^{-4} - 7x^{-8}$

15. $\dfrac{dy}{dx} = (3x^2 + 6)\dfrac{d}{dx}\left(2x - \dfrac{1}{4}\right) + \left(2x - \dfrac{1}{4}\right)\dfrac{d}{dx}(3x^2 + 6)$

$\qquad = (3x^2 + 6)(2) + \left(2x - \dfrac{1}{4}\right)(6x) = 18x^2 - \dfrac{3}{2}x + 12$

17. $\dfrac{dy}{dx} = (x^3 + 7x^2 - 8)\dfrac{d}{dx}(2x^{-3} + x^{-4}) + (2x^{-3} + x^{-4})\dfrac{d}{dx}(x^3 + 7x^2 - 8)$

$\qquad = (x^3 + 7x^2 - 8)(-6x^{-4} - 4x^{-5}) + (2x^{-3} + x^{-4})(3x^2 + 14x)$

$\qquad = -15x^{-2} - 14x^{-3} + 48x^{-4} + 32x^{-5}$

19. $12x(3x^2 + 1)$

21. $\dfrac{dy}{dx} = -\dfrac{1}{(5x - 3)^2}\dfrac{d}{dx}(5x - 3) = -\dfrac{5}{(5x - 3)^2}$

23. $\dfrac{dy}{dx} = \dfrac{(2x+1)\dfrac{d}{dx}(3x) - (3x)\dfrac{d}{dx}(2x+1)}{(2x+1)^2} = \dfrac{(2x+1)(3) - (3x)(2)}{(2x+1)^2} = \dfrac{3}{(2x+1)^2}$

25. $\dfrac{dy}{dx} = \dfrac{(x+3)\dfrac{d}{dx}(2x-1) - (2x-1)\dfrac{d}{dx}(x+3)}{(x+3)^2} = \dfrac{(x+3)(2) - (2x-1)(1)}{(x+3)^2} = \dfrac{7}{(x+3)^2}$

27. $\dfrac{dy}{dx} = \left(\dfrac{3x+2}{x}\right)\dfrac{d}{dx}\left(x^{-5}+1\right) + \left(x^{-5}+1\right)\dfrac{d}{dx}\left(\dfrac{3x+2}{x}\right)$

$= \left(\dfrac{3x+2}{x}\right)\left(-5x^{-6}\right) + \left(x^{-5}+1\right)\left[\dfrac{x(3) - (3x+2)(1)}{x^2}\right]$

$= \left(\dfrac{3x+2}{x}\right)\left(-5x^{-6}\right) + \left(x^{-5}+1\right)\left(-\dfrac{2}{x^2}\right)$

29. **(a)** $g'(x) = \sqrt{x}f'(x) + \dfrac{1}{2\sqrt{x}}f(x)$, $g'(4) = (2)(-5) + \dfrac{1}{4}(3) = -37/4$

(b) $g'(x) = \dfrac{xf'(x) - f(x)}{x^2}$, $g'(4) = \dfrac{(4)(-5) - 3}{16} = -23/16$

31. **(a)** $F'(x) = 5f'(x) + 2g'(x)$, $F'(2) = 5(4) + 2(-5) = 10$

(b) $F'(x) = f'(x) - 3g'(x)$, $F'(2) = 4 - 3(-5) = 19$

(c) $F'(x) = f(x)g'(x) + g(x)f'(x)$, $F'(2) = (-3)(-5) + (1)(4) = 19$

(d) $F'(x) = [g(x)f'(x) - f(x)g'(x)]/g^2(x)$, $F'(2) = [(1)(4) - (-3)(-5)]/(1)^2 = -11$

33. $32t$

35. $3\pi r^2$

37. $\dfrac{ds}{dt} = \dfrac{(t^3+7)\dfrac{d}{dt}(t) - t\dfrac{d}{dt}(t^3+7)}{(t^3+7)^2} = \dfrac{(t^3+7)(1) - t(3t^2)}{(t^3+7)^2} = \dfrac{7 - 2t^3}{(t^3+7)^2}$

39. $F = GmMr^{-2}$, $\dfrac{dF}{dr} = -2GmMr^{-3} = -\dfrac{2GmM}{r^3}$

41. **(a)** $dy/dx = 21x^2 - 10x + 1$, $d^2y/dx^2 = 42x - 10$

(b) $dy/dx = 24x - 2$, $d^2y/dx^2 = 24$

(c) $dy/dx = -1/x^2$, $d^2y/dx^2 = 2/x^3$

(d) $y = 35x^5 - 16x^3 - 3x$, $dy/dx = 175x^4 - 48x^2 - 3$, $d^2y/dx^2 = 700x^3 - 96x$

43. **(a)** $y' = -5x^{-6} + 5x^4$, $y'' = 30x^{-7} + 20x^3$, $y''' = -210x^{-8} + 60x^2$

(b) $y = x^{-1}$, $y' = -x^{-2}$, $y'' = 2x^{-3}$, $y''' = -6x^{-4}$

(c) $y' = 3ax^2 + b$, $y'' = 6ax$, $y''' = 6a$

45. (a) $f'(x) = 6x$, $f''(x) = 6$, $f'''(x) = 0$, $f'''(2) = 0$

(b) $\dfrac{dy}{dx} = 30x^4 - 8x$, $\dfrac{d^2y}{dx^2} = 120x^3 - 8$, $\dfrac{d^2y}{dx^2}\Big|_{x=1} = 112$

(c) $\dfrac{d}{dx}\left[x^{-3}\right] = -3x^{-4}$, $\dfrac{d^2}{dx^2}\left[x^{-3}\right] = 12x^{-5}$, $\dfrac{d^3}{dx^3}\left[x^{-3}\right] = -60x^{-6}$, $\dfrac{d^4}{dx^4}\left[x^{-3}\right] = 360x^{-7}$,

$\dfrac{d^4}{dx^4}\left[x^{-3}\right]\Big|_{x=1} = 360$

47. $y' = 3x^2 + 3$, $y'' = 6x$, and $y''' = 6$ so

$y''' + xy'' - 2y' = 6 + x(6x) - 2(3x^2 + 3) = 6 + 6x^2 - 6x^2 - 6 = 0$.

49. $F'(x) = xf'(x) + f(x)$, $F''(x) = xf''(x) + f'(x) + f'(x) = xf''(x) + 2f'(x)$

51. The graph has a horizontal tangent at points where $\dfrac{dy}{dx} = 0$, but

$\dfrac{dy}{dx} = x^2 - 3x + 2 = (x - 1)(x - 2) = 0$ if $x = 1, 2$. The corresponding values of y are $5/6$

and $2/3$ so the tangent line is horizontal at $(1, 5/6)$ and $(2, 2/3)$.

53. $y - 2 = 5(x + 3)$, $y = 5x + 17$.

55. $m_{\tan} = \dfrac{dy}{dx} = 2ax + b$ so $2a + b = 8$ when $x = 1$. Also $5 = a + b$ because $(1, 5)$ is on the curve. Solve the pair of equations $2a + b = 8$ and $a + b = 5$ to get $a = 3$ and $b = 2$.

57. If the y-intercept is -2, then the point $(0, -2)$ is on the graph so $-2 = a(0)^2 + b(0) + c$ so $c = -2$. If the x-intercept is 1, then the point $(1,0)$ is on the graph so $0 = a + b - 2$. The slope is $dy/dx = 2ax + b$; at $x = 0$ the slope is b so $b = -1$, thus $a = 3$. The function is $y = 3x^2 - x - 2$.

59. The points $(-1, 1)$ and $(2, 4)$ are on the secant line so its slope is $(4 - 1)/(2 + 1) = 1$. The slope of the tangent line to $y = x^2$ is $y' = 2x$ so $2x = 1$, $x = 1/2$.

61. $y' = -2x$, so at any point (x_0, y_0) on $y = 1 - x^2$ the tangent line is $y - y_0 = -2x_0(x - x_0)$, or $y = -2x_0x + x_0^2 + 1$. The point $(2, 0)$ is to be on the line, so $0 = -4x_0 + x_0^2 + 1$, $x_0^2 - 4x_0 + 1 = 0$. Use the quadratic formula to get $x_0 = \dfrac{4 \pm \sqrt{16 - 4}}{2} = 2 \pm \sqrt{3}$.

63. $y' = 3ax^2 + b$; the tangent line at $x = x_0$ is $y - y_0 = (3ax_0^2 + b)(x - x_0)$ where $y_0 = ax_0^3 + bx_0$. Solve with $y = ax^3 + bx$ to get

$$(ax^3 + bx) - (ax_0^3 + bx_0) = (3ax_0^2 + b)(x - x_0)$$
$$ax^3 + bx - ax_0^3 - bx_0 = 3ax_0^2 x - 3ax_0^3 + bx - bx_0$$
$$x^3 - 3x_0^2 x + 2x_0^3 = 0$$
$$(x - x_0)(x^2 + 2x_0 - 2x_0^2) = 0$$
$$(x - x_0)^2(x + 2x_0) = 0, \text{ so } x = -2x_0.$$

65. $y' = -\dfrac{1}{x^2}$; the tangent line at $x = x_0$ is $y - y_0 = -\dfrac{1}{x_0^2}(x - x_0)$, or $y = -\dfrac{x}{x_0^2} + \dfrac{2}{x_0}$. The tangent line crosses the x-axis at $2x_0$, the y-axis at $2/x_0$, so that the area of the triangle is

$$\frac{1}{2}(2/x_0)(2x_0) = 2.$$

67. If the graphs of $f(x)$ and $g(x)$ have parallel tangent lines at $x = c$, then $f'(c) = g'(c)$. If $y = f(x) - g(x)$, then $dy/dx = f'(x) - g'(x)$ so at $x = c$, $dy/dx = f'(c) - g'(c) = 0$, hence $y = f(x) - g(x)$ has a horizontal tangent line at $x = c$.

69. **(a)** $2(1 + x^{-1})(x^{-3} + 7) + (2x + 1)(-x^{-2})(x^{-3} + 7) + (2x + 1)(1 + x^{-1})(-3x^{-4})$

(b) $-5x^{-6}(x^2 + 2x)(4 - 3x)(2x^9 + 1) + x^{-5}(2x + 2)(4 - 3x)(2x^9 + 1)$
$$+x^{-5}(x^2 + 2x)(-3)(2x^9 + 1) + x^{-5}(x^2 + 2x)(4 - 3x)(18x^8)$$

(c) $(x^7 + 2x - 3)^3 = (x^7 + 2x - 3)(x^7 + 2x - 3)(x^7 + 2x - 3)$ so

$$\frac{d}{dx}(x^7 + 2x - 3)^3 = (7x^6 + 2)(x^7 + 2x - 3)(x^7 + 2x - 3)$$
$$+(x^7 + 2x - 3)(7x^6 + 2)(x^7 + 2x - 3)$$
$$+(x^7 + 2x - 3)(x^7 + 2x - 3)(7x^6 + 2)$$
$$= 3(7x^6 + 2)(x^7 + 2x - 3)^2$$

(d) $(x^2 + 1)^{50} = (x^2 + 1)(x^2 + 1)\cdots(x^2 + 1)$, where $(x^2 + 1)$ occurs 50 times so

$$\frac{d}{dx}(x^2 + 1)^{50} = [(2x)(x^2 + 1)\cdots(x^2 + 1)] + [(x^2 + 1)(2x)\cdots(x^2 + 1)]$$
$$+\cdots+ [(x^2 + 1)(x^2 + 1)\cdots(2x)]$$
$$= 2x(x^2 + 1)^{49} + 2x(x^2 + 1)^{49} + \cdots + 2x(x^2 + 1)^{49}$$
$$= 100x(x^2 + 1)^{49} \text{ because } 2x(x^2 + 1)^{49} \text{ occurs 50 times.}$$

71. $2(2x^3 - 5x^2 + 7x - 2)(6x^2 - 10x + 7)$

73. f is continuous at 1 because $\lim\limits_{x\to 1^-} f(x) \lim\limits_{x\to 1^+} = f(x) = f(1)$, also $\lim\limits_{x\to 1^-} f'(x) = \lim\limits_{x\to 1^-} 2x = 2$

and $\lim\limits_{x\to 1^+} f'(x) = \lim\limits_{x\to 1^+} \dfrac{1}{2\sqrt{x}} = \dfrac{1}{2}$ so f is not differentiable at 1.

75. If f is differentiable at $x = 1$, then f is continuous there;

$\lim\limits_{x\to 1^+} f(x) = \lim\limits_{x\to 1^-} f(x) = f(1) = 3$, $a + b = 3$; $\lim_{x\to 1^+} f'(x) = a$ and

$\lim\limits_{x\to 1^-} f'(x) = 6$ so $a = 6$ and $b = 3 - 6 = -3$.

77. (a) $f(x) = 3x - 2$ if $x \geq 2/3$, $f(x) = -3x + 2$ if $x < 2/3$ so f is differentiable everywhere except perhaps at $2/3$. f is continuous at $2/3$, also $\lim\limits_{x\to 2/3^-} f'(x) = \lim\limits_{x\to 2/3^-} (-3) = -3$

and $\lim\limits_{x\to 2/3^+} f'(x) = \lim\limits_{x\to 2/3^+} (3) = 3$ so f is not differentiable at $x = 2/3$.

(b) $f(x) = x^2 - 4$ if $|x| \geq 2$, $f(x) = -x^2 + 4$ if $|x| < 2$ so f is differentiable everywhere except perhaps at ± 2. f is continuous at -2 and 2, also $\lim\limits_{x\to 2^-} f'(x) = \lim\limits_{x\to 2^-} (-2x) = -4$

and $\lim\limits_{x\to 2^+} f'(x) = \lim\limits_{x\to 2^+} (2x) = 4$ so f is not differentiable at $x = 2$. Similarly, f is not differentiable at $x = -2$.

79. (a) $f'(x) = nx^{n-1}$, $f''(x) = n(n-1)x^{n-2}$, $f'''(x) = n(n-1)(n-2)x^{n-3}, \ldots,$

$f^{(n)}(x) = n(n-1)(n-2)\cdots 1$.

(b) From part (a), $f^{(k)}(x) = k(k-1)(k-2)\cdots 1$ so $f^{(k+1)}(x) = 0$ thus $f^{(n)}(x) = 0$ if $n > k$.

(c) From parts (a) and (b), $f^{(n)}(x) = a_n n(n-1)(n-2)\cdots 1$.

81. (a) $\dfrac{d^2}{dx^2}[cf(x)] = \dfrac{d}{dx}\left[\dfrac{d}{dx}[cf(x)]\right] = \dfrac{d}{dx}\left[c\dfrac{d}{dx}[f(x)]\right] = c\dfrac{d}{dx}\left[\dfrac{d}{dx}[f(x)]\right] = c\dfrac{d^2}{dx^2}[f(x)]$

$\dfrac{d^2}{dx^2}[f(x) + g(x)] = \dfrac{d}{dx}\left[\dfrac{d}{dx}[f(x) + g(x)]\right] = \dfrac{d}{dx}\left[\dfrac{d}{dx}[f(x)] + \dfrac{d}{dx}[g(x)]\right]$

$$= \dfrac{d^2}{dx^2}[f(x)] + \dfrac{d^2}{dx^2}[g(x)]$$

(b) yes, by repeated application of the procedure illustrated in part (a).

83. (a) The argument is based on the assumption that $\lim\limits_{h\to 0} f(x + h) = f\left(\lim\limits_{h\to 0}(x + h)\right)$ which is not necessarily true.

(b) If $h \neq 0$, then $1 + h \neq 1$ thus $f(1 + h) = 1 + h$ so $\lim\limits_{h\to 0} f(1 + h) = \lim\limits_{h\to 0}(1 + h) = 1$. But $f(1) = 3$ so $\lim\limits_{h\to 0} f(1 + h) \neq f(1)$.

85. **(a)** If a function is differentiable at a point then it is continuous at that point, thus f' is continuous on (a, b) and consequently so is f.

    (b) f and all its derivatives up to $f^{(n-1)}(x)$ are continuous on (a, b).

EXERCISE SET 3.4

1. $f'(x) = -2\sin x - 3\cos x$

3. $f'(x) = \dfrac{x(\cos x) - \sin x(1)}{x^2} = \dfrac{x\cos x - \sin x}{x^2}$

5. $f'(x) = x^3(\cos x) + (\sin x)(3x^2) - 5(-\sin x) = x^3\cos x + (3x^2 + 5)\sin x$

7. $f'(x) = \sec x \tan x - \sqrt{2}\sec^2 x$

9. $f'(x) = \sec x(\sec^2 x) + (\tan x)(\sec x \tan x) = \sec^3 x + \sec x \tan^2 x$

11. $f'(x) = 1 + 4\csc x \cot x - 2\csc^2 x$

13. $f'(x) = \dfrac{(1 + \csc x)(-\csc^2 x) - \cot x(0 - \csc x \cot x)}{(1 + \csc x)^2}$

    $\quad = \dfrac{\csc x(-\csc x - \csc^2 x + \cot^2 x)}{(1 + \csc x)^2}$

    but $1 + \cot^2 x = \csc^2 x$ (identity) thus $\cot^2 x - \csc^2 x = -1$ so

    $f'(x) = \dfrac{\csc x(-\csc x - 1)}{(1 + \csc x)^2} = -\dfrac{\csc x}{1 + \csc x}$

15. $f(x) = \sin^2 x + \cos^2 x = 1$ (identity) so $f'(x) = 0$

17. $f(x) = \dfrac{\tan x}{1 + x\tan x}$ (because $\sin x \sec x = (\sin x)(1/\cos x) = \tan x$),

    $f'(x) = \dfrac{(1 + x\tan x)(\sec^2 x) - \tan x[x(\sec^2 x) + (\tan x)(1)]}{(1 + x\tan x)^2}$

    $\quad = \dfrac{\sec^2 x - \tan^2 x}{(1 + x\tan x)^2} = \dfrac{1}{(1 + x\tan x)^2}$ (because $\sec^2 x - \tan^2 x = 1$)

19. $dy/dx = -x\sin x + \cos x$,

    $d^2y/dx^2 = -x\cos x - \sin x - \sin x = -x\cos x - 2\sin x$

21. $dy/dx = x(\cos x) + (\sin x)(1) - 3(-\sin x) = x\cos x + 4\sin x,$
$d^2y/dx^2 = x(-\sin x) + (\cos x)(1) + 4\cos x = -x\sin x + 5\cos x$

23. $dy/dx = (\sin x)(-\sin x) + (\cos x)(\cos x) = \cos^2 x - \sin^2 x,$
$d^2y/dx^2 = (\cos x)(-\sin x) + (\cos x)(-\sin x) - [(\sin x)(\cos x) + (\sin x)(\cos x)]$
$\quad = -4\sin x\cos x$

25. (a) $f'(x) = -\sin x;\ f'(x) = 0$ when $\sin x = 0$ so $x = n\pi,\ n = 0, \pm 1, \pm 2, \cdots$.
(b) $f'(x) = -\csc^2 x;\ f'(x) = 0$ when $\csc x = 0$, but $\csc x = 0$ has no solutions.
(c) $f'(x) = -\csc x\cot x;\ f'(x) = 0$ when $\cot x = 0$ so $x = \dfrac{\pi}{2} + n\pi,\ n = 0, \pm 1, \pm 2, \cdots$.

27. Let $f(x) = \tan x$, then $f'(x) = \sec^2 x$.

(a) $f(0) = 0$ and $f'(0) = 1$ so $y - 0 = (1)(x - 0),\ y = x.$
(b) $f\left(\dfrac{\pi}{4}\right) = 1$ and $f'\left(\dfrac{\pi}{4}\right) = 2$ so $y - 1 = 2\left(x - \dfrac{\pi}{4}\right),\ y = 2x - \dfrac{\pi}{2} + 1.$
(c) $f\left(-\dfrac{\pi}{4}\right) = -1$ and $f'\left(-\dfrac{\pi}{4}\right) = 2$ so $y + 1 = 2\left(x + \dfrac{\pi}{4}\right),\ y = 2x + \dfrac{\pi}{2} - 1.$

29. $x = 10\sin\theta,\ dx/d\theta = 10\cos\theta$; if $\theta = 60°$, then
$dx/d\theta = 10(1/2) = 5\ \text{ft/rad} = \pi/36\ \text{ft/}° \approx 0.087\ \text{ft/}°.$

31. $D = 50\tan\theta,\ dD/d\theta = 50\sec^2\theta$; if $\theta = 30°$, then
$dD/d\theta = 50(\sqrt{2})^2 = 100\ \text{m/rad} = 5\pi/9\ \text{m/}° \approx 1.75\ \text{m/}°.$

33. In each part, f is differentiable throughout its domain because f' can be determined there by use of the various derivative formulas that have been presented in the text; f is not differentiable elsewhere.

(a) all x
(b) all x
(c) $x \neq \pi/2 + n\pi,\ n = 0, \pm 1, \pm 2, \cdots$
(d) $x \neq n\pi,\ n = 0, \pm 1, \pm 2, \cdots$
(e) $x \neq \pi/2 + n\pi,\ n = 0, \pm 1, \pm 2, \cdots$
(f) $x \neq n\pi,\ n = 0, \pm 1, \pm 2, \cdots$
(g) $x \neq \pi + 2n\pi,\ n = 0, \pm 1, \pm 2, \cdots$
(h) $x \neq n\pi/2,\ n = 0, \pm 1, \pm 2, \cdots$
(i) all x

35. $f'(x) = -\sin x,\ f''(x) = -\cos x,\ f'''(x) = \sin x$, and $f^{(4)}(x) = \cos x$ with higher order derivatives repeating this pattern, so $f^{(n)}(x) = \sin x$ for $n = 3, 7, 11, \cdots$

37. $\displaystyle\lim_{x\to 0}\frac{\tan(x+y) - \tan y}{x} = \lim_{h\to 0}\frac{\tan(y+h) - \tan y}{h} = \frac{d}{dy}(\tan y) = \sec^2 y$

39. Let t be the radian measure, then $h = \dfrac{180}{\pi} t$ and $\cos h = \cos t$, $\sin h = \sin t$.

(a) $\displaystyle\lim_{h \to 0} \frac{\cos h - 1}{h} = \lim_{t \to 0} \frac{\cos t - 1}{180 t/\pi} = \frac{\pi}{180} \lim_{t \to 0} \frac{\cos t - 1}{t} = 0.$

(b) $\displaystyle\lim_{h \to 0} \frac{\sin h}{h} = \lim_{t \to 0} \frac{\sin t}{180 t/\pi} = \frac{\pi}{180} \lim_{t \to 0} \frac{\sin t}{t} = \frac{\pi}{180}.$

(c) $\displaystyle\frac{d}{dx}[\sin x] = \sin x \lim_{h \to 0} \frac{\cos h - 1}{h} + \cos x \lim_{h \to 0} \frac{\sin h}{h}$

$$= \sin x (0) + \cos x (\pi/180) = \frac{\pi}{180} \cos x.$$

EXERCISE SET 3.5

1. $f'(x) = 37(x^3 + 2x)^{36} \dfrac{d}{dx}(x^3 + 2x) = 37(x^3 + 2x)^{36}(3x^2 + 2)$

3. $f'(x) = -2\left(x^3 - \dfrac{7}{x}\right)^{-3} \dfrac{d}{dx}\left(x^3 - \dfrac{7}{x}\right) = -2\left(x^3 - \dfrac{7}{x}\right)^{-3}\left(3x^2 + \dfrac{7}{x^2}\right)$

5. $f(x) = 4(3x^2 - 2x + 1)^{-3},$

$\qquad f'(x) = -12(3x^2 - 2x + 1)^{-4} \dfrac{d}{dx}(3x^2 - 2x + 1)$

$\qquad\qquad = -12(3x^2 - 2x + 1)^{-4}(6x - 2) = \dfrac{24(1 - 3x)}{(3x^2 - 2x + 1)^4}$

7. $f'(x) = \dfrac{1}{2\sqrt{4 + 3\sqrt{x}}} \dfrac{d}{dx}(4 + 3\sqrt{x}) = \dfrac{3}{4\sqrt{x}\sqrt{4 + 3\sqrt{x}}}$

9. $f'(x) = \cos(x^3) \dfrac{d}{dx}(x^3) = 3x^2 \cos(x^3)$

11. $f'(x) = \sec^2(4x^2) \dfrac{d}{dx}(4x^2) = 8x \sec^2(4x^2)$

13. $f'(x) = 20 \cos^4 x \dfrac{d}{dx}(\cos x) = 20 \cos^4 x(-\sin x) = -20 \cos^4 x \sin x$

15. $f'(x) = \cos(1/x^2) \dfrac{d}{dx}(1/x^2) = -\dfrac{2}{x^3} \cos(1/x^2)$

17. $f'(x) = 4 \sec(x^7) \dfrac{d}{dx}[\sec(x^7)] = 4 \sec(x^7) \sec(x^7) \tan(x^7) \dfrac{d}{dx}(x^7) = 28x^6 \sec^2(x^7) \tan(x^7)$

19. $f'(x) = \dfrac{1}{2\sqrt{\cos(5x)}} \dfrac{d}{dx}[\cos(5x)] = -\dfrac{5\sin(5x)}{2\sqrt{\cos(5x)}}$

21. $f'(x) = -3\left[x + \csc(x^3 + 3)\right]^{-4} \dfrac{d}{dx}\left[x + \csc(x^3 + 3)\right]$

$\qquad = -3\left[x + \csc(x^3 + 3)\right]^{-4}\left[1 - \csc(x^3 + 3)\cot(x^3 + 3)\dfrac{d}{dx}(x^3 + 3)\right]$

$\qquad = -3\left[x + \csc(x^3 + 3)\right]^{-4}\left[1 - 3x^2\csc(x^3 + 3)\cot(x^3 + 3)\right]$

23. $f'(x) = x^2 \cdot \dfrac{-2x}{2\sqrt{5 - x^2}} + 2x\sqrt{5 - x^2} = \dfrac{x(10 - 3x^2)}{\sqrt{5 - x^2}}$

25. $f'(x) = x^3(2\sin 5x)\dfrac{d}{dx}(\sin 5x) + 3x^2\sin^2 5x = 10x^3\sin 5x\cos 5x + 3x^2\sin^2 5x$

27. $f'(x) = x^5\sec\left(\dfrac{1}{x}\right)\tan\left(\dfrac{1}{x}\right)\dfrac{d}{dx}\left(\dfrac{1}{x}\right) + \sec\left(\dfrac{1}{x}\right)(5x^4)$

$\qquad = x^5\sec\left(\dfrac{1}{x}\right)\tan\left(\dfrac{1}{x}\right)\left(-\dfrac{1}{x^2}\right) + 5x^4\sec\left(\dfrac{1}{x}\right)$

$\qquad = -x^3\sec\left(\dfrac{1}{x}\right)\tan\left(\dfrac{1}{x}\right) + 5x^4\sec\left(\dfrac{1}{x}\right)$

29. $f'(x) = -\sin(\cos x)\dfrac{d}{dx}(\cos x) = -\sin(\cos x)(-\sin x) = \sin(\cos x)\sin x$

31. $f'(x) = 3\cos^2(\sin 2x)\dfrac{d}{dx}[\cos(\sin 2x)]$

$\qquad = 3\cos^2(\sin 2x)[-\sin(\sin 2x)]\dfrac{d}{dx}(\sin 2x)$

$\qquad = -6\cos^2(\sin 2x)\sin(\sin 2x)\cos 2x$

33. $f'(x) = (5x + 8)^{13}12(x^3 + 7x)^{11}\dfrac{d}{dx}(x^3 + 7x) + (x^3 + 7x)^{12}13(5x + 8)^{12}\dfrac{d}{dx}(5x + 8)$

$\qquad = 12(5x + 8)^{13}(x^3 + 7x)^{11}(3x^2 + 7) + 65(x^3 + 7x)^{12}(5x + 8)^{12}$

35. $f'(x) = 3\left[\dfrac{x - 5}{2x + 1}\right]^2\dfrac{d}{dx}\left[\dfrac{x - 5}{2x + 1}\right] = 3\left[\dfrac{x - 5}{2x + 1}\right]^2 \cdot \dfrac{11}{(2x + 1)^2} = \dfrac{33(x - 5)^2}{(2x + 1)^4}$

37. $f'(x) = \dfrac{(4x^2-1)^8(3)(2x+3)^2)(2) - (2x+3)^3(8)(4x^2-1)^7(8x)}{(4x^2-1)^{16}}$

$= \dfrac{2(2x+3)^2(4x^2-1)^7[3(4x^2-1) - 32x(2x+3)]}{(4x^2-1)^{16}}$

$= -\dfrac{2(2x+3)^2(52x^2+96x+3)}{(4x^2-1)^9}$

39. $f'(x) = 5\left[x\sin 2x \tan^4(x^7)\right]^4 \dfrac{d}{dx}\left[x\sin 2x \tan^4(x^7)\right]$

$= 5\left[x\sin 2x \tan^4(x^7)\right]^4 \left[x\cos 2x \dfrac{d}{dx}(2x) + \sin 2x + 4\tan^3(x^7)\dfrac{d}{dx}\tan(x^7)\right]$

$= 5\left[x\sin 2x \tan^4(x^7)\right]^4 \left[2x\cos 2x + \sin 2x + 28x^6 \tan^3(x^7)\sec^2(x^7)\right]$

41. $\dfrac{dy}{dx} = x(-\sin(5x))\dfrac{d}{dx}(5x) + \cos(5x) - 2\sin x \dfrac{d}{dx}(\sin x)$

$= -5x\sin(5x) + \cos(5x) - 2\sin x \cos x = -5x\sin(5x) + \cos(5x) - \sin(2x),$

$\dfrac{d^2y}{dx^2} = -5x\cos(5x)\dfrac{d}{dx}(5x) - 5\sin(5x) - \sin(5x)\dfrac{d}{dx}(5x) - \cos(2x)\dfrac{d}{dx}(2x)$

$= -25x\cos(5x) - 10\sin(5x) - 2\cos(2x)$

43. $\dfrac{dy}{dx} = -3x\sin 3x + \cos 3x$; if $x = \pi$ then $y = -\pi$ and $\dfrac{dy}{dx} = -1$ so $y + \pi = -(x - \pi)$, $y = -x$.

45. $\dfrac{dy}{dx} = -3\sec^3(\pi/2 - x)\tan(\pi/2 - x)$; if $x = -\pi/2$ then $y = -1$ and $\dfrac{dy}{dx} = 0$
so $y + 1 = (0)(x + \pi/2)$, $y = -1$.

47. $y = \cot^3(\pi - \theta) = -\cot^3\theta$ so $dy/dx = 3\cot^2\theta\csc^2\theta$.

49. $\dfrac{d}{d\omega}[a\cos^2\pi\omega + b\sin^2\pi\omega] = -2\pi a\cos\pi\omega\sin\pi\omega + 2\pi b\sin\pi\omega\cos\pi\omega$

$= \pi(b-a)(2\sin\pi\omega\cos\pi\omega) = \pi(b-a)\sin 2\pi\omega$

51. **(a)** $dy/dt = -A\omega\sin\omega t, d^2y/dt^2 = -A\omega^2\cos\omega t = -\omega^2 y$
(b) One complete oscillation occurs when ωt increases over an interval of length 2π, or if t increases over an interval of length $2\pi/\omega$.
(c) $f = 1/T$
(d) amplitude $= 0.6$ cm, $\quad T = 2\pi/15$ sec/oscillation, $\quad f = 15/(2\pi)$ oscillations/sec.

53. **(a)** $p \approx 10$ lb/in^2, $dp/dh \approx -2$ lb/in^2 per mi
(b) $\dfrac{dp}{dt} = \dfrac{dp}{dh}\dfrac{dh}{dt} \approx (-2)(0.3) = -0.6$ lb/in^2 per sec

55. $3x^2y^2\dfrac{dy}{dx} + 2xy^3$

57. $\left(x\dfrac{dy}{dx} + y\right)\cos(xy)$

59. $2x\dfrac{dx}{dt} + 2y\dfrac{dy}{dt}$

61. $\dfrac{x^2}{2\sqrt{y}}\dfrac{dy}{dt} + 2x\sqrt{y}\dfrac{dx}{dt}$

63. With $u = \sin x$, $\dfrac{d}{dx}(|\sin x|) = \dfrac{d}{dx}(|u|) = \dfrac{d}{du}(|u|)\dfrac{du}{dx} = \dfrac{d}{du}(|u|)\cos x = \begin{cases} \cos x, & u > 0 \\ -\cos x, & u < 0 \end{cases}$

$= \begin{cases} \cos x, & \sin x > 0 \\ -\cos x, & \sin x < 0 \end{cases} = \begin{cases} \cos x, & 0 < x < \pi \\ -\cos x, & -\pi < x < 0 \end{cases}$.

65. **(a)** For $x \neq 0$, $f'(x) = x\left(\cos\dfrac{1}{x}\right)\left(-\dfrac{1}{x^2}\right) + \sin\dfrac{1}{x} = -\dfrac{1}{x}\cos\dfrac{1}{x} + \sin\dfrac{1}{x}$

(b) $\lim\limits_{x\to 0} x\sin\dfrac{1}{x} = 0 = f(0)$

(c) $\lim\limits_{h\to 0}\dfrac{f(0+h) - f(0)}{h} = \lim\limits_{h\to 0}\dfrac{h\sin\dfrac{1}{h}}{h} = \lim\limits_{h\to 0}\sin\dfrac{1}{h}$, which does not exist

67. **(a)** $g'(x) = 3[f(x)]^2 f'(x)$, $g'(2) = 3[f(2)]^2 f'(2) = 3(1)^2(7) = 21$

(b) $h'(x) = f'(x^3)(3x^2)$, $h'(2) = f'(8)(12) = (-3)(12) = -36$

69. $(f \circ g)'(x) = f'(g(x))g'(x)$ so $(f \circ g)'(0) = f'(g(0))g'(0) = f'(0)(3) = (2)(3) = 6.$

71. $F'(x) = f'(g(x))g'(x) = f'(\sqrt{3x-1})\dfrac{3}{2\sqrt{3x-1}} = \dfrac{\sqrt{3x-1}}{(3x-1)+1}\dfrac{3}{2\sqrt{3x-1}} = \dfrac{1}{2x}$

73. $\dfrac{d}{dx}[f(3x)] = f'(3x)\dfrac{d}{dx}(3x) = 3f'(3x) = 6x$, so $f'(3x) = 2x$. Let $u = 3x$ to get $f'(u) = \dfrac{2}{3}u$;

$\dfrac{d}{dx}[f(x)] = f'(x) = \dfrac{2}{3}x.$

75. $\dfrac{d}{dx}[f(g(h(x)))] = \dfrac{d}{dx}[f(g(u))]$, $\quad u = h(x)$

$= \dfrac{d}{du}[f(g(u))]\dfrac{du}{dx} = f'(g(u))g'(u)\dfrac{du}{dx} = f'(g(h(x)))g'(h(x))h'(x)$

EXERCISE SET 3.6

1. $y = (2x - 5)^{1/3}$; $dy/dx = \dfrac{2}{3}(2x - 5)^{-2/3}$

3. $dy/dx = \dfrac{3}{2}\left[\dfrac{x-1}{x+2}\right]^{1/2}\dfrac{d}{dx}\left[\dfrac{x-1}{x+2}\right] = \dfrac{9}{2(x+2)^2}\left[\dfrac{x-1}{x+2}\right]^{1/2}$

5. $dy/dx = x^3\left(-\dfrac{2}{3}\right)(5x^2+1)^{-5/3}(10x) + 3x^2(5x^2+1)^{-2/3} = \dfrac{1}{3}x^2(5x^2+1)^{-5/3}(25x^2+9)$

7. $dy/dx = \dfrac{5}{2}[\sin(3/x)]^{3/2}[\cos(3/x)](-3/x^2) = -\dfrac{15[\sin(3/x)]^{3/2}\cos(3/x)}{2x^2}$

9. $dy/dx = \sec^2\left[(2x-1)^{-1/3}\right]\left(-\dfrac{1}{3}\right)(2x-1)^{-4/3}(2) = -\dfrac{2}{3}(2x-1)^{-4/3}\sec^2\left[(2x-1)^{-1/3}\right]$

11. $y' = rx^{r-1}, y'' = r(r-1)x^{r-2}$ so $3x^2\left[r(r-1)x^{r-2}\right] + 4x\left(rx^{r-1}\right) - 2x^r = 0$,
$3r(r-1)x^r + 4rx^r - 2x^r = 0, (3r^2 + r - 2)x^r = 0$,
$3r^2 + r - 2 = 0, (3r-2)(r+1) = 0; r = -1, 2/3$.

13. $2x + 2y\dfrac{dy}{dx} = 0$ so $\dfrac{dy}{dx} = -\dfrac{x}{y}$

15. $x^2\dfrac{dy}{dx} + 2xy + 3x(3y^2)\dfrac{dy}{dx} + 3y^3 - 1 = 0$

$(x^2 + 9xy^2)\dfrac{dy}{dx} = 1 - 2xy - 3y^3$ so $\dfrac{dy}{dx} = \dfrac{1 - 2xy - 3y^3}{x^2 + 9xy^2}$

17. $-\dfrac{1}{y^2}\dfrac{dy}{dx} - \dfrac{1}{x^2} = 0$ so $\dfrac{dy}{dx} = -\dfrac{y^2}{x^2}$

19. $\dfrac{1}{2\sqrt{x}} + \dfrac{1}{2\sqrt{y}}\dfrac{dy}{dx} = 0$ so $\dfrac{dy}{dx} = -\dfrac{\sqrt{y}}{\sqrt{x}}$

21. $35\left(x^2 + 3y^2\right)^{34}\left(2x + 6y\dfrac{dy}{dx}\right) = 1$,

$70x(x^2 + 3y^2)^{34} + 210y(x^2 + 3y^2)^{34}\dfrac{dy}{dx} = 1$ so $\dfrac{dy}{dx} = \dfrac{1 - 70x(x^2 + 3y^2)^{34}}{210y(x^2 + 3y^2)^{34}}$

23. $3x\dfrac{dy}{dx} + 3y = \dfrac{3}{2}\left(x^3 + y^2\right)^{1/2}\left(3x^2 + 2y\dfrac{dy}{dx}\right)$,

$\left[3x - 3y(x^3 + y^2)^{1/2}\right]\dfrac{dy}{dx} = \dfrac{9}{2}x^2(x^3 + y^2)^{1/2} - 3y$ so $\dfrac{dy}{dx} = \dfrac{(3/2)x^2(x^3 + y^2)^{1/2} - y}{x - y(x^3 + y^2)^{1/2}}$

25. $\cos(x^2y^2)\left[x^2(2y)\dfrac{dy}{dx} + 2xy^2\right] = 1$, $\dfrac{dy}{dx} = \dfrac{1 - 2xy^2\cos(x^2y^2)}{2x^2y\cos(x^2y^2)}$

27. $3\tan^2(xy^2+y)\sec^2(xy^2+y)\left(2xy\dfrac{dy}{dx}+y^2+\dfrac{dy}{dx}\right)=1$

so $\dfrac{dy}{dx}=\dfrac{1-3y^2\tan^2(xy^2+y)\sec^2(xy^2+y)}{3(2xy+1)\tan^2(xy^2+y)\sec^2(xy^2+y)}$

29. $\dfrac{1}{2}\left[1+\sin^3(xy^2)\right]^{-1/2}\left[3\sin^2(xy^2)\right]\left[\cos(xy^2)\right]\left(2xy\dfrac{dy}{dx}+y^2\right)=\dfrac{dy}{dx}$,

multiply through by $2\sqrt{1+\sin^3(xy^2)}$ and solve for $\dfrac{dy}{dx}$

to get $\dfrac{dy}{dx}=\dfrac{3y^2\sin^2(xy^2)\cos(xy^2)}{2\sqrt{1+\sin^3(xy^2)}-6xy\sin^2(xy^2)\cos(xy^2)}$

31. $\dfrac{dy}{dx}=-\dfrac{3x^2y+y^3}{x^3+3y^2x};\ \dfrac{dy}{dx}\bigg|_{(1,2)}=-\dfrac{14}{13}$ **33.** $\dfrac{dy}{dx}=\dfrac{2y^{1/3}}{2x^{1/3}+3x^{1/3}y^{1/3}};\ \dfrac{dy}{dx}\bigg|_{(1,-1)}=2$

35. If $xy=8$, then $y=8/x$ so $dy/dx=-8/x^2$. By implicit differentiation we get $dy/dx=-y/x$. In both cases, $dy/dx\big|_{(2,4)}=-2$.

37. If $x^2+y^2=1$, then $y=-\sqrt{1-x^2}$ goes through the point $(1/\sqrt{2},-1/\sqrt{2})$ so $dy/dx=x/\sqrt{1-x^2}$. By implicit differentiation we get $dy/dx=-x/y$. In both cases, $dy/dx\big|_{(1/\sqrt{2},-1/\sqrt{2})}=1$.

39. Apply the quadratic formula to $y^2-3xy+2x^2-4=0$ to get $y=(3x\pm\sqrt{9x^2-4(2x^2-4)})/2=(3x\pm\sqrt{x^2+16})/2$, of which only $y=(3x-\sqrt{x^2+16})/2$ contains the point $(3,2)$ so $dy/dx=[3-x(x^2+16)^{-1/2}]/2$. By implicit differentiation we get $dy/dx=(3y-4x)/(2y-3x)$. In both cases, $dy/dx\big|_{(3,2)}=6/5$.

41. $\dfrac{dy}{dx}=-\dfrac{x^2}{y^2},\ \dfrac{d^2y}{dx^2}=-\dfrac{y^2(2x)-x^2(2y\,dy/dx)}{y^4}=-\dfrac{2xy^2-2x^2y(-x^2/y^2)}{y^4}=-\dfrac{2x(y^3+x^3)}{y^5}$,

but $x^3+y^3=1$ so $\dfrac{d^2y}{dx^2}=-\dfrac{2x}{y^5}$.

43. $\dfrac{dy}{dx}=\dfrac{y}{y-x}$,

$\dfrac{d^2y}{dx^2}=\dfrac{(y-x)(dy/dx)-y(dy/dx-1)}{(y-x)^2}=\dfrac{(y-x)\left(\dfrac{y}{y-x}\right)-y\left(\dfrac{y}{y-x}-1\right)}{(y-x)^2}$

$=\dfrac{y^2-2xy}{(y-x)^3}$ but $y^2-2xy=-3$, so $\dfrac{d^2y}{dx^2}=-\dfrac{3}{(y-x)^3}$

45. $\dfrac{dy}{dx} = \dfrac{\cos y}{1 + x \sin y}$,

$\dfrac{d^2 y}{dx^2} = \dfrac{(1 + x \sin y)(-\sin y)(dy/dx) - (\cos y)[(x \cos y)(dy/dx) + \sin y]}{(1 + x \sin y)^2}$

$= -\dfrac{2 \sin y \cos y + (x \cos y)(2 \sin^2 y + \cos^2 y)}{(1 + x \sin y)^3}$,

but $x \cos y = y$, $2 \sin y \cos y = \sin 2y$, and $\sin^2 y + \cos^2 y = 1$ so

$\dfrac{d^2 y}{dx^2} = -\dfrac{\sin 2y + y(\sin^2 y + 1)}{(1 + x \sin y)^3}$.

47. $4a^3 \dfrac{da}{dt} - 4t^3 = 6\left(a^2 + 2at\dfrac{da}{dt}\right)$, solve for $\dfrac{da}{dt}$ to get $\dfrac{da}{dt} = \dfrac{2t^3 + 3a^2}{2a^3 - 6at}$

49. $2a^2 \omega \dfrac{d\omega}{d\lambda} + 2b^2 \lambda = 0$ so $\dfrac{d\omega}{d\lambda} = -\dfrac{b^2 \lambda}{a^2 \omega}$

51. By the chain rule, $\dfrac{dy}{dx} = \dfrac{dy}{dt}\dfrac{dt}{dx}$. Use implicit differentiation on $2y^3 t + t^3 y = 1$ to get

$\dfrac{dy}{dt} = -\dfrac{2y^3 + 3t^2 y}{6ty^2 + t^3}$, but $\dfrac{dt}{dx} = \dfrac{1}{\cos t}$ so $\dfrac{dy}{dx} = -\dfrac{2y^3 + 3t^2 y}{(6ty^2 + t^3)\cos t}$.

53. $2xy \dfrac{dy}{dt} = y^2 \dfrac{dx}{dt} = 3(\cos 3x)\dfrac{dx}{dt}, \dfrac{dy}{dt} = \dfrac{3\cos 3x - y^2}{2xy}\dfrac{dx}{dt}$

55. The point $(1,1)$ is on the graph, so $1 + a = b$. The slope of the tangent line at $(1,1)$ is $-4/3$; use implicit differentiation to get $\dfrac{dy}{dx} = -\dfrac{2xy}{x^2 + 2ay}$ so at $(1,1)$, $-\dfrac{2}{1 + 2a} = -\dfrac{4}{3}$, $1 + 2a = 3/2$, $a = 1/4$ and hence $b = 1 + 1/4 = 5/4$.

57. Let $P(x_0, y_0)$ be a point where a line through the orgin is tangent to the curve $x^2 - 4x + y^2 + 3 = 0$. Implicit differentiation applied to the equation of the curve gives $dy/dx = (2 - x)/y$. At P the slope of the curve must equal the slope of the line so $(2 - x_0)/y_0 = y_0/x_0$, or $y_0^2 = 2x_0 - x_0^2$. But $x_0^2 - 4x_0 + y_0^2 + 3 = 0$ because (x_0, y_0) is on the curve, and elimination of y_0^2 in the latter two equations gives $x_0^2 - 4x_0 + (2x_0 - x_0^2) + 3 = 0$, $x_0 = 3/2$ which when substituted into $y_0^2 = 2x_0 - x_0^2$ yields $y_0^2 = 3/4$, so $y_0 = \pm\sqrt{3}/2$. The slopes of the lines are $(\pm\sqrt{3}/2)/(3/2) = \pm\sqrt{3}/3$ and their equations are $y = (\sqrt{3}/3)x$ and $y = -(\sqrt{3}/3)x$.

59. **(a)** The equation does not define y as an implicit function of x at the points where the curve crosses the x-axis. If $y = 0$ then $8x^4 = 100x^2$, $8x^2(x^2 - 25/2) = 0$ so $x = 0, \pm 5/\sqrt{2}$. The points are $(0,0)$, $(-5/\sqrt{2}, 0)$, and $(5/\sqrt{2}, 0)$.

(b) $16(x^2 + y^2)\left(2x + 2y\dfrac{dy}{dx}\right) = 100\left(2x - 2y\dfrac{dy}{dx}\right)$,

$\dfrac{dy}{dx} = \dfrac{x[25 - 4(x^2 + y^2)]}{y[25 + 4(x^2 + y^2)]}$; at $(3,1)$ $\dfrac{dy}{dx} = -9/13$ so the equation of the tangent line is

$y - 1 = (-9/13)(x - 3)$, $9x + 13y = 40$.

EXERCISE SET 3.7

1. **(a)** $\Delta y = (x + \Delta x)^2 - x^2$
$= (2 + 1)^2 - 2^2$
$= 9 - 4 = 5$

(b) $dy = 2x\,dx$
$= 2(2)(1) = 4$

(c)

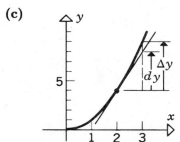

3. **(a)** $\Delta y = \dfrac{1}{x + \Delta x} - \dfrac{1}{x}$

$= \dfrac{1}{1 + 0.5} - \dfrac{1}{1}$

$= -\dfrac{1}{3}$

(b) $dy = -\dfrac{1}{x^2}\,dx$

$= -\dfrac{1}{1^2}(0.5) = -0.5$

(c)

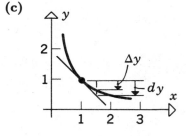

5. $dy = 3x^2\,dx$;
$\Delta y = (x + \Delta x)^3 - x^3 = x^3 3x^2\Delta x + 3x(\Delta x)^2 + (\Delta x)^3 - x^3 = 3x^2\Delta x + 3x(\Delta x)^2 + (\Delta x)^3$

7. $dy = (2x - 2)\,dx$;
$\Delta y = [(x + \Delta x)^2 - 2(x + \Delta x) + 1] - [x^2 - 2x + 1]$
$= x^2 + 2x\,\Delta x + (\Delta x)^2 - 2x - 2\Delta x + 1 - x^2 + 2x - 1 = 2x\,\Delta x + (\Delta x)^2 - 2\Delta x$

9. $dy = (12x^2 - 14x + 2)\,dx$

11. $dy = x\,d(\cos x) + \cos x\,dx = x(-\sin x)\,dx + \cos x\,dx = (-x\sin x + \cos x)\,dx$

13. $\displaystyle\lim_{\Delta x \to 0} \frac{(x + \Delta x)^2 - x^2}{\Delta x} = \frac{d}{dx}(x^2) = 2x$

15. $\displaystyle\lim_{\Delta x \to 0} \frac{\sin(\pi + \Delta x) - \sin \pi}{\Delta x} = \frac{d}{dx}(\sin x)\Big|_{x=\pi} = \cos \pi = -1$

17. $f(x) = x^4$, $f'(x) = 4x^3$, $x_0 = 3$, $\Delta x = 0.02$; $(3.02)^4 \approx 3^4 + (108)(0.02) = 81 + 2.16 = 83.16$.

19. $f(x) = \sqrt{x}$, $f'(x) = \dfrac{1}{2\sqrt{x}}$, $x_0 = 64$, $\Delta x = 1$; $\sqrt{65} \approx \sqrt{64} + \dfrac{1}{16}(1) = 8 + \dfrac{1}{16} = 8.0625$.

21. $f(x) = \sqrt{x}$, $f'(x) = \dfrac{1}{2\sqrt{x}}$, $x_0 = 81$, $\Delta x = -0.1$; $\sqrt{80.9} \approx \sqrt{81} + \dfrac{1}{18}(-0.1) \approx 8.9944$.

23. $f(x) = \sqrt[3]{x}$, $f'(x) = \dfrac{1}{3}x^{-2/3}$, $x_0 = 8$, $\Delta x = 0.06$; $\sqrt[3]{8.06} \approx \sqrt[3]{8} + \dfrac{1}{12}(0.06) = 2.005$.

25. $f(x) = \cos x$, $f'(x) = -\sin x$, $x_0 = \pi/6$, $\Delta x = \pi/180$;

$\cos 31° \approx \cos 30° + \left(-\dfrac{1}{2}\right)\left(\dfrac{\pi}{180}\right) = \dfrac{\sqrt{3}}{2} - \dfrac{\pi}{360} \approx 0.8573$.

27. $f(x) = \sin x$, $f'(x) = \cos x$, $x_0 = \pi/4$, $\Delta x = -\pi/180$;

$\sin 44° \approx \sin 45° + \dfrac{1}{\sqrt{2}}\left(-\dfrac{\pi}{180}\right) = \dfrac{1}{\sqrt{2}} - \dfrac{\pi}{180\sqrt{2}} \approx 0.6947$.

29. $dy = \dfrac{3}{2\sqrt{3x - 2}}dx$, $x = 2$, $dx = 0.03$; $\Delta y \approx dy = \dfrac{3}{4}(0.03) = 0.0225$.

31. $dy = \dfrac{1 - x^2}{(x^2 + 1)^2}dx$, $x = 2$, $dx = -0.04$; $\Delta y \approx dy = \left(-\dfrac{3}{25}\right)(-0.04) = 0.0048$.

33. **(a)** $A = x^2$ where x is the length of a side; $dA = 2x\,dx = 2(10)(\pm 0.1) = \pm 2$ ft^2.

(b) relative error in $x \approx \dfrac{dx}{x} = \dfrac{\pm 0.1}{10} = \pm 0.01$ so percentage error in $x \approx \pm 1\%$; relative error

in $A \approx \dfrac{dA}{A} = \dfrac{2x\,dx}{x^2} = 2\dfrac{dx}{x} = 2(\pm 0.01) = \pm 0.02$ so percentage error in $A \approx \pm 2\%$.

35. **(a)** $x = 10\sin\theta$, $y = 10\cos\theta$ (see figure),

$dx = 10\cos\theta\,d\theta$

$= 10\left(\cos\dfrac{\pi}{6}\right)\left(\pm\dfrac{\pi}{180}\right)$

$= 10\left(\dfrac{\sqrt{3}}{2}\right)\left(\pm\dfrac{\pi}{180}\right) \approx \pm 0.151''$,

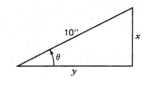

$$dy = -10(\sin\theta)d\theta = -10\left(\sin\frac{\pi}{6}\right)\left(\pm\frac{\pi}{180}\right) = -10\left(\frac{1}{2}\right)\left(\pm\frac{\pi}{180}\right) \approx \pm 0.087''.$$

(b) relative error in $x \approx \dfrac{dx}{x} = (\cot\theta)d\theta = \left(\cot\dfrac{\pi}{6}\right)\left(\pm\dfrac{\pi}{180}\right) = \sqrt{3}\left(\pm\dfrac{\pi}{180}\right) \approx \pm 0.030$

so percentage error in $x \approx \pm 3.0\%$;

relative error in $y \approx \dfrac{dy}{y} = -\tan\theta d\theta = -\left(\tan\dfrac{\pi}{6}\right)\left(\pm\dfrac{\pi}{180}\right) = -\dfrac{1}{\sqrt{3}}\left(\pm\dfrac{\pi}{180}\right) \approx \pm 0.010$

so percentage error in $y \approx \pm 1.0\%$.

37. $\dfrac{dR}{R} = \dfrac{(-2k/r^3)dr}{(k/r^2)} = -2\dfrac{dr}{r}$, but $\dfrac{dr}{r} \approx \pm 0.05$ so $\dfrac{dR}{R} \approx -2(\pm 0.05) = \pm 0.10$; percentage error in $R \approx \pm 10\%$.

39. $A = \dfrac{1}{4}(4)^2\sin 2\theta = 4\sin 2\theta$ thus $dA = 8\cos 2\theta d\theta$ so, with $\theta = 30° = \pi/6$ radians and $d\theta = \pm 15' = \pm 1/4° = \pm\pi/720$ radians, $dA = 8\cos(\pi/3)(\pm\pi/720) = \pm\pi/180 \approx \pm 0.017$ cm^2.

41. $V = x^3$ where x is the length of a side; $\dfrac{dV}{V} = \dfrac{3x^2 dx}{x^3} = 3\dfrac{dx}{x}$, but $\dfrac{dx}{x} \approx \pm 0.02$

so $\dfrac{dV}{V} \approx 3(\pm 0.02) = \pm 0.06$; percentage error in $V \approx \pm 6\%$.

43. $A = \dfrac{1}{4}\pi D^2$ where D is the diameter of the circle; $\dfrac{dA}{A} = \dfrac{(\pi D/2)dD}{\pi D^2/4} = 2\dfrac{dD}{D}$, but $\dfrac{dA}{A} \approx \pm 0.01$

so $2\dfrac{dD}{D} \approx \pm 0.01$, $\dfrac{dD}{D} \approx \pm 0.005$; maximum permissible percentage error in $D \approx \pm 0.5\%$.

45. $V = $ volume of cylindrical rod $= \pi r^2 h = \pi r^2(15) = 15\pi r^2$; approximate ΔV by dV if $r = 2.5$ and $dr = \Delta r = 0.001$. $dV = 30\pi r\, dr = 30\pi(2.5)(0.001) \approx 0.236$ cm^3.

47. **(a)** $\alpha = \Delta L/(L\Delta T) = 0.006/(40 \times 10) = 1.5 \times 10^{-5}/°$C

(b) $\Delta L = 2.3 \times 10^{-5}(180)(25) \approx 0.1$ cm, so the pole is about 180.1 cm long.

49. $f(x) = \dfrac{1}{1+x}$, $f'(x) = -\dfrac{1}{(1+x)^2}$, $x_0 = 0$, $\Delta x = x$;

$f(x_0 + \Delta x) \approx f(x_0) + f'(x_0)\Delta x$ so $f(x) \approx 1 - x$.

$x = 0.1 : 1/(1+x) = 0.9091$, $1 - x = 0.9$
$x = -0.02 : 1/(1+x) = 1.0204$, $1 - x = 1.02$

51. $y = x^k$, $dy = kx^{k-1}dx$; $\dfrac{dy}{y} = \dfrac{kx^{k-1}dx}{x^k} = k\dfrac{dx}{x}$

TECHNOLOGY EXERCISES 3

1. $f'(x) = 2x, f'(1.8) = 3.6$

3. $f'(x) = (x^2 - 4x)/(x - 2)^2, f'(3.5) \approx -0.777778$

5. 2.772589   7. 6.772589   9. 58.75 ft/sec

11. Solve $3x^2 - \cos x = 0$ to get $x = \pm 0.535428$.

CHAPTER 4
Applications of Differentiation

EXERCISE SET 4.1

1. **(a)** $A = x^2$, so $\dfrac{dA}{dt} = 2x\dfrac{dx}{dt}$

   (b) Find $\dfrac{dA}{dt}\bigg|_{x=3}$ given that $\dfrac{dx}{dt}\bigg|_{x=3} = 2$.

   From part (a), $\dfrac{dA}{dt}\bigg|_{x=3} = 2(3)(2) = 12$ ft^2/min.

3. **(a)** $V = \pi r^2 h$, so $\dfrac{dV}{dt} = \pi\left(r^2\dfrac{dh}{dt} + 2rh\dfrac{dr}{dt}\right)$.

   (b) Find $\dfrac{dV}{dt}\bigg|_{\substack{h=6,\\r=10}}$ given that $\dfrac{dh}{dt}\bigg|_{\substack{h=6,\\r=10}} = 1$ and $\dfrac{dr}{dt}\bigg|_{\substack{h=6,\\r=10}} = -1$.

   From part (a), $\dfrac{dV}{dt}\bigg|_{\substack{h=6,\\r=10}} = \pi[10^2(1) + 2(10)(6)(-1)] = -20\pi$ in^3/sec;

   the volume is decreasing.

5. **(a)** $\tan\theta = \dfrac{y}{x}$, so $\sec^2\theta\dfrac{d\theta}{dt} = \dfrac{x\dfrac{dy}{dt} - y\dfrac{dx}{dt}}{x^2}$, $\dfrac{d\theta}{dt} = \dfrac{\cos^2\theta}{x^2}\left(x\dfrac{dy}{dt} - y\dfrac{dx}{dt}\right)$.

   (b) Find $\dfrac{d\theta}{dt}\bigg|_{\substack{x=2,\\y=2}}$ given that $\dfrac{dx}{dt}\bigg|_{\substack{x=2,\\y=2}} = 1$ and $\dfrac{dy}{dt}\bigg|_{\substack{x=2,\\y=2}} = -\dfrac{1}{4}$.

   When $x = 2$ and $y = 2$, $\tan\theta = 2/2 = 1$ so $\theta = \dfrac{\pi}{4}$ and $\cos\theta = \cos\dfrac{\pi}{4} = \dfrac{1}{\sqrt{2}}$. Thus

   from part (a), $\dfrac{d\theta}{dt}\bigg|_{\substack{x=2,\\y=2}} = \dfrac{(1/\sqrt{2})^2}{2^2}\left[2\left(-\dfrac{1}{4}\right) - 2(1)\right] = -\dfrac{5}{16}$ radians/sec;

   θ is decreasing.

7. Let A be the area swept out, and θ the angle through which the minute hand has rotated. Find $\dfrac{dA}{dt}$ given that $\dfrac{d\theta}{dt} = \dfrac{\pi}{30}$ radians/min; $A = \dfrac{1}{2}r^2\theta = 8\theta$, so $\dfrac{dA}{dt} = 8\dfrac{d\theta}{dt} = \dfrac{4\pi}{15}$ in^2/min.

9. Find $\dfrac{dr}{dt}\bigg|_{A=9}$ given that $\dfrac{dA}{dt} = 6$. From $A = \pi r^2$ we get $\dfrac{dA}{dt} = 2\pi r\dfrac{dr}{dt}$ so $\dfrac{dr}{dt} = \dfrac{1}{2\pi r}\dfrac{dA}{dt}$. If

   $A = 9$ then $\pi r^2 = 9$, $r = 3/\sqrt{\pi}$ so $\dfrac{dr}{dt}\bigg|_{A=9} = \dfrac{1}{2\pi(3/\sqrt{\pi})}(6) = 1/\sqrt{\pi}$ mph.

11. Find $\dfrac{dV}{dt}\Big|_{r=9}$ given that $\dfrac{dr}{dt} = -15$. From $V = \dfrac{4}{3}\pi r^3$ we get $\dfrac{dV}{dt} = 4\pi r^2 \dfrac{dr}{dt}$ so

$\dfrac{dV}{dt}\Big|_{r=9} = 4\pi(9)^2(-15) = -4860\pi$. Air must be removed at the rate of 4860π cm^3/min.

13. Find $\dfrac{dx}{dt}\Big|_{y=5}$ given that $\dfrac{dy}{dt} = -2$. From

$x^2 + y^2 = 13^2$ we get $2x\dfrac{dx}{dt} + 2y\dfrac{dy}{dt} = 0$ so

$\dfrac{dx}{dt} = -\dfrac{y}{x}\dfrac{dy}{dt}$. Use $x^2 + y^2 = 169$ to find that

$x = 12$ when $y = 5$ so

$\dfrac{dx}{dt}\Big|_{y=5} = -\dfrac{5}{12}(-2) = \dfrac{5}{6}$ ft/sec.

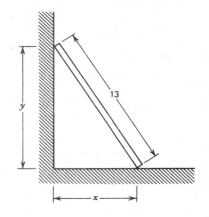

15. Let x be the length of each edge, S the surface area, and V the volume. Find $\dfrac{dS}{dt}\Big|_{x=5}$ given

that $\dfrac{dV}{dt}\Big|_{x=5} = 2$. $S = 6x^2$ and $V = x^3$, so $x = V^{1/3}$, $S = 6V^{2/3}$,

$\dfrac{dS}{dt} = 4V^{-1/3}\dfrac{dV}{dt} = \dfrac{4}{x}\dfrac{dV}{dt} = \dfrac{4}{5}(2) = 8/5$ in^2/min.

17. Find $\dfrac{dy}{dt}\Big|_{x=4000}$ given that $\dfrac{dx}{dt}\Big|_{x=4000} = 880$.

From $y^2 = x^2 + 3000^2$ we get $2y\dfrac{dy}{dt} = 2x\dfrac{dx}{dt}$ so

$\dfrac{dy}{dt} = \dfrac{x}{y}\dfrac{dx}{dt}$. If $x = 4000$, then $y = 5000$ so

$\dfrac{dy}{dt}\Big|_{x=4000} = \dfrac{4000}{5000}(880) = 704$ ft/sec.

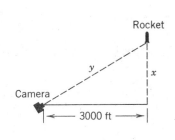

19. **(a)** $\theta = 0$ at perigee, so $r = 4995/1.12 \approx 4460$; the altitude is $4460 - 3960 = 500$ miles. $\theta = \pi$ at apogee, so $r = 4995/0.88 \approx 5676$; the altitude is $5676 - 3960 = 1716$ miles.

(b) If $\theta = 120°$, then $r = 4995/0.94 \approx 5314$; the altitude is $5314 - 3960 = 1354$ miles. The rate of change of the altitude is given by

$$\dfrac{dr}{dt} = \dfrac{4995(0.12\sin\theta)}{(1 + 0.12\cos\theta)^2}\dfrac{d\theta}{dt}.$$

Use $\theta = 120°$ and $d\theta/dt = 2.7°/$min $= (2.7)(\pi/180)$ rad/min to get $dr/dt \approx 27.7$ mi/min.

21. Find $\dfrac{dh}{dt}\Big|_{h=16}$ given that $\dfrac{dV}{dt} = 20$. The volume of water in the tank at a depth h is $V = \dfrac{1}{3}\pi r^2 h$.
Use similar triangles (see figure) to get $\dfrac{r}{h} = \dfrac{10}{24}$ so
$r = \dfrac{5}{12}h$ thus $V = \dfrac{1}{3}\pi\left(\dfrac{5}{12}h\right)^2 h = \dfrac{25}{432}\pi h^3$,
$\dfrac{dV}{dt} = \dfrac{25}{144}\pi h^2\dfrac{dh}{dt}$; $\dfrac{dh}{dt} = \dfrac{144}{25\pi h^2}\dfrac{dV}{dt}$,
$\dfrac{dh}{dt}\Big|_{h=16} = \dfrac{144}{25\pi(16)^2}(20) = \dfrac{9}{20\pi}$ ft/min.

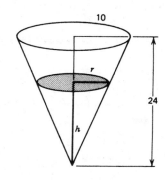

23. Find $\dfrac{dV}{dt}\Big|_{h=10}$ given that $\dfrac{dh}{dt} = 5$. $V = \dfrac{1}{3}\pi r^2 h$,
but $r = \dfrac{1}{2}h$ so $V = \dfrac{1}{3}\pi\left(\dfrac{h}{2}\right)^2 h = \dfrac{1}{12}\pi h^3$,
$\dfrac{dV}{dt} = \dfrac{1}{4}\pi h^2\dfrac{dh}{dt}$, $\dfrac{dV}{dt}\Big|_{h=10}$
$= \dfrac{1}{4}\pi(10)^2(5) = 125\pi$ ft^3/min.

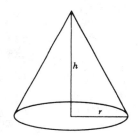

25. With s and h as shown in the figure, we want to find $\dfrac{dh}{dt}$ given that
$\dfrac{ds}{dt} = 500$. From the figure, $h = s\sin 30° = \dfrac{1}{2}s$
so $\dfrac{dh}{dt} = \dfrac{1}{2}\dfrac{ds}{dt} = \dfrac{1}{2}(500) = 250$ mph.

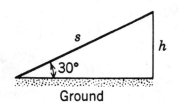

Ground

27. Find $\dfrac{dy}{dt}$ given that $\dfrac{dx}{dt}\Big|_{y=125} = -12$.
From $x^2 + 10^2 = y^2$ we get
$2x\dfrac{dx}{dt} = 2y\dfrac{dy}{dt}$ so $\dfrac{dy}{dt} = \dfrac{x}{y}\dfrac{dx}{dt}$.
Use $x^2 + 100 = y^2$ to find that
$x = \sqrt{15,525} = 15\sqrt{69}$ when $y = 125$ so $\dfrac{dy}{dt} = \dfrac{15\sqrt{69}}{125}(-12) = -\dfrac{36\sqrt{69}}{25}$. The rope must be
pulled at the rate of $\dfrac{36\sqrt{69}}{25}$ ft/min.

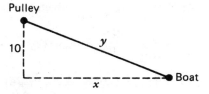

Pulley

Boat

29. Find $\dfrac{dx}{dt}\Big|_{\theta=\pi/4}$ given that $\dfrac{d\theta}{dt} = \dfrac{2\pi}{10} = \dfrac{\pi}{5}$

radians/sec. $x = 4\tan\theta$ (see figure)

so $\dfrac{dx}{dt} = 4\sec^2\theta\,\dfrac{d\theta}{dt}$,

$\dfrac{dx}{dt}\Big|_{\theta=\pi/4} = 4\left(\sec^2\dfrac{\pi}{4}\right)\left(\dfrac{\pi}{5}\right) = 8\pi/5$ kilometers/sec.

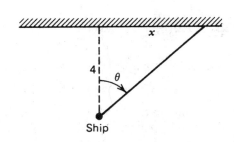

31. We wish to find $\dfrac{dz}{dt}\Big|_{\substack{x=2,\\y=4}}$ given that

$\dfrac{dx}{dt} = -600$ and $\dfrac{dy}{dt}\Big|_{\substack{x=2,\\y=4}} = -1200$

(see figure). From the law of cosines,

$z^2 = x^2 + y^2 - 2xy\cos 120°$

$\quad = x^2 + y^2 - 2xy(-1/2)$

$\quad = x^2 + y^2 + xy,$

so $2z\dfrac{dz}{dt} = 2x\dfrac{dx}{dt} + 2y\dfrac{dy}{dt} + x\dfrac{dy}{dt} + y\dfrac{dx}{dt}$,

$\dfrac{dz}{dt} = \dfrac{1}{2z}\left[(2x+y)\dfrac{dx}{dt} + (2y+x)\dfrac{dy}{dt}\right]$. When $x=2$ and $y=4$,

$z^2 = 2^2 + 4^2 + (2)(4) = 28$, so $z = \sqrt{28} = 2\sqrt{7}$, thus

$\dfrac{dz}{dt}\Big|_{\substack{x=2,\\y=4}} = \dfrac{1}{2(2\sqrt{7})}[(2(2)+4)(-600) + (2(4)+2)(-1200)] = -\dfrac{4200}{\sqrt{7}} = -600\sqrt{7}$ mph;

the distance between missile and aircraft is decreasing at the rate of $600\sqrt{7}$ mph.

33. **(a)** We want $\dfrac{dy}{dt}\Big|_{\substack{x=1,\\y=2}}$ given that $\dfrac{dx}{dt}\Big|_{\substack{x=1,\\y=2}} = 6$. For convenience, first rewrite the equation as

$xy^3 = \dfrac{8}{5} + \dfrac{8}{5}y^2$ then $3xy^2\dfrac{dy}{dt} + y^3\dfrac{dx}{dt} = \dfrac{16}{5}y\dfrac{dy}{dt}$, $\dfrac{dy}{dt} = \dfrac{y^3}{\frac{16}{5}y - 3xy^2}\dfrac{dx}{dt}$ so

$\dfrac{dy}{dt}\Big|_{\substack{x=1,\\y=2}} = \dfrac{2^3}{\frac{16}{5}(2) - 3(1)2^2}(6) = -60/7$ units/sec.

(b) falling, because $\dfrac{dy}{dt} < 0$.

35. The coordinates of P are $(x, 2x)$, so the distance between P and the point $(3,0)$ is

$D = \sqrt{(x-3)^2 + (2x-0)^2} = \sqrt{5x^2 - 6x + 9}$. Find $\left.\dfrac{dD}{dt}\right|_{x=3}$ given that $\left.\dfrac{dx}{dt}\right|_{x=3} = -2$.

$\dfrac{dD}{dt} = \dfrac{5x-3}{\sqrt{5x^2-6x+9}}\dfrac{dx}{dt}$, so $\left.\dfrac{dD}{dt}\right|_{x=3} = \dfrac{12}{\sqrt{36}}(-2) = -4$ units/sec.

37. $dy/dt = 2x\,dx/dt$, but $dy/dt = 3\,dx/dt$ so $3\,dx/dt = 2x\,dx/dt$, $(3-2x)dx/dt = 0$, $3-2x = 0$, $x = 3/2$.

39. Find $\left.\dfrac{dS}{dt}\right|_{s=10}$ given that $\left.\dfrac{ds}{dt}\right|_{s=10} = -2$. From $\dfrac{1}{s} + \dfrac{1}{S} = \dfrac{1}{6}$ we get $-\dfrac{1}{s^2}\dfrac{ds}{dt} - \dfrac{1}{S^2}\dfrac{dS}{dt} = 0$, so

$\dfrac{dS}{dt} = -\dfrac{S^2}{s^2}\dfrac{ds}{dt}$. If $s = 10$, then $\dfrac{1}{10} + \dfrac{1}{S} = \dfrac{1}{6}$ which gives $S = 15$. $\left.\dfrac{dS}{dt}\right|_{s=10} = -\dfrac{225}{100}(-2) = 4.5$

cm/sec. The image is moving away from the lens.

41. Let r be the radius, V the volume, and A the surface area of a sphere. Show that $\dfrac{dr}{dt}$ is a

constant given that $\dfrac{dV}{dt} = -kA$, where k is a positive constant. Because $V = \dfrac{4}{3}\pi r^3$,

$$\dfrac{dV}{dt} = 4\pi r^2 \dfrac{dr}{dt} \qquad\qquad (1)$$

But it is given that $\dfrac{dV}{dt} = -kA$ or, because $A = 4\pi r^2$, $\dfrac{dV}{dt} = -4r^2 k$ which when substituted

into equation (1) gives $-4\pi r^2 k = 4\pi r^2\dfrac{dr}{dt}$, $\dfrac{dr}{dt} = -k$.

43. Extend sides of cup to complete the cone and let V_0 be the volume of the portion added, then (see figure)

$V = \dfrac{1}{3}\pi r^2 h - V_0$ where

$\dfrac{r}{h} = \dfrac{4}{12} = \dfrac{1}{3}$ so $r = \dfrac{1}{3}h$ and

$V = \dfrac{1}{3}\pi\left(\dfrac{h}{3}\right)^2 h - V_0 = \dfrac{1}{27}\pi h^3 - V_0$,

$\dfrac{dV}{dt} = \dfrac{1}{9}\pi h^2\dfrac{dh}{dt}$, $\dfrac{dh}{dt} = \dfrac{9}{\pi h^2}\dfrac{dV}{dt}$,

$\left.\dfrac{dh}{dt}\right|_{h=9} = \dfrac{9}{\pi(9)^2}(20) = \dfrac{20}{9\pi}$ cm/sec.

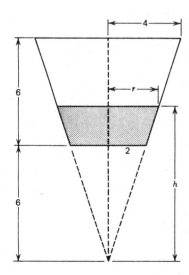

EXERCISE SET 4.2

1. **(a)** (d, f) **(b)** $(a, d), (f, g)$ **(c)** $(a, b), (c, e)$ **(d)** $(b, c), (e, g)$

3. A: $dy/dx < 0$, $d^2y/dx^2 > 0$
   B: $dy/dx > 0$, $d^2y/dx^2 < 0$
   C: $dy/dx < 0$, $d^2y/dx^2 < 0$

5. $f'(x) = 2x - 5$
   $f''(x) = 2$
   (a) $[5/2, +\infty)$
   (b) $(-\infty, 5/2]$
   (c) $(-\infty, +\infty)$
   (d) none
   (e) none

7. $f'(x) = 3(x + 2)^2$
   $f''(x) = 6(x + 2)$
   (a) $(-\infty, +\infty)$
   (b) none
   (c) $(-2, +\infty)$
   (d) $(-\infty, -2)$
   (e) -2

9. $f'(x) = 9(x^2 - 4/9)$
   $f''(x) = 18x$
   (a) $(-\infty, -2/3], [2/3, +\infty)$
   (b) $[-2/3, 2/3]$
   (c) $(0, +\infty)$
   (d) $(-\infty, 0)$
   (e) 0

11. $f'(x) = 12x^2(x - 1)$
    $f''(x) = 36x(x - 2/3)$
    (a) $[1, +\infty)$
    (b) $(-\infty, 1]$
    (c) $(-\infty, 0), (2/3, +\infty)$
    (d) $(0, 2/3)$
    (e) $0, 2/3$

13. $f'(x) = -\sin x$
    $f''(x) = -\cos x$
    (a) $[\pi, 2\pi)$
    (b) $(0, \pi]$
    (c) $(\pi/2, 3\pi/2)$
    (d) $(0, \pi/2), (3\pi/2, 2\pi)$
    (e) $\pi/2, 3\pi/2$

15. $f'(x) = \sec^2 x$
    $f''(x) = 2\sec^2 x \tan x$
    (a) $(-\pi/2, \pi/2)$
    (b) none
    (c) $(0, \pi/2)$
    (d) $(-\pi/2, 0)$
    (e) 0

17. $f'(x) = \dfrac{1}{3}(x + 2)^{-2/3}$

    $f''(x) = -\dfrac{2}{9}(x + 2)^{-5/3}$

    (a) $(-\infty, +\infty)$
    (b) none
    (c) $(-\infty, -2)$
    (d) $(-2, +\infty)$
    (e) -2

19. $f'(x) = \dfrac{4(x + 1)}{3x^{2/3}}$

    $f''(x) = \dfrac{4(x - 2)}{9x^{5/3}}$

    (a) $[-1, +\infty)$
    (b) $(-\infty, -1]$
    (c) $(-\infty, 0), (2, +\infty)$
    (d) $(0, 2)$
    (e) $0, 2$

21. $f'(x) = 1 + \cos x \geq 0$; $f'(x) = 0$ when $\cos x = -1$, $x = \pm\pi, \pm 3\pi, \pm 5\pi, \cdots$ so f is increasing on the intervals $\cdots [-3\pi, -\pi], [-\pi, \pi], [\pi, 3\pi], \cdots$ and putting all of these intervals together gives $(-\infty, +\infty)$.

23. $f'(x) = 1 - 1/x^2$ so f is increasing on $[1, +\infty)$ thus if $x > 1$, then $f(x) > f(1) = 2$, $x + 1/x > 2$.

25. $f'(x) = 1/3 - 1/[3(1+x)^{2/3}]$ so f is increasing on $[0, +\infty)$ thus if $x > 0$, then $f(x) > f(0) = 0$, $1 + x/3 - \sqrt[3]{1+x} > 0$, $\sqrt[3]{1+x} < 1 + x/3$.

27. (a)   (b)   (c)

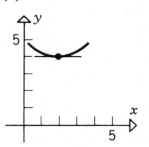
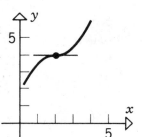
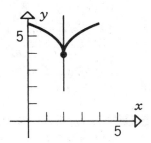

29. (a)   (b)

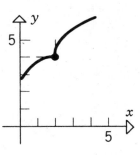
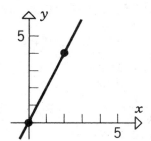

31. $f'(x) = 4(x - a)^3$, $f''(x) = 12(x - a)^2$; no inflection points.

33. $f(x_1) - f(x_2) = x_1^2 - x_2^2 = (x_1 + x_2)(x_1 - x_2) < 0$ if $x_1 < x_2$ for x_1, x_2 in $[0, +\infty)$, so $f(x_1) < f(x_2)$ and thus increasing.

35. $f(x_1) - f(x_2) = \sqrt{x_1} - \sqrt{x_2} = \dfrac{x_1 - x_2}{\sqrt{x_1} + \sqrt{x_2}} < 0$ if $x_1 < x_2$ for x_1, x_2 in $[0, +\infty)$, so

$f(x_1) < f(x_2)$ and thus increasing.

37. If $x_1 < x_2$ where x_1 and x_2 are in I, then $f(x_1) < f(x_2)$ and $g(x_1) < g(x_2)$, so
$f(x_1) + g(x_1) < f(x_2) + g(x_2)$, $(f + g)(x_1) < (f + g)(x_2)$. Thus $f + g$ is increasing on I.

39. For example, $f(x) = x$ and $g(x) = 2x$ on $(-\infty, +\infty)$.

41. $f''(x) = 6ax + 2b = 6a(x + \dfrac{b}{3a})$, $f''(x) = 0$ when $x = -\dfrac{b}{3a}$. f changes its direction of concavity
at $x = -\dfrac{b}{3a}$ so $-\dfrac{b}{3a}$ is an inflection point.

43. **(a)** Let x_1 and x_2 be any points in (a, b) where $x_1 < x_2$. If x_1 and x_2 are both in $(a, c]$ or
both in $[c, b)$, then $f(x_1) < f(x_2)$ because f is increasing on these intervals. If x_1 is in
$(a, c]$ and x_2 is in (c, b), then $f(x_1) \le f(c) < f(x_2)$ so $f(x_1) < f(x_2)$. Thus in all cases,
$f(x_1) < f(x_2)$.

(b) Similar to the proof in part (a).

EXERCISE SET 4.3

1. $f'(x) = 2x - 5$, $f'(x) = 0$ when $x = 5/2$ (stationary point).

3. $f'(x) = 3x^2 + 6x - 9 = 3(x + 3)(x - 1)$, $f'(x) = 0$ when $x = -3, 1$ (stationary points).

5. $f'(x) = 4x(x^2 - 3)$, $f'(x) = 0$ when $x = 0, \pm\sqrt{3}$ (stationary points).

7. $f'(x) = (2 - x^2)/(x^2 + 2)^2$, $f'(x) = 0$ when $x = \pm\sqrt{2}$ (stationary points).

9. $f'(x) = \dfrac{2}{3}x^{-1/3} = 2/(3x^{1/3})$, $f'(x)$ does not exist when $x = 0$.

11. $f'(x) = -3\sin 3x$, $f'(x) = 0$ when $\sin 3x = 0, 3x = n\pi, n = 0, \pm1, \pm2, \cdots$
$x = n\pi/3, n = 0, \pm1, \pm2, \cdots$ (stationary points).

13. $f'(x) = 4\sin 2x \cos 2x = 2\sin 4x$,
$f'(x) = 0$ when $4x = n\pi$, $x = n\pi/4$, $n = 1, 2, 3, \cdots, 7$ (stationary points)

15. $f'(x) = \dfrac{4(x + 1)}{3x^{2/3}}$, $f'(x) = 0$ when $x = -1$ (stationary point), $f'(x)$ does not exist when $x = 0$.

17. **(a)** $x = 2$ because $f'(x)$ changes sign from $-$ to $+$ there.
(b) $x = 0$ because $f'(x)$ changes sign from $+$ to $-$ there.
(c) $x = 1, 3$ because $f''(x)$ (the slope of the graph of $f'(x)$) changes sign at these points.

19. critical points $x = 0, \pm\sqrt{5}$; $f'(x)$:

$$
\begin{array}{ccccccc}
- & 0 & + & 0 & - & 0 & + \\
\hline
& -\sqrt{5} & & 0 & & \sqrt{5} &
\end{array}
$$

$x = 0$: relative maximum; $x = \pm\sqrt{5}$: relative minimum.

21. critical points: $\pm 3/2, -1$; $f'(x)$:

$$
\begin{array}{ccccccc}
+ & 0 & - & ? & + & 0 & - \\
\hline
& -3/2 & & -1 & & 3/2 &
\end{array}
$$

$x = \pm 3/2$: relative maximum; $x = -1$: relative minimum.

23. $f'(x) = -2(x + 2)$; critical point $x = -2$

  (a) $f'(x)$:

$$
\begin{array}{ccc}
+++ & 0 & --- \\
\hline
& -2 &
\end{array}
$$

  (b) $f''(x) = -2$; $f''(-2) < 0$, $f(-2) = 5$; relative max of 5 at $x = -2$

25. $f'(x) = 2\sin x \cos x = \sin 2x$; critical points $x = \pi/2, \pi, 3\pi/2$

  (a) $f'(x)$:

$$
\begin{array}{ccccccc}
+++ & 0 & --- & 0 & +++ & 0 & --- \\
\hline
& \pi/2 & & \pi & & 3\pi/2 &
\end{array}
$$

  (b) $f''(x) = 2\cos 2x$; $f''(\pi/2) < 0$, $f''(\pi) > 0$, $f''(3\pi/2) < 0$, $f(\pi/2) = f(3\pi/2) = 1$, $f(\pi) = 0$; relative min of 0 at $x = \pi$, relative max of 1 at $x = \pi/2, 3\pi/2$

27. $f'(x) = 3x^2 + 5$; no relative extrema because there are no critical points.

29. $f'(x) = (x - 1)(3x - 1)$; critical points $x = 1, 1/3$
$f''(x) = 6x - 4$; $f''(1) > 0$, $f''(1/3) < 0$
relative min of 0 at $x = 1$, relative max of 4/27 at $x = 1/3$

31. $f'(x) = 4x(1 - x^2)$; critical points $x = 0, 1, -1$
$f''(x) = 4 - 12x^2$; $f''(0) > 0$, $f''(1) < 0$, $f''(-1) < 0$
relative min of 0 at $x = 0$, relative max of 1 at $x = 1, -1$

33. $f'(x) = \dfrac{4}{5}x^{-1/5}$; critical point $x = 0$; relative min of 0 at $x = 0$ (first derivative test)

35. $f'(x) = 2x/(x^2 + 1)^2$; critical point $x = 0$; relative min of 0 at $x = 0$

37. $f'(x) = 2x$ if $|x| > 2$, $f'(x) = -2x$ if $|x| < 2$,
$f'(x)$ does not exist when $x = \pm 2$; critical points $x = 0, 2, -2$
relative min of 0 at $x = 2, -2$, relative max of 4 at $x = 0$

39. $f'(x) = -\sin 2x$; critical points $x = 0, \pm\pi/2, \pm\pi, \pm 3\pi/2, \cdots$
relative min of 0 at $x = \pm\pi/2, \pm 3\pi/2, \cdots$; relative max of 1 at $x = 0, \pm\pi, \pm 2\pi, \cdots$

41. $f'(x) = 2x \sec^2(x^2 + 1)$; critical point $x = 0$; relative min of $\tan 1$ at $x = 0$

43. $f'(x) = 2\cos 2x$ if $\sin 2x > 0$, $f'(x) = -2\cos 2x$ if $\sin 2x < 0$,
$f'(x)$ does not exist when $x = \pi/2, \pi, 3\pi/2$;
critical points $x = \pi/4, 3\pi/4, 5\pi/4, 7\pi/4, \pi/2, \pi, 3\pi/2$
relative min of 0 at $x = \pi/2, \pi, 3\pi/2$; relative max of 1 at $x = \pi/4, 3\pi/4, 5\pi/4, 7\pi/4$

45. Let $f(x) = x^2 + \dfrac{k}{x}$, then $f'(x) = 2x - \dfrac{k}{x^2} = \dfrac{2x^3 - k}{x^2}$. f has a relative extremum when
$2x^3 - k = 0$, so $k = 2x^3 = 2(3)^3 = 54$.

47. $f(x) = -x^4$ has a relative maximum at $x = 0$, $f(x) = x^4$ has a relative minimum at $x = 0$,
$f(x) = x^3$ has neither at $x = 0$; $f'(0) = 0$ for all three functions.

49. **(a)**

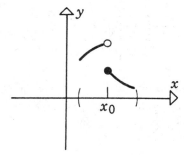

$f(x_0)$ is not an extreme value

(b)

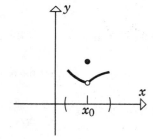

$f(x_0)$ is a relative maximum

(c)

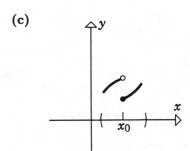

$f(x_0)$ is a relative minimum

EXERCISE SET 4.4

1. $y = x^2 - 2x - 3$
$y' = 2(x - 1)$
$y'' = 2$

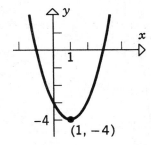

3. $y = x^3 - 3x + 1$
$y' = 3(x^2 - 1)$
$y'' = 6x$

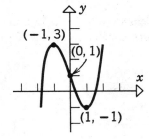

5. $y = x^3 + 3x^2 + 5$
$y' = 3x(x + 2)$
$y'' = 6(x + 1)$

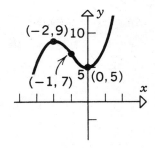

7. $y = 2x^3 - 3x^2 + 12x + 9$
$y' = 6(x^2 - x + 2)$
$y'' = 12(x - 1/2)$

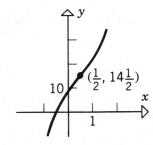

9. $y = (x-1)^4$
$y' = 4(x-1)^3$
$y'' = 12(x-1)^2$

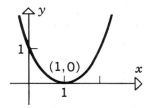

11. $y = x^4 + 2x^3 - 1$
$y' = 4x^2(x + 3/2)$
$y'' = 12x(x+1)$

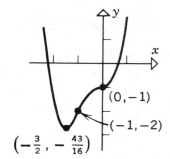

13. $y = x^4 - 3x^3 + 3x^2 + 1$
$y' = x(4x^2 - 9x + 6)$
$y'' = 12(x - 1/2)(x - 1)$

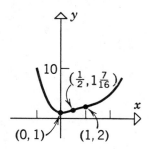

15. $y = x^3(3x^2 - 5)$
$y' = 15x^2(x^2 - 1)$
$y'' = 30x(2x^2 - 1)$

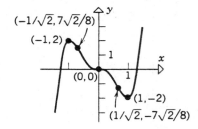

17. $y = x(x-1)^3$
$y' = (4x - 1)(x - 1)^2$
$y'' = 6(2x - 1)(x - 1)$

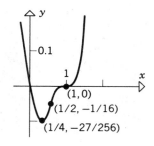

19. vertical: $x = 2$; horizontal: $y = 3$

21. vertical: $x = -\sqrt{5}$, $x = \sqrt{5}$; horizontal: none

23. vertical: $x = -1$, $x = 3$; horizontal: $y = 1$

25. $x^2/(x^2 + 2x + 5) = 1$, $2x + 5 = 0$, $x = -5/2$.

27. $(x^2 + 1)/(2x^2 - 6x) = 1/2$, $-3x = 1$, $x = -1/3$.

29. $y = 2x/(x - 3)$
$y' = -6(x - 3)^2$
$y'' = 12/(x - 3)^3$

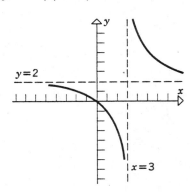

31. $y = \dfrac{x^2}{x^2 - 1}$

$y' = -\dfrac{2x}{(x^2 - 1)^2}$

$y'' = \dfrac{2(3x^2 + 1)}{(x^2 - 1)^3}$

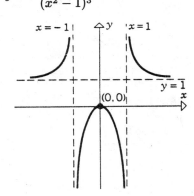

33. $y = \dfrac{x}{x^2 + 1}$

$y' = \dfrac{1 - x^2}{(x^2 + 1)^2}$

$y'' = \dfrac{2x(x^2 - 3)}{(x^2 + 1)^3}$

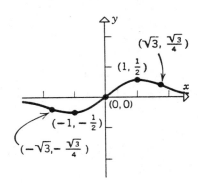

35. $y = (x - 1)/(x - 2)$
$y' = -1/(x - 2)^2$
$y'' = 2/(x - 2)^3$

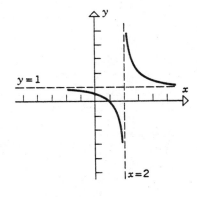

37. $y = x^2 - \dfrac{1}{x} = \dfrac{x^3 - 1}{x}$

$y' = \dfrac{2x^3 + 1}{x^2}$,

$y' = 0$ when $x = -\sqrt[3]{\dfrac{1}{2}} \approx -0.8$

$y'' = \dfrac{2(x^3 - 1)}{x^3}$

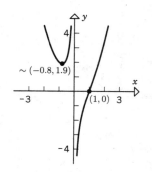

39. $y = \dfrac{1 - x}{x^2}$

$y' = \dfrac{x - 2}{x^3}$

$y'' = \dfrac{2(3 - x)}{x^4}$

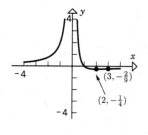

41. $y = \dfrac{x - 1}{x^2 - 4}$

$y' = -\dfrac{x^2 - 2x + 4}{(x^2 - 4)^2}$

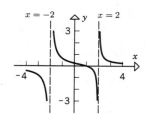

43. $y = \dfrac{(x - 1)^2}{x^2}$

$y' = \dfrac{2(x - 1)}{x^3}$

$y'' = \dfrac{2(3 - 2x)}{x^4}$

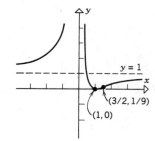

45. $y = 3 - \dfrac{4}{x} - \dfrac{4}{x^2}$

$y' = \dfrac{4(x+2)}{x^3}$

$y'' = -\dfrac{8(x+3)}{x^4}$

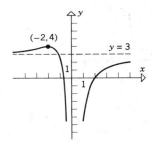

47. $y = \dfrac{x^3 - 1}{x^3 + 1}$

$y' = \dfrac{6x^2}{(x^3+1)^2}$

$y'' = \dfrac{12x(1-2x^3)}{(x^3+1)^3}$

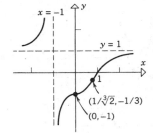

49. $y = \dfrac{x^2 - 2}{x} = x - \dfrac{2}{x}$

so $y = x$ is an oblique asymptote

$y' = \dfrac{x^2 + 2}{x^2}$

$y'' = -\dfrac{4}{x^3}$

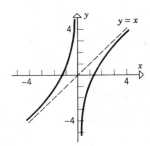

51. $y = \dfrac{(x-2)^3}{x^2} = x - 6 + \dfrac{12x - 8}{x^2}$

so $y = x - 6$ is an oblique asymptote

$y' = \dfrac{(x-2)^2(x+4)}{x^3}$

$y'' = \dfrac{24(x-2)}{x^4}$

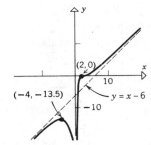

53. $y = x + 1 - \dfrac{1}{x} - \dfrac{1}{x^2} = \dfrac{(x-1)(x+1)^2}{x^2}$

$y = x + 1$ is an oblique asymptote

$y' = \dfrac{(x+1)(x^2 - x + 2)}{x^3}$

$y'' = -\dfrac{2(x+3)}{x^4}$

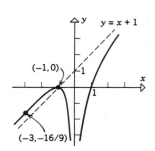

55. $\displaystyle\lim_{x \to \pm\infty} [f(x) - x^2] = \lim_{x \to \pm\infty} (1/x) = 0$

$y = x^2 + \dfrac{1}{x} = \dfrac{x^3 + 1}{x}$

$y' = 2x - \dfrac{1}{x^2} = \dfrac{2x^3 - 1}{x^2}$

$y'' = 2 + \dfrac{2}{x^3} = \dfrac{2(x^3 + 1)}{x^3}$

$y' = 0$ when $x = 1/\sqrt[3]{2} \approx 0.8$,
$\qquad\quad y = 3\sqrt[3]{2}/2 \approx 1.9$
$y'' = 0$ when $x = -1, y = 0$

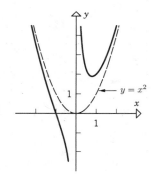

57. Let y be the length of the other side of the rectangle, then $L = 2x + 2y$ and $xy = 400$ so $y = 400/x$ and hence $L = 2x + 800/x$.

$L = 2x$ is an oblique asymptote

(see Exercise 48)

$L = 2x + \dfrac{800}{x} = \dfrac{2(x^2 + 400)}{x}$

$L' = 2 - \dfrac{800}{x^2} = \dfrac{2(x^2 - 400)}{x^2}$

$L'' = \dfrac{1600}{x^3}$

$L' = 0$ when $x = 20, L = 80$

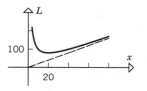

59. $y' = 0.1x^4(6x - 5)$

critical points: $x = 0, x = 5/6$

relative minimum at $x = 5/6$,

$y \approx -6.7 \times 10^{-3}$

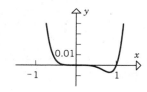

EXERCISE SET 4.5

1. $y = (x-2)^{1/3}$

$y' = \frac{1}{3}(x-2)^{-2/3}$

$y'' = -\frac{2}{9}(x-2)^{-5/3}$

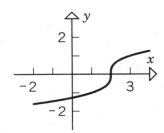

3. $y = x^{1/5}$

$y' = \frac{1}{5}x^{-4/5}$

$y'' = -\frac{4}{25}x^{-9/5}$

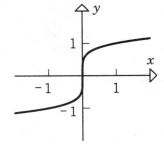

5. $y = x^{4/3}$

$y' = \frac{4}{3}x^{1/3}$

$y'' = \frac{4}{9}x^{-2/3}$

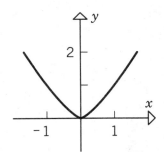

7. $y = 1 - x^{2/3}$

$y' = -\frac{2}{3}x^{-1/3}$

$y'' = \frac{2}{9}x^{-4/3}$

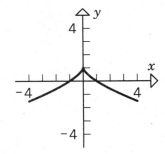

9. $y = \sqrt{x^2 - 1}$

$y' = \frac{x}{\sqrt{x^2 - 1}}$

$y'' = -\frac{1}{(x^2 - 1)^{3/2}}$

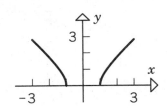

11. $y = 2x + 3x^{2/3}$

$y' = 2 + 2x^{-1/3}$,

$y'' = -\dfrac{2}{3}x^{-4/3}$

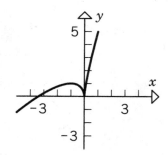

13. $y = x(3-x)^{1/2}$

$y' = \dfrac{3(2-x)}{2\sqrt{3-x}}$

$y'' = \dfrac{3(x-4)}{4(3-x)^{3/2}}$

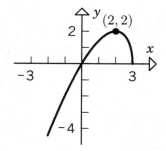

15. $y = \dfrac{8(\sqrt{x}-1)}{x}$

$y' = \dfrac{4(2-\sqrt{x})}{x^2}$

$y'' = \dfrac{2(3\sqrt{x}-8)}{x^3}$

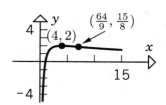

17. $y = \dfrac{\sqrt{x}}{x-3}$

$y' = -\dfrac{x+3}{2\sqrt{x}(x-3)^2}$

$y'' = \dfrac{3(x^2+6x-3)}{4x^{3/2}(x-3)^3}$

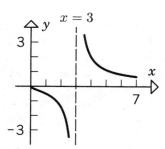

19. $y = x + \sin x$
$y' = 1 + \cos x, y' = 0$
   when $x = \pi + 2n\pi$
$y'' = -\sin x, y'' = 0$
   when $x = n\pi$

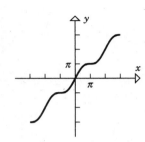

21. $y = \sin x + \cos x$
$y' = \cos x - \sin x,\ y' = 0$
   when $x = \pi/4 + n\pi$
$y'' = -\sin x - \cos x,\ y'' = 0$
   when $x = 3\pi/4 + n\pi$

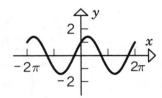

23. $y = \sin^2 x,\ 0 \le x \le 2\pi$
$y' = 2\sin x \cos x = \sin 2x$
$y'' = 2\cos 2x$

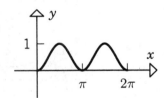

EXERCISE SET 4.6

1. $f'(x) = 8x - 4,\ f'(x) = 0$ when $x = 1/2$; $f(0) = 1,\ f(1/2) = 0,\ f(1) = 1$ so the maximum value is 1 at $x = 0, 1$ and the minimum value is 0 at $x = 1/2$.

3. $f'(x) = 3(x-1)^2,\ f'(x) = 0$ when $x = 1$; $f(0) = -1,\ f(1) = 0,\ f(4) = 27$ so the maximum value is 27 at $x = 4$ and the minimum value is -1 at $x = 0$.

5. $f'(x) = 3/(4x^2 + 1)^{3/2}$, thus there are no critical points; $f(-1) = -3/\sqrt{5},\ f(1) = 3/\sqrt{5}$ so the maximum value is $3/\sqrt{5}$ at $x = 1$ and the minimum value is $-3/\sqrt{5}$ at $x = -1$.

7. $f'(x) = \dfrac{5(8-x)}{3x^{1/3}}$, $f'(x) = 0$ when $x = 8$ and $f'(x)$ does not exist when $x = 0$; $f(-1) = 21$, $f(0) = 0$, $f(8) = 48$, $f(20) = 0$ so the maximum value is 48 at $x = 8$ and the minimum value is 0 at $x = 0, 20$.

9. $f'(x) = 1 - \sec^2 x$, $f'(x) = 0$ for x in $(-\pi/4, \pi/4)$ when $x = 0$; $f(-\pi/4) = 1 - \pi/4$, $f(0) = 0$, $f(\pi/4) = \pi/4 - 1$ so the maximum value is $1 - \pi/4$ at $x = -\pi/4$ and the minimum value is $\pi/4 - 1$ at $x = \pi/4$.

11. $f'(x) = 2\sec x \tan x - \sec^2 x = (2\sin x - 1)/\cos^2 x$, $f'(x) = 0$ for x in $(0, \pi/4)$ when $x = \pi/6$; $f(0) = 2$, $f(\pi/6) = \sqrt{3}$, $f(\pi/4) = 2\sqrt{2} - 1$ so the maximum value is 2 at $x = 0$ and the minimum value is $\sqrt{3}$ at $x = \pi/6$.

13. $f(x) = 1 + |9 - x^2| = \begin{cases} 10 - x^2, & |x| \le 3 \\ -8 + x^2, & |x| > 3 \end{cases}$, $f'(x) = \begin{cases} -2x, & |x| < 3 \\ 2x, & |x| > 3 \end{cases}$ thus $f'(x) = 0$ when $x = 0$, $f'(x)$ does not exist for x in $(-5, 1)$ when $x = -3$ because $\displaystyle\lim_{x \to -3^-} f'(x) \ne \lim_{x \to -3^+} f'(x)$ (see Theorem preceding Exercise 73, Section 3.3); $f(-5) = 17$, $f(-3) = 1$, $f(0) = 10$, $f(1) = 9$ so the maximum value is 17 at $x = -5$ and the minimum value is 1 at $x = -3$.

15. $f'(x) = 2x - 3$; critical point $x = 3/2$. Minimum value $f(3/2) = -13/4$, no maximum.

17. $f'(x) = 12x^2(1 - x)$; critical points $x = 0, 1$. Maximum value $f(1) = 1$, no minimum because $\displaystyle\lim_{x \to +\infty} f(x) = -\infty$.

19. $(x^2 - 1)^2$ can never be less than zero because it is the square of $x^2 - 1$; the minimum value is 0 for $x = \pm 1$, no maximum because $\displaystyle\lim_{x \to +\infty} f(x) = +\infty$.

21. No maximum or minimum because $\displaystyle\lim_{x \to +\infty} f(x) = +\infty$ and $\displaystyle\lim_{x \to -\infty} f(x) = -\infty$.

23. $f'(x) = -1/x^2$; no maximum or minimum because there are no critical points in $(0, +\infty)$.

25. $f'(x) = x(x + 2)/(x + 1)^2$; critical point $x = -2$ in $(-5, -1)$. Maximum value $f(-2) = -4$, no minimum.

27. $\sin 2x$ has a period of π, and $\sin 4x$ a period of $\pi/2$ so $f(x)$ is periodic with period π. Consider the interval $[0, \pi]$. $f'(x) = 4\cos 2x + 4\cos 4x$, $f'(x) = 0$ when $\cos 2x + \cos 4x = 0$, but $\cos 4x = 2\cos^2 2x - 1$ (trig identity) so

$$2\cos^2 2x + \cos 2x - 1 = 0$$
$$(2\cos 2x - 1)(\cos 2x + 1) = 0$$
$$\cos 2x = 1/2 \quad \text{or} \quad \cos 2x = -1.$$

From $\cos 2x = 1/2$, $2x = \pi/3$ or $5\pi/3$ so $x = \pi/6$ or $5\pi/6$. From $\cos 2x = -1$, $2x = \pi$ so $x = \pi/2$. $f(0) = 0$, $f(\pi/6) = 3\sqrt{3}/2$, $f(\pi/2) = 0$, $f(5\pi/6) = -3\sqrt{3}/2$, $f(\pi) = 0$. The maximum value is $3\sqrt{3}/2$ at $x = \pi/6+n\pi$ and the minimum value $is -3\sqrt{3}/2$ at $x = 5\pi/6+n\pi$, $n = 0, \pm 1, \pm 2, \cdots$.

29. $f'(x) = -[\cos(\cos x)]\sin x$; $f'(x) = 0$ if $\sin x = 0$ or if $\cos(\cos x) = 0$. If $\sin x = 0$, then $x = \pi$ is the critical point in $(0, 2\pi)$; $\cos(\cos x) = 0$ has no solutions because $-1 \le \cos x \le 1$. Thus $f(0) = \sin(1)$, $f(\pi) = \sin(-1) = -\sin(1)$, and $f(2\pi) = \sin(1)$ so the maximum value is $\sin(1) \approx 0.84147$ and the minimum value is $-\sin(1) \approx -0.84147$.

31. $f'(x) = \begin{cases} 4, & x < 1 \\ 2x - 5, & x > 1 \end{cases}$ so $f'(x) = 0$ when $x = 5/2$, and $f'(x)$ does not exist when $x = 1$ because $\lim_{x \to 1^-} f'(x) \ne \lim_{x \to 1^+} f'(x)$ (see Theorem preceding Exercise 73, Section 3.3); $f(1/2) = 0$, $f(1) = 2$, $f(5/2) = -1/4$, $f(7/2) = 3/4$ so the maximum value is 2 and the minimum value is $-1/4$.

33. $f'(x) = p(x-a)^{p-1}$; critical point $x = a$
   (a) if p is even then $p-1$ is odd and $f'(x) < 0$ when $x < a$, $f'(x) > 0$ when $x > a$ so $f(a) = 0$ is a relative minimum.
   (b) if p is odd then $p-1$ is even and $f'(x)$ does not change sign at $x = a$ so f does not have relative extrema.

35. (a) $f'(x) = -\dfrac{64\cos x}{\sin^2 x} + \dfrac{27\sin x}{\cos^2 x} = \dfrac{-64\cos^3 x + 27\sin^3 x}{\sin^2 x \cos^2 x}$, $f'(x) = 0$ when
   $27\sin^3 x = 64\cos^3 x$, $\tan^3 x = 64/27$, $\tan x = 4/3$ so the critical point is $x = x_0$ where $\tan x_0 = 4/3$ and $0 < x_0 < \pi/2$. To test x_0 first rewrite $f'(x)$ as
   $$f'(x) = \frac{27\cos^3 x(\tan^3 x - 64/27)}{\sin^2 x \cos^2 x} = \frac{27\cos x(\tan^3 x - 64/27)}{\sin^2 x};$$
   if $x < x_0$ then $\tan x < 4/3$ and $f'(x) < 0$, if $x > x_0$ then $\tan x > 4/3$ and $f'(x) > 0$ so $f(x_0)$ is the minimum value. f has no maximum because $\lim_{x \to 0^+} f(x) = +\infty$.

   (b) If $\tan x_0 = 4/3$ then (see figure)
   $\sin x_0 = 4/5$ and $\cos x_0 = 3/5$
   so $f(x_0) = 64/\sin x_0 + 27/\cos x_0$
   $\qquad = 64/(4/5) + 27/(3/5)$
   $\qquad = 80 + 45 = 125$

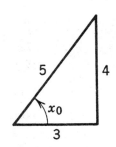

37. $f(\theta) = \sin^2\theta\cos\theta$, $f'(\theta) = \sin\theta(2\cos^2\theta - \sin^2\theta) = \sin\theta(3\cos^2\theta - 1)$; $f'(\theta) = 0$ for θ in $(0, \pi/2)$ if $\cos\theta = 1/\sqrt{3}$. $f(0) = 0$ and $f(\pi/2) = 0$, so the maximum value occurs when $\cos\theta = 1/\sqrt{3}$ where $f(\theta) = \sin^2\theta\cos\theta = (1 - \cos^2\theta)\cos\theta = (1 - 1/3)(1/\sqrt{3}) = 2/(3\sqrt{3})$.

39. $(0, 9)$ is on the graph so $f(0) = a_0 + a_1(0) + a_2(0)^2 = 9$, $a_0 = 9$ thus $f(x) = 9 + a_1 x + a_2 x^2$. $(2, 1)$ is on the graph so $f(2) = 9 + 2a_1 + 4a_2 = 1$,

$$a_1 + 2a_2 = -4 \qquad\qquad\qquad\qquad\qquad\qquad\qquad\qquad\qquad\qquad\qquad\qquad (i).$$

$f'(x) = a_1 + 2a_2 x$, but $f'(2) = 0$ because it is given that $f(2)$ is to be an extreme value, so

$$f'(2) = a_1 + 4a_2 = 0 \qquad\qquad\qquad\qquad\qquad\qquad\qquad\qquad\qquad\qquad\qquad (ii).$$

Solve (i) and (ii) to get $a_1 = -8$ and $a_2 = 2$, thus $f(x) = 9 - 8x + 2x^2$. As a check, we find that $f''(x) = 4 > 0$ so $f(2)$ is a minimum.

41. Let $f(x) = 1 - x^2/2 - \cos x$, then $f'(x) = -x + \sin x$ so $f'(x) = 0$ when $\sin x = x$ which has no solution for $0 < x < 2\pi$ thus the maximum and minimum values must occur at the endpoints of $[0, 2\pi]$. $f(0) = 0$, $f(2\pi) = -2\pi^2$, so 0 is the maximum value, thus $1 - x^2/2 - \cos x \le 0$, $1 - x^2/2 \le \cos x$ for all x in $[0, 2\pi]$.

43. By the quadratic formula, the roots are $x_1 = \dfrac{-b - \sqrt{b^2 - 4ac}}{2a}$ and $x_2 = \dfrac{-b + \sqrt{b^2 - 4ac}}{2a}$. The midpoint is $\dfrac{1}{2}(x_1 + x_2) = -\dfrac{b}{2a}$. But $f'(x) = 2ax + b$ so $f'\left(-\dfrac{b}{2a}\right) = 2a\left(-\dfrac{b}{2a}\right) + b = 0$.

45. $f'(x) = 2ax + b$; critical point is $x = -\dfrac{b}{2a}$

$f''(x) = 2a > 0$ so $f\left(-\dfrac{b}{2a}\right)$ is the minimum value of f, but

$f\left(-\dfrac{b}{2a}\right) = a\left(-\dfrac{b}{2a}\right)^2 + b\left(-\dfrac{b}{2a}\right) + c = \dfrac{-b^2 + 4ac}{4a}$ thus $f(x) \ge 0$ if and only if

$f\left(-\dfrac{b}{2a}\right) \ge 0$, $\dfrac{-b^2 + 4ac}{4a} \ge 0$, $-b^2 + 4ac \ge 0$, $b^2 - 4ac \le 0$

47. The slope of the line is -1, and the slope of the tangent to $y = -x^2$ is $-2x$ so $-2x = -1$, $x = 1/2$. The line lies above the curve so the vertical distance is given by $F(x) = 2 - x + x^2$; $F(-1) = 4$, $F(1/2) = 7/4$, $F(3/2) = 11/4$. The point $(1/2, -1/4)$ is closest, the point $(-1, -1)$ farthest.

EXERCISE SET 4.7

1. Let $x =$ one number, $y =$ the other number, and $P = xy$ where $x + y = 10$. Thus $y = 10 - x$ so $P = x(10 - x) = 10x - x^2$ for x in $[0, 10]$. $dP/dx = 10 - 2x$, $dP/dx = 0$ when $x = 5$. If $x = 0, 5, 10$ then $P = 0, 25, 0$ so P is maximum when $x = 5$ and, from $y = 10 - x$, when $y = 5$.

3. If $y = x + 1/x$ for $1/2 \leq x \leq 3/2$ then $dy/dx = 1 - 1/x^2 = (x^2 - 1)/x^2$, $dy/dx = 0$ when $x = 1$. If $x = 1/2, 1, 3/2$ then $y = 5/2, 2, 13/6$ so

   (a) y is as small as possible when $x = 1$.   (b) y is as large as possible when $x = 1/2$.

5. Let x and y be the dimensions shown in the figure and A the area, then

   $A = xy$ subject to the cost condition

   $3(2x) + 2(2y) = 6000$, or $y = 1500 - 3x/2$.

   Thus $A = x(1500 - 3x/2) = 1500x - 3x^2/2$

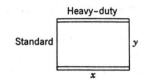

   for x in $[0, 1000]$. $dA/dx = 1500 - 3x$, $dA/dx = 0$ when $x = 500$. If $x = 0$ or 1000 then $A = 0$, if $x = 500$ then $A = 375,000$ so the area is greatest when $x = 500$ ft and (from $y = 1500 - 3x/2$) when $y = 750$ ft.

7. Let x, y, and z be as shown in the figure and A the area of the rectangle, then

   $A = xy$ and, by similar triangles, $z/10 = y/6$,

   $z = 5y/3$; also $x/10 = (8 - z)/8 = (8 - 5y/3)/8$

   thus $y = 24/5 - 12x/25$ so

   $A = x(24/5 - 12x/25) = 24x/5 - 12x^2/25$

   for x in $[0, 10]$. $dA/dx = 24/5 - 24x/25$,

   $dA/dx = 0$ when $x = 5$. If $x = 0, 5, 10$ then

   $A = 0, 12, 0$ so the area is greatest when $x = 5$ in. and $y = 12/5$ in.

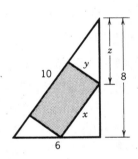

9. $A = xy$ where $x^2 + y^2 = 20^2 = 400$ so

   $y = \sqrt{400 - x^2}$ and $A = x\sqrt{400 - x^2}$ for

   $0 \leq x \leq 20$; $dA/dx = 2(200 - x^2)/\sqrt{400 - x^2}$,

   $dA/dx = 0$ when $x = \sqrt{200} = 10\sqrt{2}$. If

   $x = 0, 10\sqrt{2}, 20$ then $A = 0, 200, 0$ so

   the area is maximum when $x = 10\sqrt{2}$ and

   $y = \sqrt{400 - 200} = 10\sqrt{2}$.

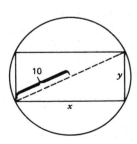

11. Let $x =$ length of each side that uses the \$1 per foot fencing,
$y =$ length of each side that uses the \$2 per foot fencing.

The cost is $C = (1)(2x) + (2)(2y) = 2x + 4y$, but $A = xy = 3200$ thus $y = 3200/x$ so

$$C = 2x + 12800/x \text{ for } x > 0,$$
$$dC/dx = 2 - 12800/x^2, \ dC/dx = 0 \text{ when } x = 80, \ d^2C/dx^2 > 0 \text{ so}$$

C is least when $x = 80, \ y = 40$.

13. Let x and y be the dimensions of a rectangle; the perimeter is $p = 2x + 2y$. But $A = xy$ thus
$y = A/x$ so $p = 2x + 2A/x$ for $x > 0$, $dp/dx = 2 - 2A/x^2 = 2(x^2 - A)/x^2$, $dp/dx = 0$ when
$x = \sqrt{A}$, $d^2p/dx^2 = 4A/x^3 > 0$ if $x > 0$ so p is a minimum when $x = \sqrt{A}$ and $y = \sqrt{A}$ and
thus the rectangle is a square.

15. The area of the window is $A = 2rh + \pi r^2/2$,

the perimeter is $p = 2r + 2h + \pi r$ thus

$h = \dfrac{1}{2}[p - (2 + \pi)r]$ so

$A = r[p - (2 + \pi)r] + \pi r^2/2$
$\quad = pr - (2 + \pi/2)r^2 \text{ for } 0 \le r \le p/(2 + \pi),$

$dA/dr = p - (4 + \pi)r, \ dA/dr = 0 \text{ when}$

$r = p/(4 + \pi).$ $d^2A/dr^2 < 0$, so A is

maximum when $r = p/(4 + \pi)$.

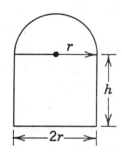

17. $V = x(12 - 2x)^2$ for $0 \le x \le 6$;
$dV/dx = 12(x - 2)(x - 6), \ dV/dx = 0$
when $x = 2$ for $0 < x < 6$. If $x = 0, 2, 6$
then $V = 0, 128, 0$ so the volume
is largest when $x = 2$ in.

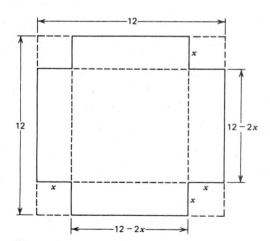

19. Let x be the length of each side of a square, then $V = x(3 - 2x)(8 - 2x) = 4x^3 - 22x^2 + 24x$ for $0 \leq x \leq 3/2$; $dV/dx = 12x^2 - 44x + 24 = 4(3x - 2)(x - 3)$, $dV/dx = 0$ when $x = 2/3$ for $0 < x < 3/2$. If $x = 0, 2/3, 3/2$ then $V = 0, 200/27, 0$ so the maximum volume is $200/27$ ft^3.

21. Let $x =$ length of each edge of base, $y =$ height, $k = \$/$cm^2 for the sides. The cost is $C = (2k)(2x^2) + (k)(4xy) = 4k(x^2 + xy)$, but $V = x^2 y = 2000$ thus $y = 2000/x^2$ so $C = 4k(x^2 + 2000/x)$ for $x > 0 . dC/dx = 4k(2x - 2000/x^2)$, $dC/dx = 0$ when $x = \sqrt[3]{1000} = 10$, $d^2 C/dx^2 > 0$ so C is least when $x = 10$, $y = 20$.

23. Let $x =$ height and width, $y =$ length. The surface area is $S = 2x^2 + 3xy$ where $x^2 y = V$, so $y = V/x^2$ and $S = 2x^2 + 3V/x$ for $x > 0$; $dS/dx = 4x - 3V/x^2$, $dS/dx = 0$ when $x = \sqrt[3]{3V/4}$, $d^2 S/dx^2 > 0$ so S is minimum when $x = \sqrt[3]{\dfrac{3V}{4}}$, $y = \dfrac{4}{3}\sqrt[3]{\dfrac{3V}{4}}$.

25. Let r and h be the dimensions shown in the figure, then the surface area is $S = 2\pi rh + 2\pi r^2$.

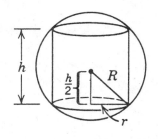

But $r^2 + \left(\dfrac{h}{2}\right)^2 = R^2$ thus $h = 2\sqrt{R^2 - r^2}$ so

$S = 4\pi r \sqrt{R^2 - r^2} + 2\pi r^2$ for $0 \leq r \leq R$,

$\dfrac{dS}{dr} = \dfrac{4\pi(R^2 - 2r^2)}{\sqrt{R^2 - r^2}} + 4\pi r$. $\dfrac{dS}{dr} = 0$ when

$$\dfrac{R^2 - 2r^2}{\sqrt{R^2 - r^2}} = -r \qquad (i)$$

$$R^2 - 2r^2 = -r\sqrt{R^2 - r^2}$$

$R^4 - 4R^2 r^2 + 4r^4 = r^2(R^2 - r^2)$

$5r^2 - 5R^2 r^2 + R^4 = 0$

and using the quadratic formula $r^2 = \dfrac{5R^2 \pm \sqrt{25R^4 - 20R^4}}{10} = \dfrac{5 \pm \sqrt{5}}{10}R^2, r = \sqrt{\dfrac{5 \pm \sqrt{5}}{10}}R$, of

which only $r = \sqrt{\dfrac{5 + \sqrt{5}}{10}}R$ satisfies (i). If $r = 0, \sqrt{\dfrac{5 + \sqrt{5}}{10}}R, 0$ then $S = 0, (5 + \sqrt{5})\pi R^2, 2\pi R^2$

so the surface area is greatest when $r = \sqrt{\dfrac{5 + \sqrt{5}}{10}}R$ and, from $h = 2\sqrt{R^2 - r^2}$,

$h = 2\sqrt{\dfrac{5 - \sqrt{5}}{10}}R.$

27. From (13), $S = 2\pi r^2 + 2\pi rh$. But $V = \pi r^2 h$ thus $h = V/(\pi r^2)$ and so $S = 2\pi r^2 + 2V/r$ for $r > 0$. $dS/dr = 4\pi r - 2V/r^2$, $dS/dr = 0$ if $r = \sqrt[3]{V/(2\pi)}$. Since $d^2 S/dr^2 = 4\pi + 4V/r^3 > 0$, the minimum surface area is achieved when $r = \sqrt[3]{V/2\pi}$ and so $h = V/(\pi r^2) = [V/(\pi r^3)]r = 2r$.

29. The surface area is $S = \pi r^2 + 2\pi r h$
where $V = \pi r^2 h = 500$ so $h = 500/(\pi r^2)$
and $S = \pi r^2 + 1000/r$ for $r > 0$;
$dS/dr = 2\pi r - 1000/r^2 = (2\pi r^3 - 1000)/r^2$,
$dS/dr = 0$ when $r = \sqrt[3]{500/\pi}$, $d^2 S/dr^2 > 0$
for $r > 0$ so S is minimum when
$r = \sqrt[3]{500/\pi}$ and
$$h = \frac{500}{\pi r^2} = \frac{500}{\pi r^3} r = \frac{500}{\pi(500/\pi)} \sqrt[3]{500/\pi}$$
$$= \sqrt[3]{500/\pi}.$$

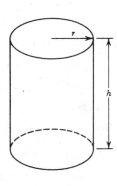

31. Let x be the length of each side of the squares and y the height of the frame, then the volume is $V = x^2 y$. The total length of the wire is L thus $8x + 4y = L$, $y = (L - 8x)/4$ so $V = x^2(L - 8x)/4 = (Lx^2 - 8x^3)/4$ for $0 \le x \le L/8$. $dV/dx = (2Lx - 24x^2)/4$, $dV/dx = 0$ for $0 < x < L/8$ when $x = L/12$. If $x = 0, L/12, L/8$ then $V = 0, L^3/1728, 0$ so the volume is greatest when $x = L/12$ and $y = L/12$.

33. Let h and r be the dimensions shown in the
figure, then the volume is $V = \frac{1}{3}\pi r^2 h$.
But $r^2 + h^2 = L^2$ thus $r^2 = L^2 - h^2$ so
$$V = \frac{1}{3}\pi(L^2 - h^2)h = \frac{1}{3}\pi(L^2 h - h^3)$$
for $0 \le h \le L$. $\dfrac{dV}{dh} = \dfrac{1}{3}\pi(L^2 - 3h^2)$.

$\dfrac{dV}{dh} = 0$ when $h = L/\sqrt{3}$. If $h = 0, L/\sqrt{3}, 0$
then $V = 0, \dfrac{2\pi}{9\sqrt{3}}L^3, 0$ so the volume is as large
as possible when $h = L/\sqrt{3}$ and $r = \sqrt{2/3}L$.

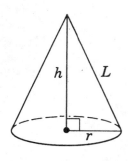

35. The area of the paper is
$A = \pi r L = \pi r \sqrt{r^2 + h^2}$, but
$V = \dfrac{1}{3}\pi r^2 h = 10$ thus $h = 30/(\pi r^2)$
so $A = \pi r \sqrt{r^2 + 900/(\pi^2 r^4)}$.
To simplify the computations let $S = A^2$,
$$S = \pi^2 r^2 \left(r^2 + \frac{900}{\pi^2 r^4}\right) = \pi^2 r^4 + \frac{900}{r^2} \text{ for } r > 0,$$

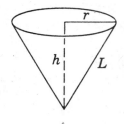

$$\frac{dS}{dr} = 4\pi^2 r^3 - \frac{1800}{r^3} = \frac{4(\pi^2 r^6 - 450)}{r^3}, \; dS/dr = 0 \text{ when } r = \sqrt[8]{450/\pi^2},$$

$$d^2 S/dr^2 > 0, \text{ so } S \text{ and hence } A \text{ is least when } r = \sqrt[8]{450/\pi^2}, \; h = \frac{30}{\pi} \sqrt[3]{\pi^2/450}.$$

37. The volume of the cone is $V = \dfrac{1}{3}\pi r^2 h$.

By similar triangles (see figure)

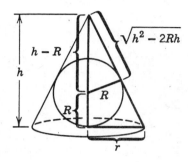

$$\frac{r}{h} = \frac{R}{\sqrt{h^2 - 2Rh}}, \; r = \frac{Rh}{\sqrt{h^2 - 2Rh}} \text{ so}$$

$$V = \frac{1}{3}\pi R^2 \frac{h^3}{h^2 - 2Rh} = \frac{1}{3}\pi R^2 \frac{h^2}{h - 2R}$$

$$\text{for } h > 2R, \; \frac{dV}{dh} = \frac{1}{3}\pi R^2 \frac{h(h - 4R)}{(h - 2R)^2},$$

$$\frac{dV}{dh} = 0 \text{ for } h > 2R \text{ when } h = 4R,$$

by the first derivative test V is minimum when $h = 4R$. If $h = 4R$ then $r = \sqrt{2}R$.

39. Let b and h be the dimensions shown in the figure, then the cross-sectional

area is $A = \dfrac{1}{2}h(5 + b)$. But $h = 5\sin\theta$

and $b = 5 + 2(5\cos\theta) = 5 + 10\cos\theta$

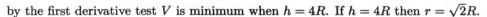

so $A = \dfrac{5}{2}\sin\theta(10 + 10\cos\theta)$

$$= 25\sin\theta(1 + \cos\theta) \text{ for } 0 \le \theta \le \pi/2.$$

$$dA/d\theta = -25\sin^2\theta + 25\cos\theta(1 + \cos\theta)$$
$$= 25(-\sin^2\theta + \cos\theta + \cos^2\theta)$$
$$= 25(-1 + \cos^2\theta + \cos\theta + \cos^2\theta)$$
$$= 25(2\cos^2\theta + \cos\theta - 1) = 25(2\cos\theta - 1)(\cos\theta + 1).$$

$dA/d\theta = 0$ for $0 < \theta < \pi/2$ when $\cos\theta = 1/2$, $\theta = \pi/3$. If $\theta = 0, \pi/3, \pi/2$ then

$A = 0, 75\sqrt{3}/4, 25$ so the cross-sectional area is greatest when $\theta = \pi/3$.

41. Let L, L_1, and L_2 be as shown in the figure, then $L = L_1 + L_2 = 8 \csc \theta + \sec \theta$,

$$\frac{dL}{d\theta} = -8 \csc \theta \cot \theta + \sec \theta \tan \theta, \ 0 < \theta < \pi/2$$

$$= -\frac{8 \cos \theta}{\sin^2 \theta} + \frac{\sin \theta}{\cos^2 \theta} = \frac{-8 \cos^3 \theta + \sin^3 \theta}{\sin^2 \theta \cos^2 \theta};$$

$\frac{dL}{d\theta} = 0$ if $\sin^3 \theta = 8 \cos^3 \theta$, $\tan^3 \theta = 8$,

$\tan \theta = 2$ which gives the absolute minimum

for L because $\lim\limits_{\theta \to 0^+} L = \lim\limits_{\theta \to \pi/2^-} L = +\infty$. If $\tan \theta = 2$, then $\csc \theta = \sqrt{5}/2$ and $\sec \theta = \sqrt{5}$ so

$L = 8(\sqrt{5}/2) + \sqrt{5} = 5\sqrt{5}$ ft.

43. **(a)** The daily profit is

$P =$ (revenue) $-$ (production cost) $= 100x - (100,000 + 50x + 0.0025x^2)$

$\quad = -100,000 + 50x - 0.0025x^2$

for $0 \le x \le 7000$, so $dP/dx = 50 - 0.005x$ and $dP/dx = 0$ when $x = 10,000$. Because 10,000 is not in the interval $[0, 7000]$, the maximum profit must occur at an endpoint. When $x = 0$, $P = -100,000$; when $x = 7000$, $P = 127,500$ so 7000 units should be manufactured and sold daily.

(b) Yes, because $dP/dx > 0$ when $x = 7000$ so profit is increasing at this production level.

45. The profit is

$P =$ (profit on nondefective) $-$ (loss on defective) $= 100(x - y) - 20y = 100x - 120y$

but $y = 0.01x + 0.00003x^2$ so $P = 100x - 120(0.01x + 0.00003x^2) = 98.8x - 0.0036x^2$ for $x > 0$, $dP/dx = 98.8 - 0.0072x$, $dP/dx = 0$ when $x = 98.8/0.0072 \approx 13,722$, $d^2P/dx^2 < 0$ so the profit is maximum at a production level of about 13,722 pounds.

47. The distance between the particles is $D = \sqrt{(1 - t - t)^2 + (t - 2t)^2} = \sqrt{5t^2 - 4t + 1}$ for $t \ge 0$. For convenience, we minimize D^2 instead, so $D^2 = 5t^2 - 4t + 1$, $dD^2/dt = 10t - 4$, which is 0 when $t = 2/5$. $d^2D^2/dt^2 > 0$ so D^2 and hence D is minimum when $t = 2/5$. The minimum distance is $D = 1/\sqrt{5}$.

49. Let $P(x, y)$ be a point on the curve $x^2 + y^2 = 1$. The distance between $P(x, y)$ and $P_0(2, 0)$ is $D = \sqrt{(x - 2)^2 + y^2}$, but $y^2 = 1 - x^2$ so $D = \sqrt{(x - 2)^2 + 1 - x^2} = \sqrt{5 - 4x}$ for $-1 \le x \le 1$,

$\frac{dD}{dx} = -\frac{2}{\sqrt{5 - 4x}}$ which has no critical points for $-1 < x < 1$. If $x = -1, 1$ then $D = 3, 1$ so the closest point occurs when $x = 1$ and $y = 0$.

51. Let (x, y) be a point on the curve, then the square of the distance between (x, y) and $(0, 2)$ is $S = x^2 + (y - 2)^2$ where $x^2 - y^2 = 1$, $x^2 = y^2 + 1$ so

$S = (y^2 + 1) + (y - 2)^2 = 2y^2 - 4y + 5$ for any y, $dS/dy = 4y - 4$, $dS/dy = 0$ when $y = 1$, $d^2S/dy^2 > 0$ so S is least when $y = 1$ and $x = \pm\sqrt{2}$.

53. If $P(x_0, y_0)$ is on the curve $y = 1/x^2$, then $y_0 = 1/x_0^2$. At P the slope of the tangent line is $-2/x_0^3$ so its equation is $y - \dfrac{1}{x_0^2} = -\dfrac{2}{x_0^3}(x - x_0)$, or $y = -\dfrac{2}{x_0^3}x + \dfrac{3}{x_0^2}$. The tangent line crosses the y-axis at $\dfrac{3}{x_0^2}$, and the x-axis at $\dfrac{3}{2}x_0$. The length of the segment then is

$L = \sqrt{\dfrac{9}{x_0^4} + \dfrac{9}{4}x_0^2}$ for $x_0 > 0$. For convenience, we minimize L^2 instead, so $L^2 = \dfrac{9}{x_0^4} + \dfrac{9}{4}x_0^2$,

$\dfrac{dL^2}{dx_0} = -\dfrac{36}{x_0^5} + \dfrac{9}{2}x_0 = \dfrac{9(x_0^6 - 8)}{2x_0^5}$, which is 0 when $x_0^6 = 8$, $x_0 = \sqrt{2}$. $\dfrac{d^2L^2}{dx_0^2} > 0$ so L^2 and hence L is minimum when $x_0 = \sqrt{2}$, $y_0 = 1/2$.

55. The area of the triangle is $A = \dfrac{1}{2}ab$. Equate slopes to get $\dfrac{b - 3}{0 - 1} = \dfrac{0 - 3}{a - 1}$, $b = \dfrac{3a}{a - 1}$ so

$A = \dfrac{3}{2}\dfrac{a^2}{a - 1}$ for $a > 1$, $\dfrac{dA}{da} = \dfrac{3a(a - 2)}{(a - 1)^2}$, $\dfrac{dA}{da} = 0$ for $a > 1$ when $a = 2$.

(a) there is no maximum for A because $\lim\limits_{a \to 1^+} A = +\infty$.

(b) by the first derivative test A is minimum when $a = 2$ so the slope is $m = \dfrac{0 - 3}{2 - 1} = -3$.

57. At each point (x, y) on the curve the slope of the tangent line is $m = \dfrac{dy}{dx} = -\dfrac{2x}{(1 + x^2)^2}$ for any x, $\dfrac{dm}{dx} = \dfrac{2(3x^2 - 1)}{(1 + x^2)^3}$, $\dfrac{dm}{dx} = 0$ when $x = \pm1/\sqrt{3}$, by the first derivative test the only relative maximum occurs at $x = -1/\sqrt{3}$, which is the absolute maximum because $\lim\limits_{x \to \pm\infty} m = 0$. The tangent line has greatest slope at the point $(-1/\sqrt{3}, 3/4)$.

59. With x and y as shown in the figure, the maximum length of pipe will be the smallest value of $L = x + y$. By similar triangles

$\dfrac{y}{8} = \dfrac{x}{\sqrt{x^2 - 16}}$, $y = \dfrac{8x}{\sqrt{x^2 - 16}}$ so

$L = x + \dfrac{8x}{\sqrt{x^2 - 16}}$ for $x > 4$,

$\dfrac{dL}{dx} = 1 - \dfrac{128}{(x^2 - 16)^{3/2}}$, $\dfrac{dL}{dx} = 0$ when

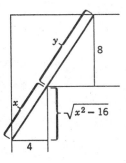

$(x^2 - 16)^{3/2} = 128$
$x^2 - 16 = 128^{2/3} = 16(2^{2/3})$
$x^2 = 16(1 + 2^{2/3})$
$x = 4(1 + 2^{2/3})^{1/2}$,

$d^2L/dx^2 = 384x/(x^2 - 16)^{5/2} > 0$ if $x > 4$ so L is smallest when $x = 4(1 + 2^{2/3})^{1/2}$. For this value of x, $L = 4(1 + 2^{2/3})^{3/2}$.

61. Let x = distance from the weaker light source, I = the intensity at that point, and k the constant of proportionality. Then

$$I = \frac{kS}{x^2} + \frac{8kS}{(90-x)^2} \text{ for } 0 < x < 90; \frac{dI}{dx} = -\frac{2kS}{x^3} + \frac{16kS}{(90-x)^3} = \frac{2kS[8x^3 - (90-x)^3]}{x^3(90-x)^3},$$

which is 0 when $8x^3 = (90-x)^3$, $2x = 90 - x$, $x = 30$. $\dfrac{dI}{dx} < 0$ if $x < 30$, and $\dfrac{dI}{dx} > 0$ if $x > 30$, so the intensity is minimum at a distance of 30 cm from the weaker source.

63. Let v = speed of light in the medium. The total time required for the light to travel from A to P to B is

$$t = \text{(total distance from } A \text{ to } P \text{ to } B)/v = \frac{1}{v}(\sqrt{(c-x)^2 + a^2} + \sqrt{x^2 + b^2}),$$

$$\frac{dt}{dx} = \frac{1}{v}\left[-\frac{c-x}{\sqrt{(c-x)^2 + a^2}} + \frac{x}{\sqrt{x^2 + b^2}}\right]$$

and $\dfrac{dt}{dx} = 0$ when $\dfrac{x}{\sqrt{x^2 + b^2}} = \dfrac{c-x}{\sqrt{(c-x)^2 + a^2}}$. But $x/\sqrt{x^2 + b^2} = \sin\theta_2$ and $(c-x)/\sqrt{(c-x)^2 + a^2} = \sin\theta_1$ thus $dt/dx = 0$ when $\sin\theta_2 = \sin\theta_1$ so $\theta_2 = \theta_1$.

65. **(a)** The rate at which the farmer walks is analogous to the speed of light in Fermat's principle.

(b) the best path occurs when $\theta_1 = \theta_2$ (see figure).

(c) by similar triangles,
$$x/(1/4) = (1-x)/(3/4)$$
$$3x = 1 - x$$
$$4x = 1$$
$$x = 1/4.$$

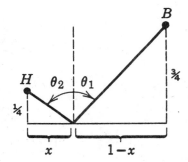

EXERCISE SET 4.8

1. $f(x) = x^2 - 2$, $f'(x) = 2x$, $x_{n+1} = x_n - \dfrac{x_n^2 - 2}{2x_n}$

$x_1 = 1$, $x_2 = 1.5$, $x_3 = 1.416666667, \cdots, x_5 = x_6 = 1.414213562$

3. $f(x) = x^3 - 6$, $f'(x) = 3x^2$, $x_{n+1} = x_n - \dfrac{x_n^3 - 6}{3x_n^2}$

$x_1 = 2$, $x_2 = 1.833333333$, $x_3 = 1.817263545, \cdots, x_5 = x_6 = 1.817120593$

5. $f(x) = x^3 - x + 3$, $f'(x) = 3x^2 - 1$, $x_{n+1} = x_n - \dfrac{x_n^3 - x_n + 3}{3x_n^2 - 1}$

$x_1 = -2$, $x_2 = -1.727272727$, $x_3 = -1.673691174, \cdots, x_5 = x_6 = -1.671699882$

7. $f(x) = x^5 + x^4 - 5$, $f'(x) = 5x^4 + 4x^3$, $x_{n+1} = x_n - \dfrac{x_n^5 + x_n^4 - 5}{5x_n^4 + 4x_n^3}$

$x_1 = 1$, $x_2 = 1.333333333$, $x_3 = 1.239420573, \cdots, x_6 = x_7 = 1.224439550$

9. $f(x) = 2x^2 + 4x - 3$, $f'(x) = 4x + 4$, $x_{n+1} = x_n - \dfrac{2x_n^2 + 4x_n - 3}{4x_n + 4}$

$x_1 = 1$, $x_2 = 0.625$, $x_3 = 0.581730769, \cdots, x_5 = x_6 = 0.581138830$

11. $f(x) = x^4 + x - 3$, $f'(x) = 4x^3 + 1$, $x_{n+1} = x_n - \dfrac{x_n^4 + x_n - 3}{4x_n^3 + 1}$

$x_1 = -2$, $x_2 = -1.645161290$, $x_3 = -1.485723955, \cdots, x_6 = x_7 = -1.452626879$

13. $f(x) = 2\sin x - x$, $f'(x) = 2\cos x - 1$, $x_{n+1} = x_n - \dfrac{2\sin x_n - x_n}{2\cos x_n - 1}$

$x_1 = 2$, $x_2 = 1.900995594$, $x_3 = 1.895511645$, $x_4 = x_5 = 1.895494267$

15. $f(x) = x - \tan x$, $f'(x) = 1 - \sec^2 x = -\tan^2 x$, $x_{n+1} = x_n + \dfrac{x_n - \tan x_n}{\tan^2 x_n}$

$x_1 = 4.5$, $x_2 = 4.493613903$, $x_3 = 4.493409655$, $x_4 = x_5 = 4.493409458$

17. **(a)** $f(x) = x^2 - a$, $f'(x) = 2x$, $x_{n+1} = \dfrac{1}{2}\left(x_n + \dfrac{a}{x_n} \right)$

(b) $a = 10$; $x_1 = 3$, $x_2 = 3.166666667$, $x_3 = 3.162280702$, $x_4 = x_5 = 3.162277660$

19. At the point of intersection, $x^3 = 0.5x - 1$, $x^3 - 0.5x + 1 = 0$. Let $f(x) = x^3 - 0.5x + 1$. By graphing $y = x^3$ and $y = 0.5x - 1$ it is evident that there is only one point of intersection and it occurs in the interval $[-2, -1]$; note that $f(-2) < 0$ and $f(-1) > 0$. $f'(x) = 3x^2 - 0.5$ so

$x_{n+1} = x_n - \dfrac{x_n^3 - 0.5x + 1}{3x_n^2 - 0.5}$; $x_1 = -1$, $x_2 = -1.2$, $x_3 = -1.166492147, \cdots,$

$x_5 = x_6 = -1.165373043$

21. The graphs of $y = x^2$ and $y = \sqrt{2x+1}$ intersect at points near $x = -0.5$ and $x = 1$; $x^2 = \sqrt{2x+1}$, $x^4 - 2x - 1 = 0$. Let $f(x) = x^4 - 2x - 1$, then $f'(x) = 4x^3 - 2$ so

$x_{n+1} = x_n - \dfrac{x_n^4 - 2x_n - 1}{4x_n^3 - 2}$.

If $x_1 = -0.5$, then $x_2 = -0.475$, $x_3 = -0.474626695$, $x_4 = x_5 = -0.474626618$;

if $x_1 = 1$, then $x_2 = 2$, $x_3 = 1.633333333, \cdots, x_8 = x_9 = 1.395336994$.

23. If $x = 1$, then $y^4 + y = 1$, $y^4 + y - 1 = 0$. Graph $z = y^4$ and $z = 1 - y$ to see that they intersect near $y = -1$ and $y = 1$. Let $f(y) = y^4 + y - 1$, then $f'(y) = 4y^3 + 1$ so $y_{n+1} = y_n - \dfrac{y_n^4 + y_n - 1}{4y_n^3 + 1}$.

If $y_1 = -1$, then $y_2 = -1.333333333$, $y_3 = -1.235807860, \cdots, y_6 = y_7 = -1.220744085$;

if $y_1 = 1$, then $y_2 = 0.8$, $y_3 = 0.731233596, \cdots, y_6 = y_7 = 0.724491959$.

25. $f'(x) = x^3 + 2x + 5$; solve $f'(x) = 0$ to find the critical points. Graph $y = x^3$ and $y = -2x - 5$ to see that they intersect at a point near $x = -1$; $f''(x) = 3x^2 + 2$ so $x_{n+1} = x_n - \dfrac{x_n^3 + 2x_n + 5}{3x_n^2 + 2}$.

$x_1 = -1$, $x_2 = -1.4$, $x_3 = -1.330964467, \cdots, x_5 = x_6 = -1.328268856$ so the minimum value of $f(x)$ occurs at $x \approx -1.328268856$ because $f''(x) > 0$; its value is approximately -4.098859132.

27. Let $f(x)$ be the square of the distance between $(1, 0)$ and any point (x, x^2) on the parabola, then $f(x) = (x-1)^2 + (x^2 - 0)^2 = x^4 + x^2 - 2x + 1$ and $f'(x) = 4x^3 + 2x - 2$. Solve $f'(x) = 0$ to find the critical points; $f''(x) = 12x^2 + 2$ so $x_{n+1} = x_n - \dfrac{4x_n^3 + 2x_n - 2}{12x_n^2 + 2} = x_n - \dfrac{2x_n^3 + x_n - 1}{6x_n^2 + 1}$.

$x_1 = 1$, $x_2 = 0.714285714$, $x_3 = 0.605168701, \cdots, x_6 = x_7 = 0.589754512$; the coordinates are approximately $(0.589754512, 0.347810385)$.

29. Let s be the arc length, and L the length of the chord, then $s = 1.5L$. But $s = r\theta$ and $L = 2r\sin(\theta/2)$ so $r\theta = 3r\sin(\theta/2)$, $\theta - 3\sin(\theta/2) = 0$. Let $f(\theta) = \theta - 3\sin(\theta/2)$, then $f'(\theta) = 1 - 1.5\cos(\theta/2)$ so $\theta_{n+1} = \theta_n - \dfrac{\theta_n - 3\sin(\theta_n/2)}{1 - 1.5\cos(\theta_n/2)}$.

$\theta_1 = 3$, $\theta_2 = 2.991592920$, $\theta_3 = 2.991563137$, $\theta_4 = \theta_5 = 2.991563136$ rad so $\theta \approx 171°$.

EXERCISE SET 4.9

1. $f(2) = f(4) = 0$, $f'(x) = 2x - 6$, $2c - 6 = 0$, $c = 3$

3. $f(\pi/2) = f(3\pi/2) = 0$, $f'(x) = -\sin x$, $-\sin c = 0$, $c = \pi$

5. $f(0) = f(4) = 0$, $f'(x) = \dfrac{1}{2} - \dfrac{1}{2\sqrt{x}}$, $\dfrac{1}{2} - \dfrac{1}{2\sqrt{c}} = 0$, $c = 1$

7. $f(-4) = 12$, $f(6) = 42$, $f'(x) = 2x + 1$, $2c + 1 = \dfrac{42 - 12}{6 - (-4)} = 3$, $c = 1$

9. $f(0) = 1$, $f(3) = 2$, $f'(x) = \dfrac{1}{2\sqrt{x+1}}$, $\dfrac{1}{2\sqrt{c+1}} = \dfrac{2-1}{3-0} = \dfrac{1}{3}$

$\sqrt{c+1} = 3/2$, $c + 1 = 9/4$, $c = 5/4$

11. $f(-5) = 0$, $f(3) = 4$, $f'(x) = -\dfrac{x}{\sqrt{25 - x^2}}$, $-\dfrac{c}{\sqrt{25 - c^2}} = \dfrac{4 - 0}{3 - (-5)} = \dfrac{1}{2}$, $-2c = \sqrt{25 - c^2}$,

$4c^2 = 25 - c^2$, $c^2 = 5$, $c = -\sqrt{5}$

(we reject $c = \sqrt{5}$ because it does not satisfy the equation $-2c = \sqrt{25 - c^2}$)

13. **(a)** $f'(x) = \sec^2 x$, $\sec^2 c = 0$ has no solution   **(b)** $\tan x$ is not continuous on $[0, \pi]$

15. Let $f(x) = \sin x$ and $x \neq y$. By the Mean-Value Theorem there is a number c between x and y such that

$\dfrac{\sin x - \sin y}{x - y} = \cos c$, $\dfrac{|\sin x - \sin y|}{|x - y|} = |\cos c| \leq 1$

so $|\sin x - \sin y| \leq |x - y|$, which also holds when $x = y$.

17. Let $f(x) = \sqrt{x}$. By the Mean-Value Theorem there is a number c between x and y such that

$\dfrac{\sqrt{y} - \sqrt{x}}{y - x} = \dfrac{1}{2\sqrt{c}} < \dfrac{1}{2\sqrt{x}}$ for c in (x, y), thus $\sqrt{y} - \sqrt{x} < \dfrac{y - x}{2\sqrt{x}}$;

multiply through and rearrange to get $\sqrt{xy} < \dfrac{1}{2}(x + y)$.

19. $f'(x) = 2a_2 x + a_1$,

$2a_2 c + a_1 = \dfrac{(a_2 b^2 + a_1 b + a_0) - (a_2 a^2 + a_1 a + a_0)}{b - a}$

$= \dfrac{a_2(b^2 - a^2) + a_1(b - a)}{b - a} = a_2(b + a) + a_1$, so $c = \dfrac{1}{2}(b + a)$.

21. $f(0) = f(1) = 0$, $f'(x) = 3ax^2 + 2bx - (a + b)$, so there is at least one number c in $(0, 1)$ where $f'(c) = 0$.

23. Let $f(x) = x^3 + 4x - 1$. Assume that $f(x) = 0$ has at least two distinct real solutions r_1 and r_2. Then $f(r_1) = f(r_2) = 0$ and so by Rolle's Theorem there is at least one number c between r_1 and r_2 where $f'(c) = 0$. But $f'(x) = 3x^2 + 4$ is never zero, so $f(x) = 0$ must have fewer than two distinct real solutions.

25. Assume that $f(x) = 0$ has at least four distinct real solutions $r_1 < r_2 < r_3 < r_4$, then by Rolle's Theorem there is at least one number in each of the intervals (r_1, r_2), (r_2, r_3) and (r_3, r_4) so that $f'(x) = 0$ at least three times. Apply Rolle's Theorem to $f'(x)$ to show that $f''(x) = 0$ at least twice; again to $f''(x)$ to show that $f'''(x) = 0$ at least once. But $f'''(x) = 60x^2 + 24ax + 6b$, and $f'''(x) = 0$ if $10x^2 + 4ax + b = 0$. Use the quadratic formula to get $x = \dfrac{-4a \pm \sqrt{16a^2 - 40b}}{20}$, which has no real solutions if $16a^2 - 40b < 0$, $16a^2 < 40b$, $2a^2 < 5b$.

27. $\dfrac{d}{dx}[f^2(x) + g^2(x)] = 2f(x)f'(x) + 2g(x)g'(x) = 2f(x)g(x) + 2g(x)[-f(x)] = 0$,

so $f^2(x) + g^2(x)$ is constant.

29. $f'(x) = 3(x-1)^2$, $g'(x) = (x^2+3) + 2x(x-3) = 3x^2 - 6x + 3 = 3(x^2 - 2x + 1) = 3(x-1)^2$, so $f'(x) = g'(x)$ and hence $f(x) - g(x) = k$. Expand $f(x)$ and $g(x)$ to get

$f(x) - g(x) = (x^3 - 3x^2 + 3x - 1) - (x^3 - 3x^2 + 3x - 9) = 8$.

31. If $f'(x) = g'(x)$, then $f(x) = g(x) + k$. Let $x = 1$,

$f(1) = g(1) + k = (1)^3 - 4(1) + 6 + k = 3 + k = 2$, so $k = -1$. $f(x) = x^3 - 4x + 5$.

33. Let $h = f - g$, then h is continuous on $[a, b]$, differentiable on (a, b), and $h(a) = f(a) - g(a) = 0$, $h(b) = f(b) - g(b) = 0$. By Rolle's Theorem there is some c in (a, b) where $h'(c) = 0$. But $h'(c) = f'(c) - g'(c)$ so $f'(c) - g'(c) = 0$, $f'(c) = g'(c)$.

35. Similar to the proof of part (a) with $f'(c) = 0$.

37. From the Mean-Value Theorem there is a point c in (a, b) where

$f'(c) = \dfrac{f(b) - f(a)}{b - a} = \dfrac{0}{b - a} = 0.$

39. Similar to proof given in text; assume that $f(x) < 0$ and replace the word "maximum" by "minimum".

41. Let $s(t)$ be the position function of the automobile for $0 \le t \le 5$, then by the Mean-Value Theorem there is at least one point c in $(0, 5)$ where

$s'(c) = v(c) = [s(5) - s(0)]/(5 - 0) = 4/5 = 0.8 \text{ mi/min} = 48 \text{ mi/hr}.$

EXERCISE SET 4.10

1. **(a)** positive, negative, slowing down   **(b)** positive, positive, speeding up
    (c) negative, positive, slowing down

3. **(a)** left because $v = ds/dt < 0$ at t_0.
    (b) negative because $a = d^2s/dt^2$ and the curve is concave down at $t_0(d^2s/dt^2 < 0)$.
    (c) speeding up because v and a have the same sign.
    (d) $v < 0$ and $a > 0$ at t_1 so the particle is slowing down because v and a have opposite signs.

5. **(a)** At 60 mi/hr the slope of the estimated tangent line is about 4.6 mi/hr per sec. Use 1 mi $= 5,280$ ft and 1 hr $= 3600$ sec to get $a = dv/dt \approx 4.6(5,280)/(3600) \approx 6.7$ ft/sec^2.

   (b) The slope of the tangent to the curve is maximum at $t = 0$.

7. $v = 3t^2 - 12t$, $a = 6t - 12$

t	s	v	$|v|$	a	direction; motion
1	-5	-9	9	-6	left; speeding up
2	-16	-12	12	0	left; neither
3	-27	-9	9	6	left; slowing down
4	-32	0	0	12	stopped
5	-25	15	15	18	right; speeding up

9. **(a)** $v = 10t - 22$, speed $= |v| = |10t - 22|$. $d|v|/dt$ does not exist at $t = 2.2$ which is the only critical point. If $t = 1, 2.2, 3$ then $|v| = 12, 0, 8$. The maximum speed is 12.

   (b) the distance from the origin is $|s| = |5t^2 - 22t| = |t(5t - 22)|$, but $t(5t - 22) < 0$ for $1 \le t \le 3$ so $|s| = -(5t^2 - 22t) = 22t - 5t^2$, $d|s|/dt = 22 - 10t$, thus the only critical point is $t = 2.2$. $d^2|s|/dt^2 < 0$ so the particle is farthest from the origin when $t = 2.2$. Its position is $s = 5(2.2)^2 - 22(2.2) = -24.2$.

11. $s = 1 + 6t - t^2$
    $v = 2(3 - t)$
    $a = -2$

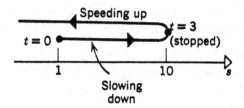

13. $s = t^3 - 9t^2 + 24t$
    $v = 3(t - 2)(t - 4)$
    $a = 6(t - 3)$

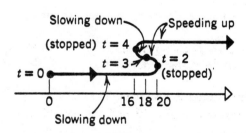

15. $s = \begin{cases} \cos t, & 0 \le t \le 2\pi \\ 1, & t > 2\pi \end{cases}$

$v = \begin{cases} -\sin t, & 0 \le t \le 2\pi \\ 0, & t > 2\pi \end{cases}$

$a = \begin{cases} -\cos t, & 0 \le t < 2\pi \\ 0, & t > 2\pi \end{cases}$

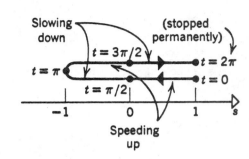

17. $s = 4t^{3/2} - 3t^2$, $v = 6t^{1/2} - 6t$, $a = 3t^{-1/2} - 6$.

    (a) $a = 0$ when $3t^{-1/2} = 6$, $t = 1/4$; $s = 5/16$, $v = 3/2$.

    (b) $v = 0$ when $6t^{1/2}(1 - t^{1/2}) = 0$ for $t > 0$, $t = 1$; $s = 1$, $a = -3$.

19. **(a)** $a = \dfrac{dv}{dt} = \dfrac{dv}{ds}\dfrac{ds}{dt} = v\dfrac{dv}{ds}$ because $v = \dfrac{ds}{dt}$.

    (b) $v = \dfrac{3}{2\sqrt{3t+7}} = \dfrac{3}{2s}$; $\dfrac{dv}{ds} = -\dfrac{3}{2s^2}$; $a = -\dfrac{9}{4s^3} = -9/500$.

21. **(a)** $s_1 = s_2$ if they collide, so $\dfrac{1}{2}t^2 - t + 3 = -\dfrac{1}{4}t^2 + t + 1$, $\dfrac{3}{4}t^2 - 2t + 2 = 0$ which has no real solution.

    (b) Find the minimum value of $D = |s_1 - s_2| = \left|\frac{3}{4}t^2 - 2t + 2\right|$. From part (a), $\dfrac{3}{4}t^2 - 2t + 2$

    is never zero, and for $t = 0$ it is positive, hence it is always positive, so $D = \dfrac{3}{4}t^2 - 2t + 2$.

    $\dfrac{dD}{dt} = \dfrac{3}{2}t - 2 = 0$ when $t = \dfrac{4}{3}$. $\dfrac{d^2D}{dt^2} > 0$ so D is minimum when $t = \frac{4}{3}$, $D = \dfrac{2}{3}$.

    (c) $v_1 = t - 1$, $v_2 = -\dfrac{1}{2}t + 1$. $v_1 < 0$ if $0 \le t < 1$, $v_1 > 0$ if $t > 1$; $v_2 < 0$ if $t > 2$, $v_2 > 0$ if $0 \le t < 2$. They are moving in opposite directions during the intervals $0 \le t < 1$ and $t > 2$.

23. **(a)** From the estimated tangent to the graph at the point where $v = 2000$, $dv/ds \approx -1.25$ ft/sec per ft.

    (b) $a = v\,dv/ds \approx (2000)(-1.25) = -2500$ ft/sec^2.

25. **(a)** $\displaystyle \lim_{t_1 \to t_0} v_{ave} = \lim_{t_1 \to t_0} \dfrac{s(t_1) - s(t_0)}{t_1 - t_0} = s'(t_0) = v(t_0)$

    (b) similar to part (a) with a in place of v.

TECHNOLOGY EXERCISES 4

1. **(c)** minimum: $(-2.111985, -0.355116)$
maximum: $(0.372591, 2.012931)$

3. relative minimum -0.232466 at $x = 0.450184$

5. relative maximum 0.355977; at $x = 1.244155$;
relative minimum -0.876839 at $x = -0.886352$

7. **(b)** relative maximum at $x = -1.414214$ and $x = 1.5$,
relative minimum at $x = 1.414214$ and $x = 2$,
inflection points at $x = -0.657920, x = 1.455739$, and $x = 1.827181$.

9. $y = x$

11. **(a)** Let x be the length of each edge of the square base, y the height of the box, and C the cost (in dollars) of one box, then $C = 0.08(2x^2 + 4xy) + 0.03(5x + y)$ where $x^2 y = 4$ so $y = 4/x^2$ and thus

$$C = 0.08(2x^2 + 16/x) + 0.03(5x + 4/x^2), \quad x > 0$$

$dC/dx = 0$ when $x = 1.508244$. This occurs when C is minimum so the dimensions are: base 1.508 ft by 1.508 ft, height 1.758 ft.

(b) The dollar cost of one box is 1.491626 so the bid should be
$1.17(75,000)(1.491626) \approx \$130,890$.

(c) If the box is a cube with x as the length of each edge, then $x^3 = 4$, $x = 4^{1/3}$ which when substituted into the formula for C in part (a) yields $C = 1.495256$ so the bid should be $1.17(75,000)(1.495256) \approx \$131,209$, which is \$319 greater than the bid in part (b).

13. **(a)** Find t so that the velocity $v = ds/dt > 0$. The particle is moving in the positive direction for $0 \le t \le 0.64$ sec.

(b) Find the maximum value of $|v|$ to obtain:
maximum speed $= 1.05$ m/sec when $t = 1.10$ sec.

15. Find t so that $N'(t)$ is maximum. The size of the population is increasing most rapidly when $t = 8.4$ years.

17. Solve $\phi - 0.0934 \sin \phi = 2\pi(1)/1.88$ to get $\phi = 3.325078$ so
$r = 228 \times 10^6 (1 - 0.0934 \cos \phi) = 248.938 \times 10^6$ km.

CHAPTER 5
Integration

EXERCISE SET 5.2

1. $x^9/9 + C$

3. $\dfrac{7}{12}x^{12/7} + C$

5. $4\displaystyle\int t^{-1/2}dt = 8\sqrt{t} + C$

7. $\displaystyle\int x^{7/2}dx = \dfrac{2}{9}x^{9/2} + C$

9. $\displaystyle\int (x^{-3} + x^{1/2} - 3x^{1/4} + x^2)dx = -\dfrac{1}{2}x^{-2} + \dfrac{2}{3}x^{3/2} - \dfrac{12}{5}x^{5/4} + \dfrac{1}{3}x^3 + C$

11. $\displaystyle\int (7y^{-3/4} - y^{1/3} + 4y^{1/2})dy = 28y^{1/4} - \dfrac{3}{4}y^{4/3} + \dfrac{8}{3}y^{3/2} + C$

13. $\displaystyle\int (x + x^4)dx = x^2/2 + x^5/5 + C$

15. $\displaystyle\int x^{1/3}(4 - 4x + x^2)dx = \int (4x^{1/3} - 4x^{4/3} + x^{7/3})dx = 3x^{4/3} - \dfrac{12}{7}x^{7/3} + \dfrac{3}{10}x^{10/3} + C$

17. $\displaystyle\int (x + 2x^{-2} - x^{-4})dx = x^2/2 - 2/x + 1/(3x^3) + C$

19. $-4\cos x + 2\sin x + C$

21. $\displaystyle\int (\sec^2 x + \sec x \tan x)dx = \tan x + \sec x + C$

23. $\displaystyle\int (\sec x \tan x + 1)dx = \sec x + x + C$

25. $\displaystyle\int \sec x \tan x \, dx = \sec x + C$

27. $\displaystyle\int (1 + \sin\theta)d\theta = \theta - \cos\theta + C$

29. $\displaystyle\int (\cos\theta - 5\sec^2\theta)d\theta = \sin\theta - 5\tan\theta + C$

31. $F(x) = \displaystyle\int x^{1/3}dx = \dfrac{3}{4}x^{4/3} + C, \; F(1) = \dfrac{3}{4} + C = 2, \; C = 5/4; \; F(x) = \dfrac{3}{4}x^{4/3} + \dfrac{5}{4}$

33. $f'(x) = \dfrac{2}{3}x^{3/2} + C_1;\ f(x) = \dfrac{4}{15}x^{5/2} + C_1 x + C_2$

35. $dy/dx = 2x + 1, y = \displaystyle\int (2x+1)dx = x^2 + x + C; y = 0$ when $x = -3$

so $(-3)^2 + (-3) + C = 0, C = -6$ thus $y = x^2 + x - 6$.

37. $dy/dx = \displaystyle\int 6x\,dx = 3x^2 + C_1$. The slope of the tangent line is -3 so $dy/dx = -3$ when $x = 1$.

Thus $3(1)^2 + C_1 = -3,\ C_1 = -6$ so $dy/dx = 3x^2 - 6,\ y = \displaystyle\int (3x^2 - 6)dx = x^3 - 6x + C_2$; If

$x = 1$, then $y = 5 - 3(1) = 2$ so $(1)^2 - 6(1) + C_2 = 2, C_2 = 7$ thus $y = x^3 - 6x + 7$.

39. $\dfrac{d}{dx}\left[\sqrt{x^3 + 5}\right] = \dfrac{3x^2}{2\sqrt{x^3+5}}$ so $\displaystyle\int \dfrac{3x^2}{2\sqrt{x^3+5}}dx = \sqrt{x^3+5} + C$.

41. $\dfrac{d}{dx}\left[\sin\left(2\sqrt{x}\right)\right] = \dfrac{\cos\left(2\sqrt{x}\right)}{\sqrt{x}}$ so $\displaystyle\int \dfrac{\cos\left(2\sqrt{x}\right)}{\sqrt{x}}dx = \sin\left(2\sqrt{x}\right) + C$

43. **(a)** $F'(x) = G'(x) = 3x + 4$.

(b) $F(x) = (9x^2 + 24x + 16)/6 = 3x^2/2 + 4x + 8/3 = G(x) + 8/3$

45. $f(x) = \dfrac{d}{dx}(5x^3 - 3x + C) = 15x^2 - 3$. **47.** $\displaystyle\int (\sec^2 x - 1)dx = \tan x - x + C$

49. $\dfrac{1}{2}\displaystyle\int (1 - \cos x)dx = \dfrac{1}{2}(x - \sin x) + C$

EXERCISE SET 5.3

1. **(a)** $\displaystyle\int u^{23}\,du = u^{24}/24 + C = (x^2 + 1)^{24}/24 + C$

(b) $-\displaystyle\int u^3\,du = -u^4/4 + C = -(\cos^4 x)/4 + C$

(c) $2\displaystyle\int \sin u\,du = -2\cos u + C = -2\cos\sqrt{x} + C$

(d) $\dfrac{3}{8}\displaystyle\int u^{-1/2}\,du = \dfrac{3}{4}u^{1/2} + C = \dfrac{3}{4}\sqrt{4x^2+5} + C$

3. (a) $-\int u\,du = -\frac{1}{2}u^2 + C = -\frac{1}{2}\cot^2 x + C$

   (b) $\int u^9 du = \frac{1}{10}u^{10} + C = \frac{1}{10}(1 + \sin t)^{10} + C$

   (c) $\int (u-1)^2 u^{1/2} du = \int (u^{5/2} - 2u^{3/2} + u^{1/2})du = \frac{2}{7}u^{7/2} - \frac{4}{5}u^{5/2} + \frac{2}{3}u^{3/2} + C$

   $$= \frac{2}{7}(1+x)^{7/2} - \frac{4}{5}(1+x)^{5/2} + \frac{2}{3}(1+x)^{3/2} + C$$

   (d) $\int \csc^2 u\,du = -\cot u + C = -\cot(\sin x) + C$

5. $u = 2 - x^2,\ du = -2x\,dx;\ -\frac{1}{2}\int u^3 du = -u^4/8 + C = -(2-x^2)^4/8 + C$

7. $u = 8x,\ du = 8dx;\ \frac{1}{8}\int \cos u\,du = \frac{1}{8}\sin u + C = \frac{1}{8}\sin 8x + C$

9. $u = 4x,\ du = 4dx;\ \frac{1}{4}\int \sec u \tan u\,du = \frac{1}{4}\sec u + C = \frac{1}{4}\sec 4x + C$

11. $u = 7t^2 + 12,\ du = 14t\,dt;\ \frac{1}{14}\int u^{1/2} du = \frac{1}{21}u^{3/2} + C = \frac{1}{21}(7t^2 + 12)^{3/2} + C$

13. $u = x^3 + 1,\ du = 3x^2 dx;\ \frac{1}{3}\int u^{-1/2} du = \frac{2}{3}u^{1/2} + C = \frac{2}{3}\sqrt{x^3 + 1} + C$

15. $u = 4x^2 + 1,\ du = 8x\,dx;\ \frac{1}{8}\int u^{-3} du = -\frac{1}{16}u^{-2} + C = -\frac{1}{16}(4x^2 + 1)^{-2} + C$

17. $u = 5/x,\ du = -(5/x^2)dx;\ -\frac{1}{5}\int \sin u\,du = \frac{1}{5}\cos u + C = \frac{1}{5}\cos(5/x) + C$

19. $u = x^3,\ du = 3x^2 dx;\ \frac{1}{3}\int \sec^2 u\,du = \frac{1}{3}\tan u + C = \frac{1}{3}\tan(x^3) + C$

21. $u = \sin 3t,\ du = 3\cos 3t\,dt;\ \frac{1}{3}\int u^5 du = \frac{1}{18}u^6 + C = \frac{1}{18}\sin^6 3t + C$

23. $u = 2 - \sin 4\theta,\ du = -4\cos 4\theta\,d\theta;\ -\frac{1}{4}\int u^{1/2} du = -\frac{1}{6}u^{3/2} + C = -\frac{1}{6}(2 - \sin 4\theta)^{3/2} + C$

25. $u = \sec 2x$, $du = 2\sec 2x \tan 2x\, dx$; $\dfrac{1}{2}\displaystyle\int u^2\, du = \dfrac{1}{6}u^3 + C = \dfrac{1}{6}\sec^3 2x + C$

27. $u = \cos 3\theta$, $du = -3\sin 3\theta\, d\theta$; $-\dfrac{1}{3}\displaystyle\int \sec^2 u\, du = -\dfrac{1}{3}\tan u + C = -\dfrac{1}{3}\tan(\cos 3\theta) + C$

29. $u = \sin(a + bx)$, $du = b\cos(a + bx)dx$

$\dfrac{1}{b}\displaystyle\int u^n\, du = \dfrac{1}{b(n+1)}u^{n+1} + C = \dfrac{1}{b(n+1)}\sin^{n+1}(a + bx) + C$

31. $u = x - 3$, $x = u + 3$, $dx = du$

$\displaystyle\int (u + 3)u^{1/2}\, du = \int (u^{3/2} + 3u^{1/2})du = \dfrac{2}{5}u^{5/2} + 2u^{3/2} + C = \dfrac{2}{5}(x - 3)^{5/2} + 2(x - 3)^{3/2} + C$

33. $u = y + 1$, $y = u - 1$, $dy = du$

$\displaystyle\int \dfrac{u - 1}{u^{1/2}}\, du = \int (u^{1/2} - u^{-1/2})du = \dfrac{2}{3}u^{3/2} - 2u^{1/2} + C = \dfrac{2}{3}(y + 1)^{3/2} - 2(y + 1)^{1/2} + C$

35. $u = 3\theta$, $du = 3\, d\theta$

$\dfrac{1}{3}\displaystyle\int \tan^2 u\, du = \dfrac{1}{3}\int (\sec^2 u - 1)du = \dfrac{1}{3}(\tan u - u) + C = \dfrac{1}{3}(\tan 3\theta - 3\theta) + C$

37. $u = \sqrt{x - 1}$, $u^2 = x - 1$, $x = u^2 + 1$, $dx = 2u\, du$

$\displaystyle\int (u^2 + 1)^2 u(2u)du = 2\int (u^4 + 2u^2 + 1)u^2\, du = 2\int (u^6 + 2u^4 + u^2)du$

$= \dfrac{2}{7}u^7 + \dfrac{4}{5}u^5 + \dfrac{2}{3}u^3 + C$

$= \dfrac{2}{7}(x - 1)^{7/2} + \dfrac{4}{5}(x - 1)^{5/2} + \dfrac{2}{3}(x - 1)^{3/2} + C$

39. **(a)** First method: $\displaystyle\int (25x^2 - 10x + 1)dx = \dfrac{25}{3}x^3 - 5x^2 + x + C$;

second method: $\dfrac{1}{5}\displaystyle\int u^2\, du = \dfrac{1}{15}u^3 + C = \dfrac{1}{15}(5x - 1)^3 + C$

(b) $\dfrac{1}{15}(5x - 1)^3 + C = \dfrac{1}{15}(125x^3 - 75x^2 + 15x - 1) + C = \dfrac{25}{3}x^3 - 5x^2 + x - \dfrac{1}{15} + C$;
the answers differ by a constant.

41. $f(x) = \displaystyle\int (6 - 5\ \sin 2x)dx = 6x + \dfrac{5}{2}\cos 2x + C$,

$f(0) = \dfrac{5}{2} + C = 3$, $C = \dfrac{1}{2}$ so $f(x) = 6x + \dfrac{5}{2}\cos 2x + \dfrac{1}{2}$

43. $u = 3x + 2$, $du = 3\,dx$; $\dfrac{1}{3}\displaystyle\int f'(u)du = \dfrac{1}{3}f(u) + C = \dfrac{1}{3}f(3x+2) + C$

45. $u = 2/x$, $du = -(2/x^2)dx$; $-\dfrac{1}{2}\displaystyle\int f'(u)du = -\dfrac{1}{2}f(u) + C = -\dfrac{1}{2}f(2/x) + C$

EXERCISE SET 5.4

1. **(a)** $1 + 8 + 27 = 36$ **(b)** $5 + 8 + 11 + 14 + 17 = 55$
    (c) $20 + 12 + 6 + 2 + 0 + 0 = 40$ **(d)** $1 + 1 + 1 + 1 + 1 + 1 = 6$

3. $\displaystyle\sum_{k=1}^{10} k$ **5.** $\displaystyle\sum_{k=1}^{49} k(k+1)$ **7.** $\displaystyle\sum_{k=1}^{10} 2k$

9. $\displaystyle\sum_{k=1}^{6} (-1)^{k+1}(2k-1)$ **11.** $\displaystyle\sum_{k=1}^{5} (-1)^k \dfrac{1}{k}$

13. $\displaystyle\sum_{k=1}^{4} \sin \dfrac{(2k-1)\pi}{8}$ **15.** $\displaystyle\sum_{k=1}^{5} \dfrac{k}{k+1}$

17. **(a)** $\displaystyle\sum_{k=1}^{5} (-1)^{k+1} a_k$ **(b)** $\displaystyle\sum_{k=0}^{5} (-1)^{k+1} b_k$ **(c)** $\displaystyle\sum_{k=0}^{n} a_k x^k$ **(d)** $\displaystyle\sum_{k=0}^{5} a^{5-k} b^k$

19. $\displaystyle\sum_{k=1}^{100} k - \sum_{k=1}^{2} k = \dfrac{1}{2}(100)(100+1) - (1+2) = 5050 - 3 = 5047$

21. $\dfrac{1}{6}(20)(21)(41) = 2{,}870$

23. $4\displaystyle\sum_{k=1}^{6} k^3 - 2\sum_{k=1}^{6} k + \sum_{k=1}^{6} 1 = 4\left[\dfrac{1}{4}(6)^2(7)^2\right] - 2\left[\dfrac{1}{2}(6)(7)\right] + 6 = 1728$

25. $\displaystyle\sum_{k=1}^{30} k(k^2 - 4) = \sum_{k=1}^{30}(k^3 - 4k) = \sum_{k=1}^{30} k^3 - 4\sum_{k=1}^{30} k = \dfrac{1}{4}(30)^2(31)^2 - 4 \times \dfrac{1}{2}(30)(31) = 214{,}365$

27. $\displaystyle\sum_{k=1}^{n} \dfrac{3k}{n} = \dfrac{3}{n}\sum_{k=1}^{n} k = \dfrac{3}{n} \times \dfrac{1}{2}n(n+1) = \dfrac{3}{2}(n+1)$

29. $\displaystyle\sum_{k=1}^{n-1}\frac{k^3}{n^2} = \frac{1}{n^2}\sum_{k=1}^{n-1}k^3 = \frac{1}{n^2}\times\frac{1}{4}(n-1)^2n^2 = \frac{1}{4}(n-1)^2$

31. $\displaystyle\frac{1+2+3+\cdots+n}{n^2} = \sum_{k=1}^{n}\frac{k}{n^2} = \frac{1}{n^2}\sum_{k=1}^{n}k = \frac{1}{n^2}\times\frac{1}{2}n(n+1) = \frac{n+1}{2n};\ \lim_{n\to+\infty}\frac{n+1}{2n} = \frac{1}{2}.$

33. $\displaystyle\sum_{k=1}^{n}\frac{5k}{n^2} = \frac{5}{n^2}\sum_{k=1}^{n}k = \frac{5}{n^2}\times\frac{1}{2}n(n+1) = \frac{5(n+1)}{2n};\ \lim_{n\to+\infty}\frac{5(n+1)}{2n} = \frac{5}{2}.$

35. $\displaystyle\sum_{k=1}^{n-1}\left(\frac{9}{n}-\frac{k}{n^2}\right) = \frac{9}{n}\sum_{k=1}^{n-1}1 - \frac{1}{n^2}\sum_{k=1}^{n-1}k = \frac{9}{n}(n-1) - \frac{1}{n^2}\times\frac{1}{2}(n-1)(n) = \frac{17}{2}\left(\frac{n-1}{n}\right);$

$\displaystyle\lim_{n\to+\infty}\frac{17}{2}\left(\frac{n-1}{n}\right) = \frac{17}{2}.$

37. $\displaystyle 1+3+5+\cdots+(2n-1) = \sum_{k=1}^{n}(2k-1) = 2\sum_{k=1}^{n}k - \sum_{k=1}^{n}1 = 2\times\frac{1}{2}n(n+1) - n = n^2.$

39. $(3^5 - 3^4) + (3^6 - 3^5) + \cdots + (3^{17} - 3^{16}) = 3^{17} - 3^4$

41. $\displaystyle\left(\frac{1}{2^2}-\frac{1}{1^2}\right) + \left(\frac{1}{3^2}-\frac{1}{2^2}\right) + \cdots + \left(\frac{1}{20^2}-\frac{1}{19^2}\right) = \frac{1}{20^2} - 1 = -\frac{399}{400}$

43. $(a_1 - a_0) + (a_2 - a_1) + \cdots + (a_n - a_{n-1}) = a_n - a_0$

45. **(a)** $\displaystyle\sum_{k=1}^{n}\frac{1}{(2k-1)(2k+1)} = \frac{1}{2}\sum_{k=1}^{n}\left(\frac{1}{2k-1}-\frac{1}{2k+1}\right)$

$\displaystyle = \frac{1}{2}\left[\left(1-\frac{1}{3}\right) + \left(\frac{1}{3}-\frac{1}{5}\right) + \left(\frac{1}{5}-\frac{1}{7}\right) + \cdots + \left(\frac{1}{2n-1}-\frac{1}{2n+1}\right)\right]$

$\displaystyle = \frac{1}{2}\left[1-\frac{1}{2n+1}\right] = \frac{n}{2n+1}$

(b) $\displaystyle\lim_{n\to+\infty}\frac{n}{2n+1} = \frac{1}{2}$

47. **(a)** $n+n+\cdots+n = n^2$
(n terms)

(b) -3

(c) $\displaystyle x\sum_{k=1}^{n}k = \frac{1}{2}n(n+1)x$

(d) $c+c+\cdots+c = (n-m+1)c$
($n-m+1$ terms)

49. (a) $\displaystyle\sum_{k=0}^{14}(k+4)(k+1)$ (b) $\displaystyle\sum_{k=5}^{19}(k-1)(k-4)$

51. $\displaystyle\sum_{k=1}^{18}k\sin\frac{\pi}{k}$ **53.** both are valid

55. (a) $\displaystyle\sum_{k=0}^{19}3^{k+1}=\sum_{k=0}^{19}3(3^k)=\frac{3(1-3^{20})}{1-3}=\frac{3}{2}(3^{20}-1)$

(b) $\displaystyle\sum_{k=0}^{25}2^{k+5}=\sum_{k=0}^{25}2^5 2^k=\frac{2^5(1-2^{26})}{1-2}=2^{31}-2^5$

(c) $\displaystyle\sum_{k=0}^{100}(-1)\left(\frac{-1}{2}\right)^k=\frac{(-1)(1-(-1/2)^{101})}{1-(-1/2)}=-\frac{2}{3}(1+1/2^{101})$

57. $\displaystyle\sum_{i=1}^{4}\left[\sum_{j=1}^{5}i+\sum_{j=1}^{5}j\right]\sum_{i=1}^{4}\left[5i+\frac{1}{2}(5)(6)\right]=5\sum_{i=1}^{4}i+\sum_{i=1}^{4}15=5\cdot\frac{1}{2}(4)(5)+(4)(15)=110$

59. $\displaystyle\sum_{k=1}^{n}(a_k-b_k)=(a_1-b_1)+(a_2-b_2)+\cdots+(a_n-b_n)$

$$=(a_1+a_2+\cdots+a_n)-(b_1+b_2+\cdots+b_n)=\sum_{k=1}^{n}a_k-\sum_{k=1}^{n}b_k$$

EXERCISE SET 5.5

1. $\Delta x=\dfrac{6-2}{4}=1,\ f(x)=3x+1$

(a) $x_k^*=2,3,4,5;\ \displaystyle\sum_{k=1}^{4}f(x_k^*)\Delta x=(7+10+13+16)(1)=46$

(b) $x_k^*=3,4,5,6;\ \displaystyle\sum_{k=1}^{4}f(x_k^*)\Delta x=(10+13+16+19)(1)=58$

3. $\Delta x = \pi/4$, $f(x) = \cos x$

(a) $x_k^* = -\pi/2, -\pi/4, 0, \pi/4$

$$\sum_{k=1}^{4} f(x_k^*)\Delta x = (0 + 1/\sqrt{2} + 1 + 1/\sqrt{2})(\pi/4) = (1 + \sqrt{2})\pi/4 \approx 1.896$$

(b) $x_k^* = -\pi/4, 0, \pi/4, \pi/2$

$$\sum_{k=1}^{4} f(x_k^*)\Delta x = (1/\sqrt{2} + 1 + 1/\sqrt{2} + 0)(\pi/4) = (1 + \sqrt{2})\pi/4 \approx 1.896$$

5. $\Delta x = \dfrac{3}{n}$, $x_k^* = 1 + \dfrac{3}{n}k$; $f(x_k^*)\Delta x = \dfrac{1}{2}x_k^*\Delta x = \dfrac{1}{2}\left(1 + \dfrac{3}{n}k\right)\dfrac{3}{n} = \dfrac{3}{2}\left[\dfrac{1}{n} + \dfrac{3}{n^2}k\right]$

$$\sum_{k=1}^{n} f(x_k^*)\Delta x = \dfrac{3}{2}\left[\sum_{k=1}^{n}\dfrac{1}{n} + \sum_{k=1}^{n}\dfrac{3}{n^2}k\right] = \dfrac{3}{2}\left[1 + \dfrac{3}{n^2}\cdot\dfrac{1}{2}n(n+1)\right] = \dfrac{3}{2}\left[1 + \dfrac{3}{2}\dfrac{n+1}{n}\right]$$

$$A = \lim_{n\to+\infty}\dfrac{3}{2}\left[1 + \dfrac{3}{2}\left(1 + \dfrac{1}{n}\right)\right] = \dfrac{3}{2}\left(1 + \dfrac{3}{2}\right) = \dfrac{15}{4}$$

7. $\Delta x = \dfrac{1}{n}$, $x_k^* = \dfrac{k}{n}$; $f(x_k^*)\Delta x = x_k^*\Delta x = \dfrac{k^2}{n^2}\dfrac{1}{n} = \dfrac{k^2}{n^3}$

$$\sum_{k=1}^{n} f(x_k^*)\Delta x = \sum_{k=1}^{n}\dfrac{k^2}{n^3} = \dfrac{1}{n^3}\dfrac{1}{6}n(n+1)(2n+1) = \dfrac{1}{6}\dfrac{(n+1)(2n+1)}{n^2}$$

$$A = \lim_{n\to+\infty}\dfrac{1}{6}\left(1 + \dfrac{1}{n}\right)\left(2 + \dfrac{1}{n}\right) = \dfrac{1}{3}$$

9. $\Delta x = \dfrac{4}{n}$, $x_k^* = 2 + k\dfrac{4}{n}$

$$f(x_k^*)\Delta x = (x_k^*)^3\Delta x = \left[2 + \dfrac{4}{n}k\right]^3\dfrac{4}{n} = \dfrac{32}{n}\left[1 + \dfrac{2}{n}k\right]^3 = \dfrac{32}{n}\left[1 + \dfrac{6}{n}k + \dfrac{12}{n^2}k^2 + \dfrac{8}{n^3}k^3\right]$$

$$\sum_{k=1}^{n} f(x_k^*)\Delta x = \dfrac{32}{n}\left[\sum_{k=1}^{n}1 + \dfrac{6}{n}\sum_{k=1}^{n}k + \dfrac{12}{n^2}\sum_{k=1}^{n}k^2 + \dfrac{8}{n^3}\sum_{k=1}^{n}k^3\right]$$

$$= \dfrac{32}{n}\left[n + \dfrac{6}{n}\cdot\dfrac{1}{2}n(n+1) + \dfrac{12}{n^2}\cdot\dfrac{1}{6}n(n+1)(2n+1) + \dfrac{8}{n^3}\cdot\dfrac{1}{4}n^2(n+1)^2\right]$$

$$= 32\left[1 + 3\dfrac{n+1}{n} + 2\dfrac{(n+1)(2n+1)}{n^2} + 2\dfrac{(n+1)^2}{n^2}\right]$$

$$A = \lim_{n\to+\infty}32\left[1 + 3\left(1 + \dfrac{1}{n}\right) + 2\left(1 + \dfrac{1}{n}\right)\left(2 + \dfrac{1}{n}\right) + 2\left(1 + \dfrac{1}{n}\right)^2\right]$$

$$= 32[1 + 3(1) + 2(1)(2) + 2(1)^2] = 320$$

11. $\Delta x = \dfrac{3}{n}, \; x_k^* = 1 + (k-1)\dfrac{3}{n}$

$f(x_k^*)\Delta x = \dfrac{1}{2}x_k^*\Delta x = \dfrac{1}{2}\left[1 + (k-1)\dfrac{3}{n}\right]\dfrac{3}{n} = \dfrac{1}{2}\left[\dfrac{3}{n} + (k-1)\dfrac{9}{n^2}\right]$

$\displaystyle\sum_{k=1}^{n} f(x_k^*)\Delta x = \dfrac{1}{2}\left[\sum_{k=1}^{n}\dfrac{3}{n} + \dfrac{9}{n^2}\sum_{k=1}^{n}(k-1)\right] = \dfrac{1}{2}\left[3 + \dfrac{9}{n^2}\cdot\dfrac{1}{2}(n-1)n\right] = \dfrac{3}{2} + \dfrac{9}{4}\dfrac{n-1}{n}$

$A = \displaystyle\lim_{n\to+\infty}\left[\dfrac{3}{2} + \dfrac{9}{4}\left(1-\dfrac{1}{n}\right)\right] = \dfrac{3}{2} + \dfrac{9}{4} = \dfrac{15}{4}$

13. $\Delta x = \dfrac{1}{n}, \; x_k^* = (k-1)\dfrac{1}{n}; \; f(x_k^*)\Delta x = (x_k^*)^2\Delta x = (k-1)^2\dfrac{1}{n^2}\cdot\dfrac{1}{n} = \dfrac{1}{n^3}(k-1)^2$

$\displaystyle\sum_{k=1}^{n} f(x_k^*)\Delta x = \dfrac{1}{n^3}\sum_{k=1}^{n}(k-1)^2 = \dfrac{1}{n^3}\sum_{k=1}^{n-1}k^2 = \dfrac{1}{n^3}\cdot\dfrac{1}{6}(n-1)(n)(2n-1) = \dfrac{1}{6}\dfrac{(n-1)(2n-1)}{n^2}$

$A = \displaystyle\lim_{n\to+\infty}\dfrac{1}{6}\left(1-\dfrac{1}{n}\right)\left(2-\dfrac{1}{n}\right) = \dfrac{1}{6}(1)(2) = \dfrac{1}{3}$

15. **(a)** With x_k^* as the right endpoint, $\Delta x = \dfrac{b}{n}, \; x_k^* = \dfrac{b}{n}k$

$f(x_k^*)\Delta x = (x_k^*)^3\Delta x = \dfrac{b^4}{n^4}k^3, \; \displaystyle\sum_{k=1}^{n} f(x_k^*)\Delta x = \dfrac{b^4}{n^4}\sum_{k=1}^{n}k^3 = \dfrac{b^4}{4}\dfrac{(n+1)^2}{n^2}$

$A = \displaystyle\lim_{n\to+\infty}\dfrac{b^4}{4}\left(1+\dfrac{1}{n}\right)^2 = b^4/4$

(b) $\Delta x = \dfrac{b-a}{n}, \; x_k^* = a + \dfrac{b-a}{n}k$

$f(x_k^*)\Delta x = (x_k^*)^3\Delta x = \left[a + \dfrac{b-a}{n}k\right]^3\dfrac{b-a}{n}$

$= \dfrac{b-a}{n}\left[a^3 + \dfrac{3a^2(b-a)}{n}k + \dfrac{3a(b-a)^2}{n^2}k^2 + \dfrac{(b-a)^3}{n^3}k^3\right]$

$\displaystyle\sum_{k=1}^{n} f(x_k^*)\Delta x = (b-a)\left[a^3 + \dfrac{3}{2}a^2(b-a)\dfrac{n+1}{n} + \dfrac{1}{2}a(b-a)^2\dfrac{(n+1)(2n+1)}{n^2}\right.$

$\left. + \dfrac{1}{4}(b-a)^3\dfrac{(n+1)^2}{n^2}\right]$

$A = \displaystyle\lim_{n\to+\infty}\sum_{k=1}^{n} f(x_k^*)\Delta x$

$= (b-a)\left[a^3 + \dfrac{3}{2}a^2(b-a) + a(b-a)^2 + \dfrac{1}{4}(b-a)^3\right] = \dfrac{1}{4}(b^4 - a^4).$

17. (a) $\Delta x = \dfrac{b-a}{n}$, $x_k^* = a + \dfrac{b-a}{n}(k-1)$; $f(x_k^*)\Delta x = \dfrac{b-a}{n}\left[a + \dfrac{b-a}{n}(k-1)\right]$

$$\sum_{k=1}^{n} f(x_k^*)\Delta x = (b-a)\left[a + \dfrac{b-a}{2}\cdot\dfrac{n-1}{n}\right]$$

$$A = \lim_{n\to+\infty}(b-a)\left[a + \dfrac{b-a}{2}\left(1-\dfrac{1}{n}\right)\right] = \dfrac{1}{2}(b^2 - a^2)$$

(b) $\Delta x = \dfrac{b-a}{n}$, $x_k^* = a + \dfrac{b-a}{n}k$; $f(x_k^*)\Delta x = \dfrac{b-a}{n}\left[a + \dfrac{b-a}{n}k\right]$

$$\sum_{k=1}^{n} f(x_k^*)\Delta x = (b-a)\left[a + \dfrac{b-a}{2}\cdot\dfrac{n+1}{n}\right]$$

$$A = \lim_{n\to+\infty}(b-a)\left[a + \dfrac{b-a}{2}\left(1+\dfrac{1}{n}\right)\right] = \dfrac{1}{2}(b^2 - a^2)$$

(c) The region is enclosed by a trapezoid so its area is

$$A = \dfrac{1}{2}(b-a)(b+a)$$

$$= \dfrac{1}{2}(b^2 - a^2).$$

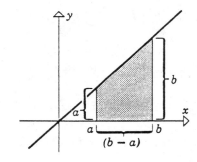

19. (a) 0.761923639, 0.712712753, 0.684701150
    (b) 0.584145862, 0.623823864, 0.649145594
    (c) 0.663501867, 0.665867079, 0.666538346

21. (a) 0.919403170, 0.960215997, 0.984209789
    (b) 1.076482803, 1.038755813, 1.015625715
    (c) 1.001028824, 1.000257067, 1.000041125

EXERCISE SET 5.6

1. (a) $(4/3)(1) + (5/2)(1) + (4)(2) = 71/6$
   (b) 2

3. **(a)** $(-9/4)(1) + (3)(2) + (63/16)(1) + (-5)(3) = -117/16$
   (b) 3

5. For the smallest, find x_k^* so that $f(x_k^*)$ is minimum on each subinterval: $x_1^* = 1$, $x_2^* = 3/2$, $x_3^* = 3$ so $(2)(1) + (7/4)(2) + (4)(1) = 9.5$. For the largest, find x_k^* so that $f(x_k^*)$ is maximum on each subinterval: $x_1^* = 0$, $x_2^* = 3$, $x_3^* = 4$ so $(4)(1) + (4)(2) + (8)(1) = 20$.

7. $\displaystyle\int_{-3}^{3} 4x(1 - 3x)\,dx$

9. $\displaystyle\lim_{\max \Delta x_k \to 0} \sum_{k=1}^{n} 2x_k^* \Delta x_k$; $a = 1$, $b = 2$

11. $\displaystyle\lim_{\max \Delta x_k \to 0} \sum_{k=1}^{n} \frac{x_k^*}{x_k^* + 1} \Delta x_k$; $a = 0$, $b = 1$

13. **(a)** 0.8   **(b)** −2.6   **(c)** −1.8   **(d)** −0.3

15. **(a)** $A = \dfrac{1}{2}(1)(2) = 1$ **(b)** $A = \dfrac{1}{2}(2)(3/2 + 1/2) = 2$

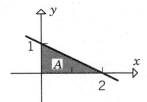
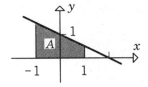

(c) $-A = -\dfrac{1}{2}(1/2)(1) = -1/4$ **(d)** $A_1 - A_2 = 1 - 1/4 = 3/4$

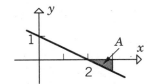
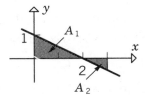

17. $A_1 - A_2 = \dfrac{1}{2}(2)(1/2) - \dfrac{1}{2}(6)(3/2)$

$\qquad = -4$

19. $A_1 + A_2 = \dfrac{1}{2}(5)(5/2) + \dfrac{1}{2}(1)(1/2)$

$\qquad = 13/2$

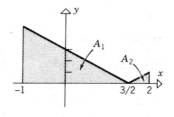

21. $\dfrac{1}{2}[\pi(1)^2] = \pi/2$

23. $\sqrt{10x - x^2} = \sqrt{25 - (x-5)^2};\ \displaystyle\int_0^{10} \sqrt{10x - x^2}\, dx = \dfrac{1}{2}[\pi(5)^2] = 25\pi/2$

25. $A_1 - A_2 = 0$ because
$A_1 = A_2$ by symmetry

27. $A_1 + A_2 = (3)(2) + \dfrac{1}{2}(2)(5+3) = 14$

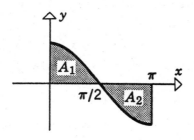

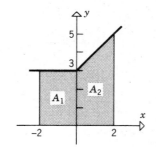

29. $\displaystyle\int_{-1}^2 f(x)dx + 2\int_{-1}^2 g(x)dx = 5 + 2(-3) = -1$

31. $\displaystyle\int_1^5 f(x)dx = \int_0^5 f(x)dx - \int_0^1 f(x)dx = 1 - (-2) = 3.$

33. negative, because $\sqrt{x}/(1-x) < 0$ for $2 \le x \le 3$.

35. positive, because $x^2/(3 - \cos x) > 0$ for $0 < x \le 4$.

37. negative, because $\int_2^0 x^2 \sin\sqrt{x}\,dx = -\int_0^2 x^2 \sin\sqrt{x}\,dx$ and $x^2 \sin\sqrt{x} > 0$ for $0 < x \le 2$.

39. f is not bounded on $[0,1]$ because $\lim_{x\to 0+} f(x) = +\infty$, so f is not integrable on $[0,1]$.

41. Suppose that L_1 and L_2 satisfy (7). Let $S_n = \sum_{k=1}^n f(x_k^*)\Delta x_k$. From definition 5.6.10, for any $\epsilon > 0$ there are numbers $\delta_1 > 0$ and $\delta_2 > 0$ such that $|S_n - L_1| < \epsilon$ for max $\Delta x_k < \delta_1$ and $|S_n - L_2| < \epsilon$ for max $\Delta x_k < \delta_2$.

If $\delta = \min(\delta_1, \delta_2)$ then $|S_n - L_1| < \epsilon$ and $|S_n - L_2| < \epsilon$ for max $\Delta x_k < \delta$ thus

$$|L_1 - L_2| = |L_1 - S_n + S_n - L_2| = |(L_1 - S_n) + (S_n - L_2)| \le |S_n - L_1| + |S_n - L_2| < \epsilon + \epsilon = 2\epsilon$$

so $|L_1 - L_2| < 2\epsilon$ for max $\Delta x_k < \delta$. Suppose $L_1 \ne L_2$ and let $\epsilon = \frac{1}{2}|L_1 - L_2|$ then $|L_1 - L_2| < 2\epsilon$ yields $|L_1 - L_2| < |L_1 - L_2|$ which is false so $L_1 \ne L_2$ is impossible.

43. Let $R_n = \sum_{k=1}^n f(x_k^*)\Delta x_k$, $S_n = \sum_{k=1}^n g(x_k^*)\Delta x_k$, $T_n = \sum_{k=1}^n [f(x_k^*) + g(x_k^*)]\Delta x_k$, $R = \int_a^b f(x)dx$, and $S = \int_a^b g(x)dx$ then $T_n = R_n + S_n$ and we want to prove that $\lim_{\max \Delta x_k \to 0} T_n = R + S$.

$$|T_n - (R+S)| = |(R_n - R) + (S_n - S)| \le |R_n - R| + |S_n - S|$$

so for any $\epsilon > 0$ $|T_n - (R+S)| < \epsilon$ if $|R_n - R| + |S_n - S| < \epsilon$.

Because f and g are integrable on $[a, b]$, there are numbers δ_1 and δ_2 such that $|R_n - R| < \epsilon/2$ for max $\Delta x_k < \delta_1$ and $|S_n - S| < \epsilon/2$ for max $\Delta x_k < \delta_2$.

If $\delta = \min(\delta_1, \delta_2)$ then $|R_n - R| < \epsilon/2$ and $|S_n - S| < \epsilon/2$ for max $\Delta x_k < \delta$ thus $|R_n - R| + |S_n - S| < \epsilon$ and so $|T_n - (R+S)| < \epsilon$ for max $\Delta x_k < \delta$ which shows that $\lim_{\max \Delta x_k \to 0} T_n = R + S$.

45. (b), (c) are always valid. Use $f(x) = g(x) = 1$ to show that (a), (d), and (e) are false.

EXERCISE SET 5.7

1. $\left.\frac{1}{4}x^4\right]_2^3 = 65/4$

3. $\left.\frac{1}{2}x^2 + \frac{1}{5}x^5\right]_{-1}^2 = 81/10$

5. $\left.\frac{1}{3}t^3 - t^2 + 8t\right]_1^2 = 22/3$

7. $\int_1^3 x^{-2}dx = \left.-\frac{1}{x}\right]_1^3 = 2/3$

9. $-\dfrac{1}{2x^2} + \dfrac{2}{x} - \dfrac{1}{3x^3}\Big]_1^2 = -1/3$

11. $\dfrac{2}{3}x^{3/2}\Big]_1^9 = 52/3$

13. $\dfrac{4}{5}y^{5/2}\Big]_4^9 = 844/5$

15. $6\sqrt{x} - \dfrac{10}{3}x^{3/2} + \dfrac{2}{\sqrt{x}}\Big]_1^4 = -55/3$

17. $-\cos\theta\Big]_{-\pi/2}^{\pi/2} = 0$

19. $\displaystyle\int_{-\pi/4}^{\pi/4} \cos x\, dx = \sin x\Big]_{-\pi/4}^{\pi/4} = \sqrt{2}$

21. $\dfrac{1}{2}x^2 - 2\cot x\Big]_{\pi/6}^{\pi/2} = \pi^2/9 + 2\sqrt{3}$

23. $\displaystyle\int_0^{3/2}(3-2x)dx + \int_{3/2}^2 (2x-3)dx = (3x-x^2)\Big]_0^{3/2} + (x^2-3x)\Big]_{3/2}^2 = 9/4 + 1/4 = 5/2$

25. $\displaystyle\int_0^{\pi/2}\cos x\, dx + \int_{\pi/2}^{3\pi/4}(-\cos x)dx = \sin x\Big]_0^{\pi/2} - \sin x\Big]_{\pi/2}^{3\pi/4} = 2 - \sqrt{2}/2$

27. $\displaystyle\int_{-2}^0 x^2 dx + \int_0^3(-x)dx = \dfrac{1}{3}x^3\Big]_{-2}^0 - \dfrac{1}{2}x^2\Big]_0^3 = -11/6$

29. $0.665867079;\ \displaystyle\int_1^3 \dfrac{1}{x^2}\, dx = -\dfrac{1}{x}\Big]_1^3 = 2/3$

31. $\dfrac{\sqrt{1}+\sqrt{2}+\sqrt{3}+\cdots+\sqrt{n}}{n^{3/2}} = \displaystyle\sum_{k=1}^n \dfrac{\sqrt{k}}{n^{3/2}} = \sum_{k=1}^n \sqrt{\dfrac{k}{n}}\dfrac{1}{n} = \sum_{k=1}^n f(x_k^*)\Delta x$, where

$f(x) = \sqrt{x}, x_k^* = k/n$, and $\Delta x = 1/n$, so $\displaystyle\lim_{n\to+\infty}\sum_{k=1}^n f(x_k^*)\Delta x = \int_0^1 \sqrt{x}\, dx = 2/3$.

33. $A = \displaystyle\int_0^3 (x^2+1)dx = \dfrac{1}{3}x^3 + x\Big]_0^3 = 12$

35. $A = \displaystyle\int_0^{2\pi/3} 3\sin x\, dx = -3\cos x\Big]_0^{2\pi/3} = 9/2$

37. $A_1 = \int_{-3}^{-2} (x^2 - 3x - 10)dx$

$= \dfrac{1}{3}x^3 - \dfrac{3}{2}x^2 - 10x \Big]_{-3}^{-2} = 23/6,$

$A_2 = -\int_{-2}^{5} (x^2 - 3x - 10)dx = 343/6,$

$A_3 = \int_{5}^{8} (x^2 - 3x - 10)dx = 243/6,$

$A = A_1 + A_2 + A_3 = 203/2$

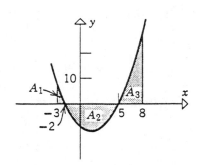

39. $\int_{1}^{2} \dfrac{x}{x^4 + 1}\,dx < \int_{1}^{2} \dfrac{1}{x^3}\,dx = -\dfrac{1}{2x^2}\Big]_{1}^{2} = 3/8.$

41. $m = \int_{1}^{3} \dfrac{1}{x^{1.01}}\,dx = -\dfrac{100}{x^{0.01}}\Big]_{1}^{3} = 1.092599583;$

$M = \int_{1}^{3} \dfrac{1}{x^{0.99}}\,dx = 100x^{0.01}\Big]_{1}^{3} = 1.104669194$

43. $f_{\text{ave}} = \dfrac{1}{3-1}\int_{1}^{3} 3x\,dx = \dfrac{3}{4}x^2\Big]_{1}^{3} = 6.$

45. $f_{\text{ave}} = \dfrac{1}{\pi - 0}\int_{0}^{\pi} \sin x\,dx = -\dfrac{1}{\pi}\cos x\Big]_{0}^{\pi} = 2/\pi.$

47. $f_{\text{ave}} = \dfrac{1}{4-0}\int_{0}^{4} \sqrt{x}\,dx = \dfrac{1}{6}x^{3/2}\Big]_{0}^{4} = 4/3.$

49. $f_{\text{ave}} = \dfrac{1}{2-(-2)}\int_{-2}^{2} \sqrt{4 - x^2}\,dx = \dfrac{1}{4}\times\dfrac{1}{2}\pi(2)^2 = \pi/2.$

51. **(a)** $f_{\text{ave}} = \dfrac{1}{2-0}\int_{0}^{2} x^2\,dx = 4/3$

(b) $(x^*)^2 = 4/3, x^* = \pm 2/\sqrt{3},$
but only $2/\sqrt{3}$ is in $[0,2]$.

(c)

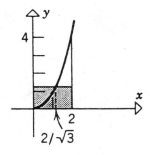

53. $f_{ave} = \dfrac{1}{9}\displaystyle\int_0^9 x^{1/2}dx = 2;\ \sqrt{x^*} = 2,\ x^* = 4$

55. $f_{ave} = \dfrac{1}{x_1 - x_0}\displaystyle\int_{x_0}^{x_1} (\alpha x + \beta)dx = \dfrac{1}{2}\alpha(x_0 + x_1) + \beta;$

$\alpha x^* + \beta = \dfrac{1}{2}\alpha(x_0 + x_1) + \beta,\ x^* = (x_0 + x_1)/2.$

57. $a_{ave} = \dfrac{v(t_1) - v(t_0)}{t_1 - t_0} = \dfrac{1}{t_1 - t_0}v(t)\Big]_{t_0}^{t_1} = \dfrac{1}{t_1 - t_0}\displaystyle\int_{t_0}^{t_1} a(t)dt$

59. time to fill tank = (volume of tank)/(rate of filling) = $[\pi(3)^2 5]/(1) = 45\pi$, force on bottom at
time t = weight of water in tank at time t = (62.4) (rate of filling)(time) = 62.4t,

$force_{ave} = \dfrac{1}{45\pi}\displaystyle\int_0^{45\pi} 62.4t\,dt = 1404\pi$ lb.

61. (a) $\displaystyle\int_a^b [f(x) - f_{ave}]\,dx = \int_a^b f(x)dx - \int_a^b f_{ave}dx = \int_a^b f(x)dx - f_{ave}(b - a) = 0$

because $f_{ave}(b - a) = \displaystyle\int_a^b f(x)dx.$

(b) No, because if $\displaystyle\int_a^b [f(x) - c]dx = 0$ then $\displaystyle\int_a^b f(x)dx - c(b - a) = 0$ so

$c = \dfrac{1}{b - a}\displaystyle\int_a^b f(x)dx = f_{ave}$ is the only value.

EXERCISE SET 5.8

1. (a) $\displaystyle\int_1^3 u^7\,du$ **(b)** $-\dfrac{1}{2}\displaystyle\int_7^4 u^{1/2}\,du$ **(c)** $\dfrac{1}{\pi}\displaystyle\int_{-\pi}^{\pi} \sin u\,du$

(d) $\displaystyle\int_0^1 u^2\,du$ **(e)** $\dfrac{1}{2}\displaystyle\int_3^4 (u - 3)u^{1/2}\,du$ **(f)** $\displaystyle\int_{-3}^0 (u + 5)u^{20}\,du$

3. $u = 2x + 1,\ \dfrac{1}{2}\displaystyle\int_1^3 u^4\,du = \dfrac{1}{10}u^5\Big]_1^3 = 121/5,$ or $\dfrac{1}{10}(2x + 1)^5\Big]_0^1 = 121/5$

5. $u = 1 - 2x$, $-\dfrac{1}{2}\displaystyle\int_3^1 u^3\,du = -\dfrac{1}{8}u^4\Big]_3^1 = 10$, or $-\dfrac{1}{8}(1-2x)^4\Big]_{-1}^0 = 10$

7. $u = 1 + x$, $\displaystyle\int_1^9 (u-1)u^{1/2}\,du = \int_1^9 (u^{3/2} - u^{1/2})\,du = \dfrac{2}{5}u^{5/2} - \dfrac{2}{3}u^{3/2}\Big]_1^9 = 1192/15,$

or $\dfrac{2}{5}(1+x)^{5/2} - \dfrac{2}{3}(1+x)^{3/2}\Big]_0^8 = 1192/15$

9. $u = x/2$, $8\displaystyle\int_0^{\pi/4} \sin u\,du = -8\cos u\Big]_0^{\pi/4} = 8 - 4\sqrt{2}$, or $-8\cos(x/2)\Big]_0^{\pi/2} = 8 - 4\sqrt{2}$

11. $u = x^2 + 2$, $\dfrac{1}{2}\displaystyle\int_6^3 u^{-3}\,du = -\dfrac{1}{4u^2}\Big]_6^3 = -1/48$, or $-\dfrac{1}{4}\dfrac{1}{(x^2+2)^2}\Big]_{-2}^{-1} = -1/48$

13. $\dfrac{2}{3}(3u+1)^{1/2}\Big]_0^1 = 2/3$ **15.** $\dfrac{2}{3}(x^3+9)^{1/2}\Big]_{-1}^1 = \dfrac{2}{3}(\sqrt{10} - 2\sqrt{2})$

17. $u = x^2 + 4x + 7$, $\dfrac{1}{2}\displaystyle\int_{12}^{28} u^{-1/2}\,du = u^{1/2}\Big]_{12}^{28} = \sqrt{28} - \sqrt{12} = 2(\sqrt{7} - \sqrt{3})$

19. $\dfrac{1}{2}\sin^2 x\Big]_{-3\pi/4}^{-\pi/4} = 0$ **21.** $\dfrac{5}{2}\sin(x^2)\Big]_0^{\sqrt{\pi}} = 0$

23. $u = \sqrt{x}$, $2\displaystyle\int_\pi^{2\pi} \sin u\,du = -2\cos u\Big]_\pi^{2\pi} = -4$

25. $u = \sin 3x$, $\dfrac{1}{3}\displaystyle\int_0^{-1} u^2\,du = \dfrac{1}{9}u^3\Big]_0^{-1} = -1/9$

27. $u = 3\theta$, $\dfrac{1}{3}\displaystyle\int_{\pi/4}^{\pi/3} \sec^2 u\,du = \dfrac{1}{3}\tan u\Big]_{\pi/4}^{\pi/3} = (\sqrt{3} - 1)/3$

29. $u = 4 - 3y$, $y = \dfrac{1}{3}(4-u)$, $dy = -\dfrac{1}{3}du$

$-\dfrac{1}{27}\displaystyle\int_4^1 \dfrac{16 - 8u + u^2}{u^{1/2}}\,du = \dfrac{1}{27}\int_1^4 (16u^{-1/2} - 8u^{1/2} + u^{3/2})\,du$

$= \dfrac{1}{27}\left[32u^{1/2} - \dfrac{16}{3}u^{3/2} + \dfrac{2}{5}u^{5/2}\right]_1^4 = 106/405$

31. $A = \displaystyle\int_0^{\pi/8} 3\,\cos 2x\,dx = \frac{3}{2}\sin 2x\Big]_0^{\pi/8} = 3\sqrt{2}/4$

33. $V_p/\sqrt{2} = 120, V_p = 120\sqrt{2} \approx 169.7$ volts

35. 0.692835360, 0.693069098, 0.693134682

37. 3.142425985, 3.141800987, 3.141625987

39. $\displaystyle\int_{-2}^{2} \sqrt{4-u^2}\,du = \frac{1}{2}[\pi(2)^2] = 2\pi$

41. $-\dfrac{1}{2}\displaystyle\int_1^0 \sqrt{1-u^2}\,du = \dfrac{1}{2}\displaystyle\int_0^1 \sqrt{1-u^2}\,du = \dfrac{1}{2}\cdot\dfrac{1}{4}[\pi(1)^2] = \pi/8$

43. $u = 1/x, \; -\displaystyle\int_2^1 f(u)du = \displaystyle\int_1^2 f(u)du = 3$

45. $u = 3x+1, \; \dfrac{1}{3}\displaystyle\int_1^4 f(u)du = \dfrac{5}{3}$

47. $\sin x = \cos(\pi/2 - x)$,

$$\int_0^{\pi/2} \sin^n x\,dx = \int_0^{\pi/2} \cos^n(\pi/2 - x)dx = -\int_{\pi/2}^0 \cos^n u\,du \;(u = \pi/2 - x)$$

$$= \int_0^{\pi/2} \cos^n u\,du = \int_0^{\pi/2} \cos^n x\,dx \;(\text{by replacing } u \text{ by } x)$$

49. $2k$, because $f(-x) = f(x)$ on $[-2,2]$.   **51.** 0, because $f(-x) = -f(x)$ on $[-1,1]$.

53. $\displaystyle\sum_{k=1}^{n} \sin[\pi(k/n)](1/n) = \sum_{k=1}^{n} f(x_k^*)\Delta x$, where $f(x) = \sin(\pi x)$, $x_k^* = k/n$, and $\Delta x = 1/n$ so

$$\lim_{n\to+\infty} \sum_{k=1}^{n} f(x_k^*)\Delta x = \int_0^1 \sin(\pi x)dx = -\frac{1}{\pi}\cos(\pi x)\Big]_0^1 = 2/\pi.$$

55. **(a)** $I = -\displaystyle\int_a^0 \frac{f(a-u)}{f(a-u)+f(u)}\,du = \int_0^a \frac{f(a-u)+f(u)-f(u)}{f(a-u)+f(u)}\,du$

$\quad\quad x = \displaystyle\int_0^a du - \int_0^a \frac{f(u)}{f(a-u)+f(u)}\,du, I = a - I \text{ so } 2I = a, I = a/2$

(b) $3/2$ **(c)** $\pi/4$

EXERCISE SET 5.9

1. (a) 3   (b) 3

3. (a) $x^3 + 1$ (b) $F(x) = \frac{1}{4}t^4 + t \Big]_1^x = \frac{1}{4}x^4 + x - \frac{5}{4}$; $F'(x) = x^3 + 1$

5. $\sin \sqrt{x}$ 7. $|x|$

9. $\displaystyle\int_2^x \frac{1}{t-1}\,dt$ 11. $\displaystyle\int_0^x \frac{1}{t-1}\,dt$

13. (a) $(0, +\infty)$ because f is continuous there and 1 is in $(0, +\infty)$.
     (b) at $x = 1$ because $F(1) = 0$

15. $F'(x) = \dfrac{\cos x}{x^2 + 3}$, $F''(x) = \dfrac{-(x^2 + 3)\sin x - 2x \cos x}{(x^2 + 3)^2}$
     (a) 0   (b) 1/3   (c) 0

17. The domain is $(-\infty, +\infty)$; $F(x)$ is 0 if $x = 1$, positive if $x > 1$, and negative if $x < 1$.

19. The domain is $[-2, 2]$; $F(x)$ is 0 if $x = -1$, positive if $-1 < x \le 2$, and negative if $-2 \le x < -1$.

21. $x < 0 : F(x) = \displaystyle\int_{-1}^x (-t)\,dt = -\frac{1}{2}t^2 \Big]_{-1}^x = \frac{1}{2}(1 - x^2)$,

     $x \ge 0 : F(x) = \displaystyle\int_{-1}^0 (-t)\,dt + \int_0^x t\,dt = \frac{1}{2} + \frac{1}{2}x^2$; $F(x) = \begin{cases} (1 - x^2)/2, & x < 0 \\ (1 + x^2)/2, & x \ge 0 \end{cases}$

23. $x \le 0 : F(x) = \displaystyle\int_{-1}^x t^2\,dt = \frac{1}{3}t^3 \Big]_{-1}^x = \frac{1}{3}(x^3 + 1)$,

     $x > 0 : F(x) = \displaystyle\int_{-1}^0 t^2\,dt + \int_0^x 2t\,dt = \frac{1}{3} + x^2$; $F(x) = \begin{cases} (x^3 + 1)/3, & x \le 0 \\ x^2 + 1/3, & x > 0 \end{cases}$

25. $\dfrac{1}{x^3}(3x^2) = \dfrac{3}{x}$

27. $F'(x) = \dfrac{1}{1 + x^2} + \dfrac{1}{1 + (1/x)^2}(-1/x^2) = 0$ so F is constant on $(0, +\infty)$.

29. (a) $\sin^2(x^3)(3x^2) - \sin^2(x^2)(2x) = 3x^2\sin^2(x^3) - 2x\sin^2(x^2)$

    (b) $\dfrac{1}{1+x}(1) - \dfrac{1}{1-x}(-1) = \dfrac{2}{1-x^2}$

31. $\displaystyle\int_x^b f(t)dt = -\int_b^x f(t)dt$ so $\dfrac{d}{dx}\int_x^b f(t)dt = -\dfrac{d}{dx}\int_b^x f(t)dt = -f(x)$

TECHNOLOGY EXERCISES 5

1. $\dfrac{1}{3}\sqrt{5 + 2\sin 3x} + C$ **3.** $-\dfrac{1}{3a(ax^3 + b)} + C$

5. $f(x) = -\dfrac{\sqrt{9 + x^2}}{9x} + 1.138889$

7. $f(x) = \dfrac{1}{3}x^2\sin 3x - \dfrac{2}{27}\sin 3x + \dfrac{2}{9}x\cos 3x - 0.251607$

9. 1.137630   **11.** 4.368876

13. With $b = 1.618034$, area $= \displaystyle\int_0^b (x + x^2 - x^3)dx = 1.007514$.

15. Solve $\dfrac{1}{4}k^4 - k - k^2 + \dfrac{7}{4} = 0$ to get $k = 2.073948$.

17. Solve $\dfrac{\sqrt{1 + 16k}}{k} - \dfrac{\sqrt{1 + k}}{k} = 2.5$ to get $k = 1.210563$.

19. (a)

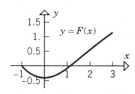

    (b) $x = 1.128780$ (by trial and error or some other method)

    (c) $F'(x) = f(x) = 0$ if $x = 0$ (relative minimum);

          $F''(x) = f'(x) = 0$ if $x = 1.587401$ (inflection point).

21.

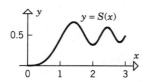

$S'(x) = \sin(\pi x^2/2) = 0$ if $\pi x^2/2 = \pi$ so $x^2 = 2$, $x = \sqrt{2}$; $S(\sqrt{2}) = 0.713972$

23.

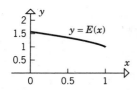

$E(x) = 1.5$ if $x = 0.174139$

25. $\displaystyle\lim_{n \to +\infty} \sum_{k=1}^{n} \left(-1 + \frac{4k}{n}\right)^3 \frac{4}{n} = 20$

CHAPTER 6
Applications of the Definite Integral

EXERCISE SET 6.1

1. $A = \int_{-1}^{2}(x^2 + 1 - x)dx = (x^3/3 + x - x^2/2)\Big]_{-1}^{2} = 9/2$

3. $A = \int_{1}^{2}(y - 1/y^2)dy = (y^2/2 + 1/y)\Big]_{1}^{2} = 1$

5. (a) $A = \int_{0}^{4}(4x - x^2)dx = 32/3$

   (b) $A = \int_{0}^{16}(\sqrt{y} - y/4)dy = 32/3$

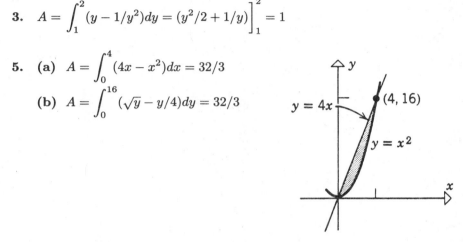

7. Eliminate x to get $y^2 = y + 2$, $y^2 - y - 2 = 0, (y+1)(y-2) = 0$, $y = -1$ and 2 with corresponding values of $x = 1/2$ and 2.

   (a) $A = \int_{0}^{1/2}[\sqrt{2x} - (-\sqrt{2x})]dx$

   $+ \int_{1/2}^{2}[\sqrt{2x} - (2x - 2)]dx$

   $= \int_{0}^{1/2}2\sqrt{2}x^{1/2}dx + \int_{1/2}^{2}(\sqrt{2}x^{1/2} - 2x + 2)dx$

   $= 2/3 + 19/12 = 9/4$

   (b) $A = \int_{-1}^{2}[(y/2 + 1) - y^2/2]dy = 9/4$

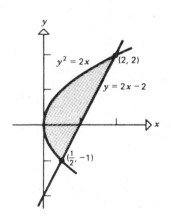

9. $A = \int_{1/4}^{1} (\sqrt{x} - x^2)dx = 49/192$

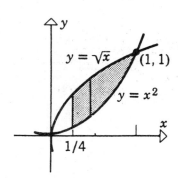

11. $A = \int_{\pi/4}^{\pi/2} (0 - \cos 2x)dx$

$= -\int_{\pi/4}^{\pi/2} \cos 2x\, dx = 1/2$

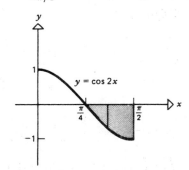

13. $A = \int_{0}^{4} [0 - (y^2 - 4y)]dy$

$= \int_{0}^{4} (4y - y^2)dy = 32/3$

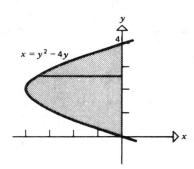

15. Equate $\sec^2 x$ and 2 to get $\sec^2 x = 2$, $\sec x = \pm\sqrt{2}$, $x = \pm\pi/4$

$A = \int_{-\pi/4}^{\pi/4} (2 - \sec^2 x)dx = \pi - 2$

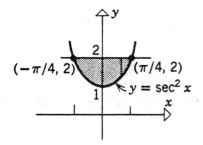

17. Eliminate y to get $6 - x = x^2 + 4$,

$x^2 + x - 2 = 0$, $(x + 2)(x - 1) = 0$,

$x = -2, 1$ with corresponding values of $y = 8, 5$.

$A = \int_{-2}^{1} [(6 - x) - (x^2 + 4)]dx$

$= \int_{-2}^{1} (2 - x - x^2)dx = 9/2$

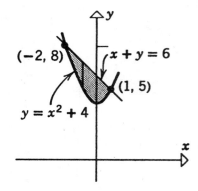

19. $A = \int_{-1}^{4} [(y+6) - (-y^2)]dy$

$\qquad = \int_{-1}^{4} (y + 6 + y^2)dy$

$\qquad = 355/6$

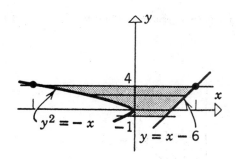

21. $y = 2 + |x - 1| = \begin{cases} 3 - x, & x \le 1 \\ 1 + x, & x \ge 1 \end{cases}$,

$A = \int_{-5}^{1} \left[\left(-\frac{1}{5}x + 7\right) - (3 - x) \right] dx$

$\qquad + \int_{1}^{5} \left[\left(-\frac{1}{5}x + 7\right) - (1 + x) \right] dx$

$\qquad = \int_{-5}^{1} \left(\frac{4}{5}x + 4\right) dx + \int_{1}^{5} \left(6 - \frac{6}{5}x\right) dx$

$\qquad = 72/5 + 48/5 = 24$

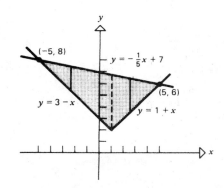

23. $A = \int_{-1}^{0} (y^3 - y)dy + \int_{0}^{1} -(y^3 - y)dy$

$\qquad = 1/2$

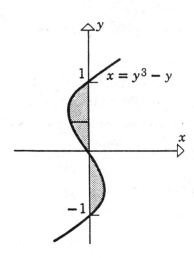

25. From the symmetry of the region

$$A = 2 \int_{\pi/4}^{5\pi/4} (\sin x - \cos x)dx = 4\sqrt{2}$$

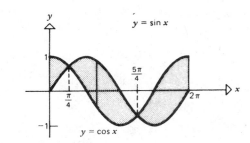

27. $A = \int_1^4 \left(y - \dfrac{1}{\sqrt{y}} \right) dy = 11/2$

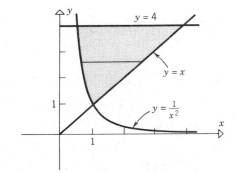

29. The tangent line at $(4, 2)$ is

$$y = \frac{1}{4}x + 1;$$

$$A = \int_0^4 \left[\left(\frac{1}{4}x + 1 \right) - \sqrt{x} \right] dx = 2/3$$

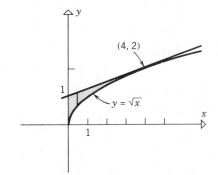

31. **(a)** $A = \int_1^b x^{-1/2}dx = 2(\sqrt{b} - 1)$ **(b)** $\lim\limits_{b \to +\infty} A = +\infty$

33. $\displaystyle\int_0^k 2\sqrt{y}\,dy = \int_k^9 2\sqrt{y}\,dy$

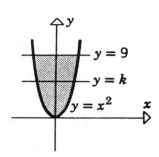

$\displaystyle\int_0^k y^{1/2}\,dy = \int_k^9 y^{1/2}\,dy$

$\displaystyle\frac{2}{3}k^{3/2} = \frac{2}{3}(27 - k^{3/2})$

$k^{3/2} = 27/2$

$k = (27/2)^{2/3} = 9/\sqrt[3]{4}$

35. Solve for y to get $y = (b/a)\sqrt{a^2 - x^2}$ for the upper half of the ellipse; make use of symmetry to get $A = 4\displaystyle\int_0^a \frac{b}{a}\sqrt{a^2 - x^2}\,dx = \frac{4b}{a}\int_0^a \sqrt{a^2 - x^2}\,dx = \frac{4b}{a} \times \frac{1}{4}\pi a^2 = \pi ab.$

37. **(a)** It gives the area of the region that is between f and g when $f(x) > g(x)$ <u>minus</u> the area of the region between f and g when $f(x) < g(x)$, for $a \le x \le b$.

(b) It gives the area of the region that is between f and g for $a \le x \le b$.

39. The curves intersect at $x = 0$ and, by Newton's Method, at $x \approx 2.595739080 = b$, so

$$A \approx \int_0^b (\sin x - 0.2x)\,dx = -\cos x - 0.1x^2 \Big]_0^b \approx 1.180898334$$

EXERCISE SET 6.2

1. $V = \pi\displaystyle\int_{-1}^3 (3 - x)\,dx = 8\pi$

3. $V = \pi\displaystyle\int_0^2 \frac{1}{4}(3 - y)^2\,dy = 13\pi/6$

5. $V = \pi\displaystyle\int_0^2 x^4\,dx = 32\pi/5$

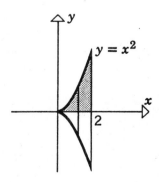

7. $V = \pi \int_1^2 (1+x^3)^2 dx$

$= \pi \int_1^2 (1 + 2x^3 + x^6) dx$

$= 373\pi/14$

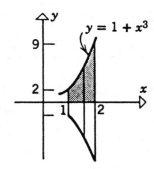

9. $V = \pi \int_{-3}^3 (9 - x^2)^2 dx$

$= \pi \int_{-3}^3 (81 - 18x^2 + x^4) dx$

$= 1296\pi/5$

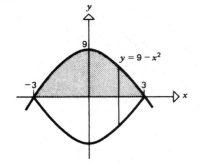

11. $V = \pi \int_0^4 [(4x)^2 - (x^2)^2] dx$

$= \pi \int_0^4 (16x^2 - x^4) dx = 2048\pi/15$

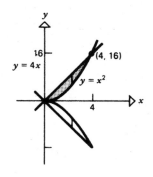

13. $V = \pi \int_0^{\pi/4} (\cos^2 x - \sin^2 x) dx$

$= \pi \int_0^{\pi/4} \cos 2x \, dx = \pi/2$

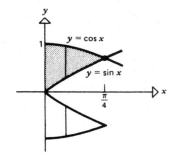

15. $V = \pi \int_0^1 [(\sqrt{x})^2 - x^2] dx$

$ = \pi \int_0^1 (x - x^2) dx = \pi/6$

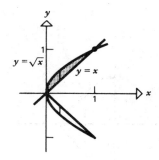

17. $V = \pi \int_0^1 y^{2/3} dy = 3\pi/5$

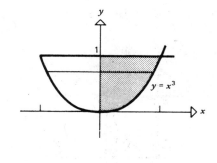

19. $V = \pi \int_{-1}^3 (1+y) dy = 8\pi$

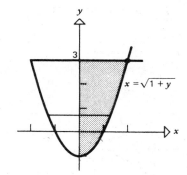

21. $V = \pi \int_{\pi/4}^{3\pi/4} \csc^2 y \, dy = 2\pi$

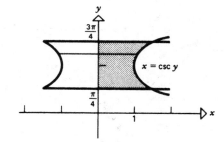

23. $V = \pi \int_1^3 (9 - y^2)dy = 28\pi/3$

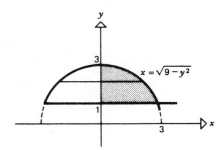

25. $V = \pi \int_2^9 [(y-1)^{2/3} - 1]dy$

$= 58\pi/5$

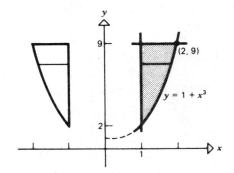

27. $V = \pi \int_{-1}^2 [(y+2)^2 - y^4]dy$

$= 72\pi/5$

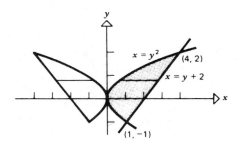

29. $V = \pi \int_{-4}^4 [(25 - x^2) - 9]dx$

$= 2\pi \int_0^4 (16 - x^2)dx$

$= 256\pi/3$

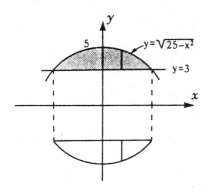

31. **(a)** $V(b) = \pi \int_1^b \frac{1}{x^2}dx = \pi(1 - 1/b)$ **(b)** $\lim_{b \to +\infty} \pi(1 - 1/b) = \pi$

33. $V = \pi \int_0^b x\,dx = \pi b^2/2; \pi b^2/2 = 2, b^2 = 4/\pi, b = 2/\sqrt{\pi}$

35. $V = \pi \int_0^3 (9 - y^2)^2 dy$

$= \pi \int_0^3 (81 - 18y^2 + y^4) dy$

$= 648\pi/5$

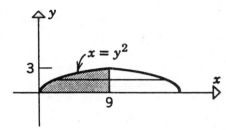

37. $V = \pi \int_0^1 [(\sqrt{x} + 1)^2 - (x+1)^2] dx$

$= \pi \int_0^1 (2\sqrt{x} - x - x^2) dx = \pi/2$

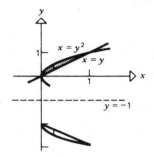

39. $V = \pi \int_{-a}^a \frac{b^2}{a^2}(a^2 - x^2) dx$

$= 4\pi ab^2/3$

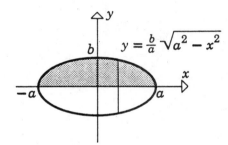

41. $V = \pi \int_{-1}^0 (x+1) dx$

$+ \pi \int_0^1 [(x+1) - 2x] dx$

$= \pi/2 + \pi/2 = \pi$

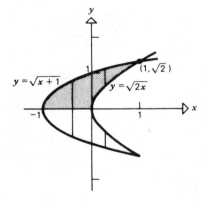

43. By similar triangles, $R/r = y/h$ so
$R = ry/h$ and $A(y) = \pi r^2 y^2/h^2$.

$$V = (\pi r^2/h^2) \int_0^h y^2 \, dy = \pi r^2 h/3$$

45. $V = 2\pi \int_0^{L/2} [(r^2 - y^2) - (r^2 - L^2/4)] \, dy$

$$= 2\pi \int_0^{L/2} (L^2/4 - y^2) \, dy$$

$$= \pi L^3/6$$

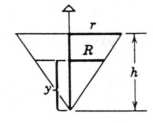

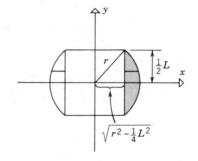

47. (a)

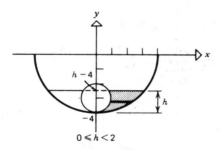

(b)

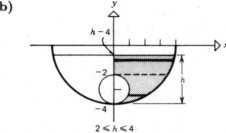

If the cherry is partially submerged then $0 \le h < 2$ as shown in Figure (a); if it is totally submerged then $2 \le h \le 4$ as shown in Figure (b). The radius of the glass is 4 cm and that of the cherry is 1 cm so points on the sections shown in the figures satisfy the equations $x^2 + y^2 = 16$ and $x^2 + (y+3)^2 = 1$. We will find the volumes of the solids that are generated when the shaded regions are revolved about the y-axis.

For $0 \le h < 2$,

$$V = \pi \int_{-4}^{h-4} [(16 - y^2) - (1 - (y+3)^2)] \, dy = 6\pi \int_{-4}^{h-4} (y+4) \, dy = 3\pi h^2;$$

for $2 \le h \le 4$,

$$V = \pi \int_{-4}^{-2} [(16 - y^2) - (1 - (y+3)^2)]dy + \pi \int_{-2}^{h-4} (16 - y^2)dy$$

$$= 6\pi \int_{-4}^{-2} (y+4)dy + \pi \int_{-2}^{h-4} (16 - y^2)dy = 12\pi + \frac{1}{3}\pi(12h^2 - h^3 - 40)$$

$$= \frac{1}{3}\pi(12h^2 - h^3 - 4)$$

so

$$V = \begin{cases} 3\pi h^2 & \text{if } 0 \le h < 2 \\ \frac{1}{3}\pi(12h^2 - h^3 - 4) & \text{if } 2 \le h \le 4 \end{cases}$$

49. $A(x) = \pi(x^2/4)^2 = \pi x^4/16$, $V = \int_0^{20} (\pi x^4/16)dx = 40,000\pi$ ft^3

51. $V = \int_0^1 (x - x^2)^2 dx$

$= \int_0^1 (x^2 - 2x^3 + x^4)dx$

$= 1/30$

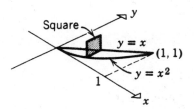

53. With $y = \sqrt{9 - x^2}$, which is the upper half of the circle, $A(x)$ is the area of an equilateral triangle whose sides are each of length $2y$ so

$$A(x) = \frac{\sqrt{3}}{4}(2y)^2 = \sqrt{3}y^2 = \sqrt{3}(9 - x^2),$$

$$V = \int_{-3}^{3} \sqrt{3}(9 - x^2)dx$$

$$= 2\sqrt{3}\int_0^3 (9 - x^2)dx = 36\sqrt{3}.$$

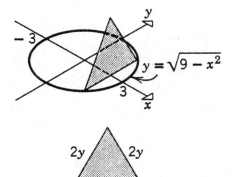

55. $V = \int_{\pi/4}^{3\pi/4} \sin^2 x \, dx$

$= \dfrac{1}{2} \int_{\pi/4}^{3\pi/4} (1 - \cos 2x) dx$

$= (\pi + 2)/4$

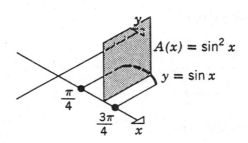

$A(x) = \sin^2 x$

$y = \sin x$

57. $\tan \theta = h/x$ so $h = x \tan \theta$,

$A(y) = \dfrac{1}{2} hx = \dfrac{1}{2} x^2 \tan \theta = \dfrac{1}{2}(r^2 - y^2) \tan \theta$

because $x^2 = r^2 - y^2$,

$V = \dfrac{1}{2} \tan \theta \int_{-r}^{r} (r^2 - y^2) dy$

$= \tan \theta \int_0^r (r^2 - y^2) dy = \dfrac{2}{3} r^3 \tan \theta$

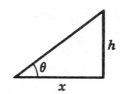

59. Each cross section perpendicular to the y-axis is a square so

$A(y) = x^2 = r^2 - y^2$,

$\dfrac{1}{8} V = \int_0^r (r^2 - y^2) dy$

$V = 8(2r^3/3) = 16r^3/3$

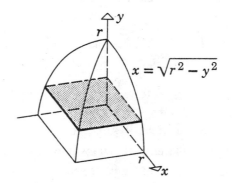

$x = \sqrt{r^2 - y^2}$

61. (a) If $V = \int_0^h A(x) dx$, then $\dfrac{dV}{dt} = \dfrac{dV}{dh}\dfrac{dh}{dt} = A(h)\dfrac{dh}{dt}$ so $A(h)\dfrac{dh}{dt} = -kA(h)$, $\dfrac{dh}{dt} = -k$.

(b) If $dh/dt = -k$, then $h = -kt + C$. But $h = h_0$ when $t = 0$ so $C = h_0$, thus $h = h_0 - kt$; $h = 0$ when $t = h_0/k$.

EXERCISE SET 6.3

1. $V = \int_1^2 2\pi x (x^2) dx = 2\pi \int_1^2 x^3 dx = 15\pi/2$

3. $V = \int_0^1 2\pi y (2y - 2y^2) dy = 4\pi \int_0^1 (y^2 - y^3) dy = \pi/3$

5. $V = \int_0^1 2\pi(x)(x^3) dx$

   $= 2\pi \int_0^1 x^4 dx = 2\pi/5$

7. $V = \int_1^3 2\pi x (1/x) dx$

   $= 2\pi \int_1^3 dx = 4\pi$

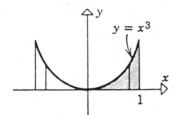

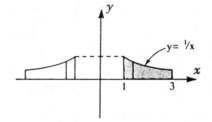

9. $V = \int_1^2 2\pi x[(2x-1) - (-2x+3)] dx$

   $= 8\pi \int_1^2 (x^2 - x) dx = 20\pi/3$

11. $V = \int_0^2 2\pi x (4 - x^2)^{1/3} dx$

    $+ \int_2^4 2\pi x[-(4 - x^2)^{1/3}] dx$

    $= 3\pi \sqrt[3]{4}(1 + 3\sqrt[3]{3})$

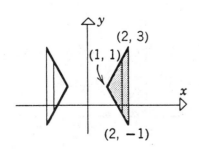

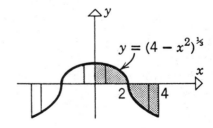

13. $V = \int_0^1 2\pi y^3 dy = \pi/2$

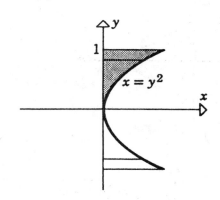

15. $V = \int_0^1 2\pi y(1 - \sqrt{y})dy$

$= 2\pi \int_0^1 (y - y^{3/2})dy = \pi/5$

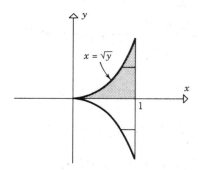

17. **(b)** $V = 2\pi \int_0^\pi x \sin x dx = 2\pi(\sin x - x \cos x)\Big]_0^\pi = 2\pi^2$

19. **(a)** $V = \int_0^1 2\pi x(x^3 - 3x^2 + 2x)dx = 7\pi/30$

(b) much easier; the method of slicing would require that x be expressed in terms of y.

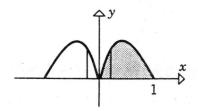

21. $V = \int_0^1 2\pi(1 - y)y^{1/3}dy$

$= 2\pi \int_0^1 (y^{1/3} - y^{4/3})dy = 9\pi/14$

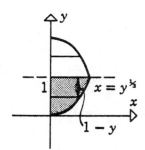

23. $x = \dfrac{h}{r}(r-y)$ is an equation of

line through $(0, r)$ and $(h, 0)$ so

$$V = \int_0^r 2\pi y \left[\frac{h}{r}(r-y)\right] dy$$

$$= \frac{2\pi h}{r} \int_0^r (ry - y^2) dy = \pi r^2 h/3$$

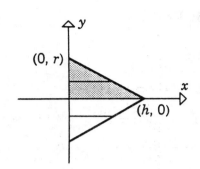

25. $V = \displaystyle\int_0^a 2\pi x (2\sqrt{r^2 - x^2}) dx$

$$= 4\pi \int_0^a x(r^2 - x^2)^{1/2} dx$$

$$= -\frac{4\pi}{3}(r^2 - x^2)^{3/2}\Big]_0^a$$

$$= \frac{4\pi}{3}\left[r^3 - (r^2 - a^2)^{3/2}\right]$$

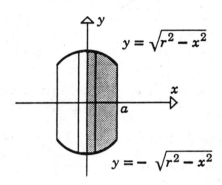

27. $V_x = \pi \displaystyle\int_{1/2}^b \frac{1}{x^2} dx = \pi(2 - 1/b)$, $V_y = 2\pi \displaystyle\int_{1/2}^b dx = \pi(2b - 1)$;

$V_x = V_y$ if $2 - 1/b = 2b - 1$, $2b^2 - 3b + 1 = 0$, solve to get $b = 1/2$ (reject) or $b = 1$.

EXERCISE SET 6.4

1. **(a)** $L = \int_1^2 \sqrt{1+2^2}\,dx = \sqrt{5}\int_1^2 dx = \sqrt{5}$

    (b) $L = \int_2^4 \sqrt{1+(1/2)^2}\,dy = \frac{1}{2}\sqrt{5}\int_2^4 dy = \sqrt{5}$

    (c) $L = \sqrt{(2-1)^2+(4-2)^2} = \sqrt{5}$ ($y = 2x$ is a line).

3. $f'(x) = \frac{9}{2}x^{1/2}$, $1 + [f'(x)]^2 = 1 + \frac{81}{4}x$,

    $L = \int_0^1 \sqrt{1+81x/4}\,dx = \frac{8}{243}\left(1+\frac{81}{4}x\right)^{3/2}\Bigg]_0^1 = (85\sqrt{85}-8)/243$

5. $\dfrac{dy}{dx} = \dfrac{2}{3}x^{-1/3}$, $1 + \left(\dfrac{dy}{dx}\right)^2 = 1 + \dfrac{4}{9}x^{-2/3} = \dfrac{9x^{2/3}+4}{9x^{2/3}}$,

    $L = \int_1^8 \dfrac{\sqrt{9x^{2/3}+4}}{3x^{1/3}}\,dx = \dfrac{1}{18}\int_{13}^{40} u^{1/2}\,du$, $u = 9x^{2/3}+4$

    $\qquad = \dfrac{1}{27}u^{3/2}\Bigg]_{13}^{40} = \dfrac{1}{27}(40\sqrt{40}-13\sqrt{13}) = \dfrac{1}{27}(80\sqrt{10}-13\sqrt{13})$

    or (alternate solution)

    $x = y^{3/2}$, $\dfrac{dx}{dy} = \dfrac{3}{2}y^{1/2}$, $1 + \left(\dfrac{dx}{dy}\right)^2 = 1 + \dfrac{9}{4}y = \dfrac{4+9y}{4}$,

    $L = \dfrac{1}{2}\int_1^4 \sqrt{4+9y}\,dy = \dfrac{1}{18}\int_{13}^{40} u^{1/2}\,du = \dfrac{1}{27}(80\sqrt{10}-13\sqrt{13})$.

7. $x = g(y) = \dfrac{1}{24}y^3 + 2y^{-1}$, $g'(y) = \dfrac{1}{8}y^2 - 2y^{-2}$,

    $1 + [g'(y)]^2 = 1 + \left(\dfrac{1}{64}y^4 - \dfrac{1}{2} + 4y^{-4}\right) = \dfrac{1}{64}y^4 + \dfrac{1}{2} + 4y^{-4} = \left(\dfrac{1}{8}y^2 + 2y^{-2}\right)^2$,

    $L = \int_2^4 \left(\dfrac{1}{8}y^2 + 2y^{-2}\right) dy = 17/6$

9. **(a)**

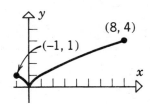

(b) dy/dx does not exist at $x = 0$

(c) $x = g(y) = y^{3/2}$, $g'(y) = \dfrac{3}{2}y^{1/2}$,

$$L = \int_0^1 \sqrt{1 + 9y/4}\,dy \quad \text{(portion for } -1 \le x \le 0\text{)}$$

$$+ \int_0^4 \sqrt{1 + 9y/4}\,dy \quad \text{(portion for } 0 \le x \le 8\text{)}$$

$$= \frac{8}{27}\left(\frac{13}{8}\sqrt{13} - 1\right) + \frac{8}{27}(10\sqrt{10} - 1) = (13\sqrt{13} + 80\sqrt{10} - 16)/27$$

11.
$$m \le f'(x) \le M$$
$$m^2 \le [f'(x)]^2 \le M^2$$
$$1 + m^2 \le 1 + [f'(x)]^2 \le 1 + M^2$$
$$\sqrt{1 + m^2} \le \sqrt{1 + [f'(x)]^2} \le \sqrt{1 + M^2}$$
$$\int_a^b \sqrt{1 + m^2}\,dx \le \int_a^b \sqrt{1 + [f'(x)]^2}\,dx \le \int_a^b \sqrt{1 + M^2}\,dx$$
$$(b - a)\sqrt{1 + m^2} \le L \le (b - a)\sqrt{1 + M^2}$$

13. 4.645975301

EXERCISE SET 6.5

1. $S = \displaystyle\int_0^1 2\pi(7x)\sqrt{1 + 49}\,dx = 70\pi\sqrt{2}\int_0^1 x\,dx = 35\pi\sqrt{2}$

3. $f'(x) = -x/\sqrt{4 - x^2}$, $1 + [f'(x)]^2 = 1 + \dfrac{x^2}{4 - x^2} = \dfrac{4}{4 - x^2}$,

$S = \displaystyle\int_{-1}^1 2\pi\sqrt{4 - x^2}(2/\sqrt{4 - x^2})\,dx = 4\pi\int_{-1}^1 dx = 8\pi$

5. $f'(x) = \frac{1}{2}x^{-1/2} - \frac{1}{2}x^{1/2}$, $1 + [f'(x)]^2 = 1 + \frac{1}{4}x^{-1} - \frac{1}{2} + \frac{1}{4}x = \left(\frac{1}{2}x^{-1} + \frac{1}{2}x\right)^2$,

$S = \int_1^3 2\pi\left(x^{1/2} - \frac{1}{3}x^{3/2}\right)\left(\frac{1}{2}x^{-1} + \frac{1}{2}x\right)dx = \frac{\pi}{3}\int_1^3 (3 + 2x - x^2)dx = 16\pi/9$

7. $S = \int_0^2 2\pi(9y + 1)\sqrt{82}dy = 2\pi\sqrt{82}\int_0^2 (9y + 1)dy = 40\pi\sqrt{82}$

9. $g'(y) = -y/\sqrt{9 - y^2}$, $1 + [g'(y)]^2 = \frac{9}{9 - y^2}$,

$S = \int_{-2}^2 2\pi\sqrt{9 - y^2} \cdot \frac{3}{\sqrt{9 - y^2}}dy = 6\pi\int_{-2}^2 dy = 24\pi$

11. $x = g(y) = \frac{1}{4}y^4 + \frac{1}{8}y^{-2}$, $g'(y) = y^3 - \frac{1}{4}y^{-3}$,

$1 + [g'(y)]^2 = 1 + \left(y^6 - \frac{1}{2} + \frac{1}{16}y^{-6}\right) = \left(y^3 + \frac{1}{4}y^{-3}\right)^2$

$S = \int_1^2 2\pi\left(\frac{1}{4}y^4 + \frac{1}{8}y^{-2}\right)\left(y^3 + \frac{1}{4}y^{-3}\right)dy = \frac{\pi}{16}\int_1^2 (8y^7 + 6y + y^{-5})dy = 16,911\pi/1024$

13. Revolve the line segment joining the points $(0, 0)$ and (h, r) about the x-axis. An equation of the line segment is $y = (r/h)x$ for $0 \le x \le h$ so

$S = \int_0^h 2\pi(r/h)x\sqrt{1 + r^2/h^2}dx = \frac{2\pi r}{h^2}\sqrt{r^2 + h^2}\int_0^h x\,dx = \pi r\sqrt{r^2 + h^2}$

15. $f(x) = \sqrt{r^2 - x^2}$, $f'(x) = -x/\sqrt{r^2 - x^2}$, $1 + [f'(x)]^2 = r^2/(r^2 - x^2)$,

$S = \int_a^{a+h} 2\pi\sqrt{r^2 - x^2}(r/\sqrt{r^2 - x^2})dx = 2\pi r\int_a^{a+h} dx = 2\pi rh$

17. (a) length of arc of sector = circumference of base of cone, $\ell\theta = 2\pi r$, $\theta = 2\pi r/\ell$;

$\qquad$ $S = $ area of sector $= \frac{1}{2}\ell^2(2\pi r/\ell) = \pi r\ell$

$\quad$ **(b)** $S = \pi r_2\ell_2 - \pi r_1\ell_1$
$\qquad = \pi r_2(\ell_1 + \ell) - \pi r_1\ell_1$
$\qquad = \pi[(r_2 - r_1)\ell_1 + r_2\ell]$
$\qquad$ Using similar triangles
$\qquad\qquad \ell_2/r_2 = \ell_1/r_1$
$\qquad\qquad r_1\ell_2 = r_2\ell_1$
$\qquad\quad r_1(\ell_1 + \ell) = r_2\ell_1$
$\qquad\quad (r_2 - r_1)\ell_1 = r_1\ell$
$\qquad$ so $S = \pi(r_1\ell + r_2\ell) = \pi(r_1 + r_2)\ell$

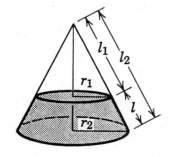

19. (a) $1 \le \sqrt{1 + [f'(x)]^2}$ so $2\pi f(x) \le 2\pi f(x)\sqrt{1 + [f'(x)]^2}$,

$$\int_a^b 2\pi f(x)dx \le \int_a^b 2\pi f(x)\sqrt{1 + [f'(x)]^2}dx, \; 2\pi \int_a^b f(x)dx \le S, 2\pi A \le S$$

   (b) $2\pi A = S$ if $f'(x) = 0$ for all x in $[a, b]$ so $f(x)$ is constant on $[a, b]$.

EXERCISE SET 6.6

1. $s(t) = \displaystyle\int (2t - 3)dt = t^2 - 3t + C$, $s(1) = (1)^2 - 3(1) + C = 5$, $C = 7$, $s(t) = t^2 - 3t + 7$.

3. $s(t) = \displaystyle\int (t^3 - 2t^2 + 1)dt = \frac{1}{4}t^4 - \frac{2}{3}t^3 + t + C$,

   $s(0) = \frac{1}{4}(0)^4 - \frac{2}{3}(0)^3 + 0 + C = 1$, $C = 1$, $s(t) = \frac{1}{4}t^4 - \frac{2}{3}t^3 + t + 1$.

5. $v(t) = \displaystyle\int 4\, dt = 4t + C_1$, $v(0) = 4(0) + C_1 = 1$, $C_1 = 1$, $v(t) = 4t + 1$,

   $s(t) = \displaystyle\int (4t + 1)dt = 2t^2 + t + C_2$, $s(0) = 2(0)^2 + 0 + C_2 = 0$, $C_2 = 0$, $s(t) = 2t^2 + t$.

7. $v(t) = \displaystyle\int 4\cos 2t\, dt = 2\sin 2t + C_1$, $v(0) = 2\sin 0 + C_1 = -1$, $C_1 = -1$,

   $v(t) = 2\sin 2t - 1$, $s(t) = \displaystyle\int (2\sin 2t - 1)dt = -\cos 2t - t + C_2$,

   $s(0) = -\cos 0 - 0 + C_2 = -3$, $C_2 = -2$, $s(t) = -\cos 2t - t - 2$.

9. (a) $s = \displaystyle\int \sin\frac{1}{2}\pi t\, dt = -\frac{2}{\pi}\cos\frac{1}{2}\pi t + C$

   $s = 0$ when $t = 0$ which gives $C = \dfrac{2}{\pi}$ so $s = -\dfrac{2}{\pi}\cos\dfrac{1}{2}\pi t + \dfrac{2}{\pi}$.

   $a = \dfrac{dv}{dt} = \dfrac{\pi}{2}\cos\dfrac{1}{2}\pi t$. When $t = 1 : s = 2/\pi$, $v = 1$, $|v| = 1$, $a = 0$.

   (b) $v = -3\displaystyle\int t\, dt = -\frac{3}{2}t^2 + C_1$, $v = 0$ when $t = 0$ which gives $C_1 = 0$ so $v = -\frac{3}{2}t^2$.

   $s = -\dfrac{3}{2}\displaystyle\int t^2 dt = -\frac{1}{2}t^3 + C_2$, $s = 1$ when $t = 0$ which gives $C_2 = 1$ so $s = -\frac{1}{2}t^3 + 1$.

   When $t = 1 : s = 1/2$, $v = -3/2$, $|v| = 3/2$, $a = -3$.

11. If $a(t) = k$ then $v(t) = \int k\,dt = kt + C_1$, but $v(0) = 60$ mph $= 88$ ft/sec so $k(0) + C_1 = 88$, $C_1 = 88$, $v(t) = kt + 88$; $s(t) = \int (kt + 88)dt = \frac{1}{2}kt^2 + 88t + C_2$, $s(0) = 0$ so $C_2 = 0$, $s(t) = \frac{1}{2}kt^2 + 88t$. $v(t) = 0$ at the instant when the car comes to a stop so $kt + 88 = 0$, $t = -88/k$; $s(t) = 180$ at this instant so $\frac{1}{2}k(-88/k)^2 + 88(-88/k) = 180$, $-3872/k = 180$, $k \approx -21.5$ ft/sec^2.

13. $s = 0$ and $v = 112$ when $t = 0$ so $v(t) = -32t + 112$, $s(t) = -16t^2 + 112t$.

    (a) $v(3) = 16$ ft/sec, $v(5) = -48$ ft/sec.

    (b) $v = 0$ when the projectile is at its maximum height so $-32t + 112 = 0$, $t = 7/2$ sec, $s(7/2) = -16(7/2)^2 + 112(7/2) = 196$ ft.

    (c) $s = 0$ when it reaches the ground so $-16t^2 + 112t = 0$, $-16t(t - 7) = 0$, $t = 0, 7$ of which $t = 7$ is when it is at ground level on its way down. $v(7) = -112$, $|v| = 112$ ft/sec.

15.   **(a)** $s(t) = 0$ when it hits the ground, $s(t) = -16t^2 + 16t = -16t(t - 1) = 0$ when $t = 1$ sec.

    (b) The projectile moves upward until it gets to its highest point where $v(t) = 0$, $v(t) = -32t + 16 = 0$ when $t = 1/2$ sec.

17.   **(a)** $s(t) = 0$ when the package hits the ground, $s(t) = -16t^2 + 20t + 200 = 0$ when $t = (5 + 5\sqrt{33})/8$ sec.

    (b) $v(t) = -32t + 20$, $v[(5 + 5\sqrt{33})/8] = -20\sqrt{33}$, the speed at impact is $20\sqrt{33}$ ft/sec.

19. $s(t) = -4.9t^2 + 49t + 150$ and $v(t) = -9.8t + 49$.

    (a) The projectile reaches its maximum height when $v(t) = 0$, $-9.8t + 49 = 0$, $t = 5$ sec.

    (b) $s(5) = -4.9(5)^2 + 49(5) + 150 = 272.5$ m.

    (c) The projectile reaches its starting point when $s(t) = 150$, $-4.9t^2 + 49t + 150 = 150$, $-4.9t(t - 10) = 0$, $t = 10$ sec.

    (d) $v(10) = -9.8(10) + 49 = -49$ m/sec.

    (e) $s(t) = 0$ when the projectile hits the ground, $-4.9t^2 + 49t + 150 = 0$ when (use the quadratic formula) $t \approx 12.46$ sec.

    (f) $v(12.46) = -9.8(12.46) + 49 \approx -73.1$, the speed at impact is about 73.1 m/sec.

21. $s(t) = -16t^2 + v_0 t$, $v(t) = -32t + v_0$; $v = 0$ when it reaches maximum height so $-32t + v_0 = 0$, $t = v_0/32$ is the time it takes to get there, thus $s(v_0/32) = -16(v_0/32)^2 + v_0(v_0/32) = 1,000$, $v_0^2 = 64,000$, $v_0 = 80\sqrt{10}$ ft/sec (positive because fired upward).

23. $s = -16t^2 + v_0 t + s_0$, but $s_0 = 0$ so $s = -16t^2 + v_0 t$. $s = 0$ when $t = 8$ so $0 = -16(8)^2 + v_0(8)$, $v_0 = 128$ ft/sec. $v = 0$ at its highest point so $-32t + 128 = 0$, $t = 4$, $s = -16(4)^2 + 128(4) = 256$ ft.

25. **(a)** negative, because v is decreasing
    (b) increasing, because the graph of $v(t)$ is concave up
    (c) negative, because the area between the graph of $v(t)$ and the t-axis appears to be greater where $v < 0$ compared to where $v > 0$.

27. displacement $= \int_0^2 (t^2 + t - 2)dt = 2/3$

    distance $= \int_0^2 |t^2 + t - 2|dt = \int_0^1 -(t^2 + t - 2)dt + \int_1^2 (t^2 + t - 2)dt = 7/6 + 11/6 = 3$

29. displacement $= \int_0^\pi \cos t \, dt = 0$

    distance $= \int_0^\pi |\cos t|dt = \int_0^{\pi/2} \cos t \, dt + \int_{\pi/2}^\pi -\cos t \, dt = 1 + 1 = 2$

31. $v(t) = t^3 - 3t^2 + 2t = t(t-1)(t-2)$

    displacement $= \int_0^3 (t^3 - 3t^2 + 2t)dt = 9/4$

    distance $= \int_0^3 |v(t)|dt = \int_0^1 v(t)dt + \int_1^2 -v(t)dt + \int_2^3 v(t)dt = 11/4$

33. $v(t) = \frac{1}{2}t^2 - 2t$

    displacement $= \int_1^5 (\frac{1}{2}t^2 - 2t)dt = -10/3$

    distance $= \int_1^5 |\frac{1}{2}t^2 - 2t|dt = \int_1^4 -(\frac{1}{2}t^2 - 2t)dt + \int_4^5 (\frac{1}{2}t^2 - 2t)dt = 17/3$

35. $v(t) = \frac{2}{5}\sqrt{5t + 1} + \frac{8}{5}$

    displacement $= \int_0^3 \left(\frac{2}{5}\sqrt{5t + 1} + \frac{8}{5}\right) dt = \frac{4}{75}(5t + 1)^{3/2} + \frac{8}{5}t \Big]_0^3 = 204/25$

    distance $= \int_0^3 |v(t)|dt = \int_0^3 v(t)dt = 204/25$

EXERCISE SET 6.7

1. **(a)** $W = 30[5 - (-2)] = 210$ ft·lb   **(b)** $W = \int_1^6 x^{-2} dx = 5/6$ ft·lb

3. $F(x) = kx$, $F(0.2) = 0.2k = 100$, $k = 500$ N/m, $W = \int_0^{0.8} 500x\,dx = 160$ J

5. $W = \int_0^1 kx\,dx = k/2 = 10$, $k = 20$ lb/ft   7. $W = \int_0^6 (9-x)\rho(25\pi)\,dx = 900\pi\rho$ ft·lb

9. $w/4 = x/3$, $w = 4x/3$

$$W = \int_0^2 (3-x)(9810)(4x/3)(6)\,dx$$

$$= 78480 \int_0^2 (3x - x^2)\,dx$$

$$= 261,600 \text{ J}$$

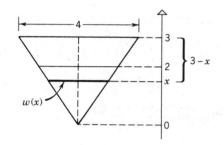

11. **(a)** $W = \int_0^9 (10-x)62.4(300)\,dx$

$$= 18,720 \int_0^9 (10-x)\,dx$$
$$= 926,640 \text{ ft·lb}$$

   (b) to empty the pool in one hour
   would require $926,640/3600 = 257.4$
   ft·lb of work per second so
   hp of motor $= 257.4/550 = 0.468$

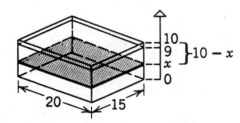

13. $W = \displaystyle\int_0^{100} 15(100 - x)\,dx$

$= 75,000 \text{ ft·lb}$

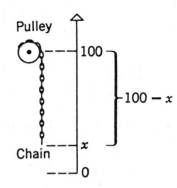

15. **(a)** $F(4000) = k/(4000)^2 = 6000,\ k = 9.6 \times 10^{10}$

(b) $W = \displaystyle\int_{4000}^{5000} 9.6 \times 10^{10} x^{-2}\,dx = 4,800,000 \text{ mi·lb}$

EXERCISE SET 6.8

1. **(a)** $F = \rho h A = (62.4)(5)(9) = 2,808 \text{ lb}$ **(b)** $F = (40)(10)(9) = 3,600 \text{ lb}$

3. $F = \displaystyle\int_1^3 9810 x(4)\,dx$

$= 39240 \displaystyle\int_1^3 x\,dx$

$= 156,960 \text{ N}$

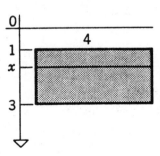

5. by similar triangles

$$\frac{w(x)}{6} = \frac{10 - x}{8}$$

$$w(x) = \frac{3}{4}(10 - x),$$

$$F = \int_2^{10} 62.4x \left[\frac{3}{4}(10 - x) \right] dx$$

$$= 46.8 \int_2^{10} (10x - x^2) dx = 6988.8 \, \text{lb}$$

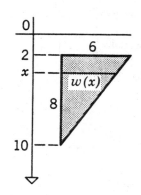

7. $$F = \int_0^5 9810x(2\sqrt{25 - x^2}) dx$$

$$= 19,620 \int_0^5 x(25 - x^2)^{1/2} dx$$

$$= 8.175 \times 10^5 \, \text{N}$$

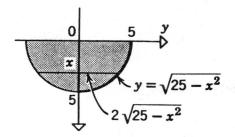

9. Find the forces on the upper and lower halves and add them:

$$\frac{w_1(x)}{\sqrt{2}a} = \frac{x}{\sqrt{2}a/2}, \quad w_1(x) = 2x$$

$$F_1 = \int_0^{\sqrt{2}a/2} \rho x(2x) dx$$

$$= 2\rho \int_0^{\sqrt{2}a/2} x^2 dx = \sqrt{2}\rho a^3/6,$$

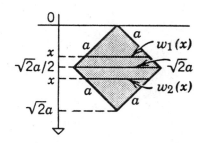

$$\frac{w_2(x)}{\sqrt{2}a} = \frac{\sqrt{2}a - x}{\sqrt{2}a/2}, \quad w_2(x) = 2(\sqrt{2}a - x)$$

$$F_2 = \int_{\sqrt{2}a/2}^{\sqrt{2}a} \rho x[2(\sqrt{2}a - x)]dx = 2\rho \int_{\sqrt{2}a/2}^{\sqrt{2}a} (\sqrt{2}ax - x^2)dx = \sqrt{2}\rho a^3/3,$$

$$F = F_1 + F_2 = \sqrt{2}\rho a^3/6 + \sqrt{2}\rho a^3/3 = \rho a^3/\sqrt{2}$$

11. $\sqrt{16^2 + 4^2} = \sqrt{272} = 4\sqrt{17}$ is the
other dimension of the bottom.
$(h(x) - 4)/4 = x/(4\sqrt{17})$
$h(x) = x/\sqrt{17} + 4$,

$$F = \int_0^{4\sqrt{17}} 62.4(x/\sqrt{17} + 4)10\,dx$$

$$= 624 \int_0^{4\sqrt{17}} (x/\sqrt{17} + 4)\,dx$$

$$= 14{,}976\sqrt{17}\,\text{lb}$$

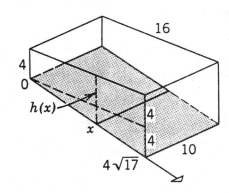

13. (a) From Exercise 12, $F = 4\rho_0(h + 1)$ so (assuming that ρ_0 is constant) $dF/dt = 4\rho_0(dh/dt)$ which is a positive constant if dh/dt is a positive constant.
(b) If $dh/dt = 20$ then $dF/dt = 80\rho_0$ lb/min from part (a).

TECHNOLOGY EXERCISES 6

1. The x-coordinates of the points of intersection are $a = -0.423028$ and $b = 1.725171$; the area is $\int_a^b (2\sin x - x^2 + 1)\,dx = 2.542696$.

3. The area is given by $\int_0^k \left(\frac{1}{2}x + 1 - \sqrt{x}\right)\,dx = \frac{1}{4}k^2 + k - \frac{2}{3}k^{3/2} = 2$; solve for k to get $k = 3.268669$.

5. The curve intersects the x-axis at $a = 1.465571$ so the volume is
$$2\pi \int_0^a x(1 + x^2 - x^3)\,dx = 5.498059.$$

7. The volume is given by $\pi \int_0^k \left[\left(\frac{1}{2}x + 1\right)^2 - (\sqrt{x})^2\right]\,dx = \pi\left(\frac{1}{12}k^3 + k\right) = 10$; solve for k to get $k = 2.242872$.

9. The arc length is given by $\int_0^k \sqrt{1 + 1/x^4}\,dx = 2$; find k (by trial and error, or some other method) to get $k = 2.854368$.

11. The distance is $\int_0^{4.6} \sqrt{1 + (2.09 - 0.82x)^2}\, dx = 6.65$ meters

13. **(a)** $v < 0$ for t between 0 and $T = 1.895494$ sec; the maximum distance to the left of its starting position is $-\int_0^T v(t)dt = 4.21$ cm.

(b) Let $t = k$ be the time to arrive back at the point where it started, then
$$\int_0^k (5t - 10\sin t)dt = \frac{5}{2}k^2 + 10\cos k - 10 = 0; \text{ solve for } k \text{ to get } k = 2.78 \text{ sec.}$$

15. The work is given by $\int_0^2 \frac{x}{\sqrt{1 + kx}}dx = \frac{4}{3}\left[\frac{(k-1)\sqrt{1+2k}}{k^2} + \frac{1}{k^2}\right] = 0.8$; solve for k to get $k = 4.425822$.

CHAPTER 7
Logarithmic and Exponential Functions

EXERCISE SET 7.1

1. (a) -4 (b) 4 (c) $1/4$

3. (a) 2.9690 (b) 0.0341

5. (a) $\log_2 16 = \log_2(2^4) = 4$

    (b) $\log_2\left(\dfrac{1}{32}\right) = \log_2(2^{-5}) = -5$

    (c) $\log_4 4 = 1$

    (d) $\log_9 3 = \log_9(9^{1/2}) = 1/2$

7. (a) 1.3655 (b) -0.3011

9. (a) $2\ln a + \dfrac{1}{2}\ln b + \dfrac{1}{2}\ln c = 2r + s/2 + t/2.$

    (b) $\ln b - 3\ln a - \ln c = s - 3r - t.$

11. (a) $1 + \log x + \dfrac{1}{2}\log(x-3)$

    (b) $2\ln|x| + 3\ln\sin x - \dfrac{1}{2}\ln(x^2+1)$

13. $\log\dfrac{2^4(16)}{3} = \log(256/3)$

15. $\ln\dfrac{\sqrt[3]{x}(x+1)^2}{\cos x}$

17. $\sqrt{x} = 10^{-1} = 0.1,\ x = 0.01$

19. $1/x = e^{-2},\ x = e^2$

21. $2x = 8,\ x = 4$

23. $\log_{10} x = 5,\ x = 10^5$

25. $\ln 2x^2 = \ln 3,\ 2x^2 = 3,\ x^2 = 3/2,\ x = \sqrt{3/2}$ (we discard $-\sqrt{3/2}$ because it does not satisfy the original equation).

27. $\ln 5^{-2x} = \ln 3,\ -2x\ln 5 = \ln 3,\ x = -\dfrac{\ln 3}{2\ln 5}$

29. $e^{3x} = 7/2,\ 3x = \ln(7/2),\ x = \dfrac{1}{3}\ln(7/2)$

31. $e^{-x}(x+2) = 0$ so $e^{-x} = 0$ (impossible) or $x + 2 = 0,\ x = -2$

33. $e^{-2x} - 3e^{-x} + 2 = 0,\ (e^{-x}-1)(e^{-x}-2) = 0$; if $e^{-x} - 1 = 0$, then $x = 0$, if $e^{-x} - 2 = 0$, then $x = -\ln 2.$

35. $4 - 12e^{2x} = 1$, $12e^{2x} = 3$, $e^{2x} = 1/4$, $x = \frac{1}{2}\ln(1/4) = -\ln 2$

37. $\log(1/2) < 0$ so $3\log(1/2) < 2\log(1/2)$

39. $\log_2 7.35 = (\log 7.35)/(\log 2) = (\ln 7.35)/(\ln 2) \approx 2.8777$;
$\log_5 0.6 = (\log 0.6)/(\log 5) = (\ln 0.6)/(\ln 5) \approx -0.3174$

41. $75e^{-t/125} = 15$, $t = -125\ln(1/5) = 125\ln 5 \approx 201$ days.

43. **(a)** 7.4; basic **(b)** 4.2; acidic **(c)** 6.4; acidic **(d)** 5.9; basic

45. **(a)** 140 dB; damage **(b)** 120 dB; damage
(c) 80 dB; no damage **(d)** 75 dB; no damage

47. Let I_A and I_B be the intensities of the automobile and blender, respectively. Then
$\log_{10} I_A/I_0 = 7$ and $\log_{10} I_B/I_0 = 9.3$, $I_A = 10^7 I_0$ and $I_B = 10^{9.3} I_0$, so $I_B/I_A = 10^{2.3} \approx 200$.

49. **(a)** $\log E = 4.4 + 1.5(8.2) = 16.7$, $E = 10^{16.7} \approx 5 \times 10^{16}$ J
(b) Let M_1 and M_2 be the magnitudes of earthquakes with energies of E and $10E$,
respectively. Then $1.5(M_2 - M_1) = \log(10E) - \log E = \log 10 = 1$,
$M_2 - M_1 = 1/1.5 = 2/3 \approx 0.67$.

51. If $t = -2x$, then $x = -t/2$ and $\lim_{x \to 0}(1 - 2x)^{1/x} = \lim_{t \to 0}(1 + t)^{-2/t} = \lim_{t \to 0}[(1 + t)^{1/t}]^{-2} = e^{-2}$.

EXERCISE SET 7.2

1. $\dfrac{1}{2x}(2) = 1/x$

3. $2(\ln x)\left(\dfrac{1}{x}\right) = \dfrac{2\ln x}{x}$

5. $\dfrac{1}{\tan x}(\sec^2 x) = \dfrac{\sec^2 x}{\tan x}$

7. $\dfrac{1}{x/(1+x^2)}\left[\dfrac{(1+x^2)(1) - x(2x)}{(1+x^2)^2}\right] = \dfrac{1 - x^2}{x(1+x^2)}$

9. $\dfrac{3x^2 - 14x}{x^3 - 7x^2 - 3}$

11. $\dfrac{1}{2}(\ln x)^{-1/2}\left(\dfrac{1}{x}\right) = \dfrac{1}{2x\sqrt{\ln x}}$

13. $\cos(5/\ln x)\dfrac{d}{dx}[5(\ln x)^{-1}] = \cos(5/\ln x)[-5(\ln x)^{-2}(1/x)] = -\dfrac{5\cos(5/\ln x)}{x(\ln x)^2}$

15. $-\dfrac{2x^3}{3 - 2x} + 3x^2\ln(3 - 2x)$

17. $2(x^2+1)[\ln(x^2+1)]\dfrac{2x}{x^2+1} + 2x[\ln(x^2+1)]^2 = 4x\ln(x^2+1) + 2x[\ln(x^2+1)]^2$

19. $\dfrac{(1+\ln x)(2x) - x^2(0+1/x)}{(1+\ln x)^2} = \dfrac{x(1+2\ln x)}{(1+\ln x)^2}$

21. $\dfrac{dy}{dx} + \dfrac{1}{xy}\left(x\dfrac{dy}{dx}+y\right) = 0,\; \dfrac{dy}{dx} = -\dfrac{y}{x(y+1)}$

23. $\dfrac{1}{2}\ln|x| + C$

25. $u = x^3 - 4,\; du = 3x^2 dx,\; \dfrac{1}{3}\displaystyle\int \dfrac{1}{u}du = \dfrac{1}{3}\ln|x^3 - 4| + C$

27. $u = \tan x,\; du = \sec^2 x\, dx,\; \displaystyle\int \dfrac{1}{u}du = \ln|\tan x| + C$

29. $u = 1 + \cos 3\theta,\; du = -3\sin 3\theta d\theta$

$-\dfrac{1}{3}\displaystyle\int \dfrac{1}{u}du = -\dfrac{1}{3}\ln|1+\cos 3\theta| + C = -\dfrac{1}{3}\ln(1+\cos 3\theta) + C$ because $1 + \cos 3\theta \geq 0$

31. divide $x^2 + 1$ into x^3 to get

$\displaystyle\int \dfrac{x^3}{x^2+1}dx = \int \left[x - \dfrac{x}{x^2+1}\right]dx = \int x\,dx - \int \dfrac{x}{x^2+1}dx = \dfrac{1}{2}x^2 - \dfrac{1}{2}\ln(x^2+1) + C$

33. $u = \ln y,\; du = \dfrac{1}{y}dy,\; \displaystyle\int u^3 du = \dfrac{1}{4}(\ln y)^4 + C$

35. $u = 3x + 2,\; \dfrac{1}{3}\displaystyle\int_2^5 \dfrac{1}{u}du = \dfrac{1}{3}\ln|u|\Big]_2^5 = \dfrac{1}{3}(\ln 5 - \ln 2) = \dfrac{1}{3}\ln\dfrac{5}{2}$

37. $u = x^2 + 5,\; \dfrac{1}{2}\displaystyle\int_6^5 \dfrac{1}{u}du = \dfrac{1}{2}\ln|u|\Big]_6^5 = \dfrac{1}{2}(\ln 5 - \ln 6) = \dfrac{1}{2}\ln\dfrac{5}{6}$

39. $\dfrac{d}{dx}\left[\ln\cos x - \dfrac{1}{2}\ln(4-3x^2)\right] = -\tan x + \dfrac{3x}{4-3x^2}$

41. $\dfrac{d}{dx}\left[\dfrac{1}{2}\ln x + \dfrac{1}{3}\ln(x+3) + \dfrac{1}{5}\ln(3x-2)\right] = \dfrac{1}{2x} + \dfrac{1}{3(x+3)} + \dfrac{3}{5(3x-2)}$

43. $\ln|y| = \ln|x| + \dfrac{1}{3}\ln|1+x^2|,\; \dfrac{dy}{dx} = x^3\sqrt{1+x^2}\left[\dfrac{1}{x} + \dfrac{2x}{3(1+x^2)}\right]$

45. $\ln|y| = \dfrac{1}{3}\ln|x^2 - 8| + \dfrac{1}{2}\ln|x^3 + 1| - \ln|x^6 - 7x + 5|$

$\dfrac{dy}{dx} = \dfrac{(x^2 - 8)^{1/3}\sqrt{x^3 + 1}}{x^6 - 7x + 5}\left[\dfrac{2x}{3(x^2 - 8)} + \dfrac{3x^2}{2(x^3 + 1)} - \dfrac{6x^5 - 7}{x^6 - 7x + 5}\right]$

47. (a) $\log_x e = \dfrac{\ln e}{\ln x} = \dfrac{1}{\ln x}, \dfrac{d}{dx}[\log_x e] = -\dfrac{1}{x(\ln x)^2}$

(b) $\log_x 2 = \dfrac{\ln 2}{\ln x}, \dfrac{d}{dx}[\log_x 2] = -\dfrac{\ln 2}{x(\ln x)^2}$

49. $\dfrac{k}{n^2 + k^2} = \dfrac{k/n}{1 + k^2/n^2}\dfrac{1}{n}$ so $\displaystyle\sum_{k=1}^{n}\dfrac{k}{n^2 + k^2} = \sum_{k=1}^{n} f(x_k^*)\Delta x$ where $f(x) = \dfrac{x}{1 + x^2}, x_k^* = \dfrac{k}{n}$, and

$\Delta x = \dfrac{1}{n}, 0 \le x \le 1;\ \displaystyle\lim_{n\to+\infty}\sum_{k=1}^{n}\dfrac{k}{n^2 + k^2} = \lim_{n\to+\infty}\sum_{k=1}^{n} f(x_k^*)\Delta x = \int_0^1 \dfrac{x}{1 + x^2}\,dx = \dfrac{1}{2}\ln 2.$

51. Let $f(x) = x^2 - \ln x$, then $f'(x) = 2x - 1/x = (2x^2 - 1)/x;\ f'(x) = 0$ if $x = 1/\sqrt{2}$ at which there is a relative minimum and hence the absolute minimum on $(0, +\infty)$. The minimum value is $1/2 - \ln(1/\sqrt{2}) = (1 + \ln 2)/2.$

53. Let $f(x) = (\ln x)^2/x$, then $f'(x) = [2\ln x - (\ln x)^2]/x^2 = (2 - \ln x)(\ln x)/x^2;\ f'(x) = 0$ if $x = 1$ or $x = e^2$, at which there is a relative minimum and a relative maximum, respectively. The relative minimum value is $f(1) = 0$; the relative maximum value is $f(e^2) = 4/e^2.$

55. Let $f(x) = x - 1 - \ln x$, then $f'(x) = 1 - 1/x = (x - 1)/x, f'(x) = 0$ when $x = 1$ where the minimum value occurs, so $f(x) \ge f(1) = 0, x - 1 - \ln x \ge 0, \ln x \le x - 1.$

57. $p = c/v$ so the average pressure with respect to volume is

$p_{\text{ave}} = \dfrac{1}{v_1 - v_0}\displaystyle\int_{v_0}^{v_1}\dfrac{c}{v}\,dv = \dfrac{c}{v_1 - v_0}(\ln v_1 - \ln v_0) = \dfrac{c}{v_1 - v_0}\ln(v_1/v_0)$

59. $A = \displaystyle\int_0^{\pi/3}\tan x\,dx = \int_0^{\pi/3}\dfrac{\sin x}{\cos x}\,dx = -\ln|\cos x|\Big]_0^{\pi/3} = \ln 2$

61. $V = \pi\displaystyle\int_1^4 \dfrac{1}{x}\,dx = \pi\ln|x|\Big]_1^4 = \pi\ln 4$

63. $f(x) = \ln x - x + 2, x_{n+1} = x_n - \dfrac{\ln x_n - x_n + 2}{1/x_n - 1};$

$x_1 = 0.2, x_2 = 0.152359478, x_3 = 0.158447821, \cdots, x_5 = x_6 = 0.158594340;$

$x_1 = 3, x_2 = 3.147918433, x_3 = 3.146193441, x_4 = x_5 = 3.146193221.$

65. $-10xe^{-5x^2}$ **67.** $x^3 e^x + 3x^2 e^x = x^2 e^x (x+3)$

69. $\dfrac{dy}{dx} = \dfrac{(e^x + e^{-x})(e^x + e^{-x}) - (e^x - e^{-x})(e^x - e^{-x})}{(e^x + e^{-x})^2}$

$\qquad = \dfrac{(e^{2x} + 2 + e^{-2x}) - (e^{2x} - 2 + e^{-2x})}{(e^x + e^{-x})^2} = 4/(e^x + e^{-x})^2$

71. $(x \sec^2 x + \tan x)e^{x \tan x}$ **73.** $(1 - 3e^{3x})e^{(x - e^{3x})}$ **75.** $\dfrac{(x-1)e^{-x}}{1 - xe^{-x}} = \dfrac{x-1}{e^x - x}$

77. $e^{ax}(a \cos bx - b \sin bx)$ **79.** $y = e^{\ln(x^3 + 1)} = x^3 + 1,\ dy/dx = 3x^2$

81. $f'(x) = -3^{-x} \ln 3;\ y = 3^{-x},\ \ln y = -x \ln 3,\ \dfrac{1}{y}y' = -\ln 3,\ y' = -y \ln 3 = -3^{-x} \ln 3$

83. $f'(x) = \pi^{x \tan x}(\ln \pi)(x \sec^2 x + \tan x);$

$\qquad y = \pi^{x \tan x},\ \ln y = (x \tan x) \ln \pi,\ \dfrac{1}{y}y' = (\ln \pi)(x \sec^2 x + \tan x)$

$\qquad y' = \pi^{x \tan x}(\ln \pi)(x \sec^2 x + \tan x)$

85. Let $y = u^v$ then $\ln y = v \ln u,\ \dfrac{1}{y}\dfrac{dy}{dx} = v\left(\dfrac{1}{u}\dfrac{du}{dx}\right) + (\ln u)\dfrac{dv}{dx},$

$\qquad \dfrac{dy}{dx} = u^v\left[\dfrac{v}{u}\dfrac{du}{dx} + \ln u\dfrac{dv}{dx}\right] = vu^{v-1}\dfrac{du}{dx} + u^v \ln u\dfrac{dv}{dx}$

$\qquad$ when u is constant, $d(u^v)/dx = u^v \ln u\, dv/dx$; when v is constant, $d(u^v)/dx = vu^{v-1}du/dx$

87. $\ln y = (\ln x) \ln(x^3 - 2x),\ \dfrac{1}{y}\dfrac{dy}{dx} = \dfrac{3x^2 - 2}{x^3 - 2x}\ln x + \dfrac{1}{x}\ln(x^3 - 2x),$

$\qquad \dfrac{dy}{dx} = (x^3 - 2x)^{\ln x}\left[\dfrac{3x^2 - 2}{x^3 - 2x}\ln x + \dfrac{1}{x}\ln(x^3 - 2x)\right]$

89. $\ln y = (\tan x) \ln(\ln x),\ \dfrac{1}{y}\dfrac{dy}{dx} = \dfrac{1}{x \ln x}\tan x + (\sec^2 x) \ln(\ln x),$

$\qquad \dfrac{dy}{dx} = (\ln x)^{\tan x}\left[\dfrac{\tan x}{x \ln x} + (\sec^2 x) \ln(\ln x)\right]$

91. $\ln y = e^x \ln x,\ \dfrac{1}{y}\dfrac{dy}{dx} = \dfrac{e^x}{x} + e^x \ln x,\ \dfrac{dy}{dx} = x^{(e^x)}\left[\dfrac{e^x}{x} + e^x \ln x\right]$.

93. $y = Ae^{2x} + Be^{-4x},\ y' = 2Ae^{2x} - 4Be^{-4x},\ y'' = 4Ae^{2x} + 16Be^{-4x}$ so

$\qquad y'' + 2y' - 8y = (4Ae^{2x} + 16Be^{-4x}) + 2(2Ae^{2x} - 4Be^{-4x}) - 8(Ae^{2x} + Be^{-4x}) = 0.$

95. (a) $f'(x) = ke^{kx}$, $f''(x) = k^2 e^{kx}$, $f'''(x) = k^3 e^{kx}, \ldots, f^{(n)}(x) = k^n e^{kx}$
(b) $f'(x) = -ke^{-kx}$, $f''(x) = k^2 e^{-kx}$, $f'''(x) = -k^3 e^{-kx}, \ldots, f^{(n)}(x) = (-1)^n k^n e^{-kx}$

97. $f'(x) = \dfrac{1}{\sqrt{2\pi}\sigma} \exp\left[-\dfrac{1}{2}\left(\dfrac{x-\mu}{\sigma}\right)^2\right] \dfrac{d}{dx}\left[-\dfrac{1}{2}\left(\dfrac{x-\mu}{\sigma}\right)^2\right]$

$= \dfrac{1}{\sqrt{2\pi}\sigma} \exp\left[-\dfrac{1}{2}\left(\dfrac{x-\mu}{\sigma}\right)^2\right]\left[-\left(\dfrac{x-\mu}{\sigma}\right)\left(\dfrac{1}{\sigma}\right)\right]$

$= -\dfrac{1}{\sqrt{2\pi}\sigma^3}(x-\mu)\exp\left[-\dfrac{1}{2}\left(\dfrac{x-\mu}{\sigma}\right)^2\right]$

99. $-\dfrac{1}{5}e^{-5x} + C$ **101.** $e^{\sin x} + C$

103. $-\dfrac{1}{6}\displaystyle\int e^{-2x^3}(-6x^2)dx = -\dfrac{1}{6}e^{-2x^3} + C$

105. $u = 1 + e^x$, $du = e^x dx$, $\displaystyle\int \dfrac{1}{u}du = \ln(1+e^x) + C$

107. $u = 1 + e^{2t}$, $du = 2e^{2t}dt$, $\dfrac{1}{2}\displaystyle\int u^{1/2}du = \dfrac{1}{3}(1+e^{2t})^{3/2} + C$

109. $-\exp(\cos x) + C$

111. $u = 2 - e^{-x}$, $du = e^{-x}dx$, $\displaystyle\int \sec^2 u\, du = \tan(2 - e^{-x}) + C$

113. $\dfrac{\pi^{\sin x}}{\ln \pi} + C$ **115.** $\dfrac{1}{2}x^2 \ln 3 - 4\pi e^2 \sin x + C$

117. $\ln(e^x) + \ln(e^{-x}) = \ln(e^x e^{-x}) = \ln 1 = 0$ so $\displaystyle\int [\ln(e^x) + \ln(e^{-x})]dx = C$

119. $u = \sqrt{y}$, $du = \dfrac{1}{2\sqrt{y}}dy$, $2\displaystyle\int e^u du = 2e^{\sqrt{y}} + C$

121. $u = 3 - 4e^x$, $du = -4e^x dx$, $u = -1$ when $x = 0$, $u = -17$ when $x = \ln 5$

$-\dfrac{1}{4}\displaystyle\int_{-1}^{-17} u\, du = -\dfrac{1}{8}u^2\Big|_{-1}^{-17} = -36$

123. $3x - e^x]_1^2 = 3 + e - e^2$

125. $u = e^x + 4$, $du = e^x dx$, $u = e^{-\ln 3} + 4 = \dfrac{1}{3} + 4 = \dfrac{13}{3}$ when $x = -\ln 3$,

$u = e^{\ln 3} + 4 = 3 + 4 = 7$ when $x = \ln 3$, $\displaystyle\int_{13/3}^{7} \dfrac{1}{u} du = \ln u \Big]_{13/3}^{7} = \ln(7) - \ln(13/3) = \ln(21/13)$

127. $g(x) = e^{-x} f(x)$, $g'(x) = e^{-x} f'(x) - e^{-x} f(x) = e^{-x}[f'(x) - f(x)] = 0$ if $f'(x) = f(x)$ thus $g(x) = k$ because $g'(x) = 0$ so $e^{-x} f(x) = k$, $f(x) = ke^x$.

129. $2^x = 3^{x+1}$, $\ln(2^x) = \ln(3^{x+1})$, $x \ln 2 = (x+1)\ln 3$, $x \ln 2 = x \ln 3 + \ln 3$,

$x(\ln 2 - \ln 3) = \ln 3$, $x = \dfrac{\ln 3}{\ln 2 - \ln 3} = \dfrac{\ln 3}{\ln(2/3)}$

131. $f'(x) = e x^{e-1}$

133. $\dfrac{dk}{dT} = k_0 \exp\left[-\dfrac{q(T - T_0)}{2T_0 T}\right]\left(-\dfrac{q}{2T^2}\right) = -\dfrac{qk_0}{2T^2}\exp\left[-\dfrac{q(T - T_0)}{2T_0 T}\right]$

135. Divide $e^x + 3$ into e^{2x} to get $\dfrac{e^{2x}}{e^x + 3} = e^x - \dfrac{3e^x}{e^x + 3}$ so

$\displaystyle\int \dfrac{e^{2x}}{e^x + 3} dx = \int e^x dx - 3 \int \dfrac{e^x}{e^x + 3} dx = e^x - 3\ln(e^x + 3) + C$

137. Let $f(x) = x^3 e^{-2x}$, then $f'(x) = -2x^3 e^{-2x} + 3x^2 e^{-2x} = x^2 e^{-2x}(3 - 2x)$; $f'(x) = 0$ if $x = 0$ or $x = 3/2$. The maximum value occurs at $x = 3/2$ so the maximum value is $f(3/2) = (27/8)e^{-3}$.

139. $A = \displaystyle\int_0^{\ln 3} (3 - e^x)\,dx$

$= (3x - e^x)]_0^{\ln 3} = 3\ln 3 - 2$

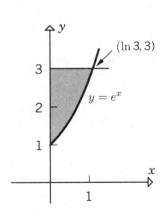

141. $V = \pi \int_0^{\ln 3} e^{2x} dx = \frac{\pi}{2} e^{2x} \Big]_0^{\ln 3} = 4\pi.$

143. The area of the region shown in the diagram is $A = \int_1^5 \ln x \, dx.$

If $y = \ln x$, then $x = e^y$ so

$A = \int_0^{\ln 5} (5 - e^y) dy = (5y - e^y) \Big]_0^{\ln 5}$

$= 5\ln 5 - 4.$

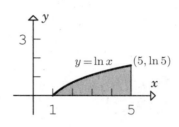

145. $f(x) = e^x + x^2 - 2, x_{n+1} = x_n - \dfrac{e^{x_n} + x_n^2 - 2}{e^{x_n} + 2x_n}$

$x_1 = -1.5, \cdots, x_5 = x_6 = -1.315973778;$

$x_1 = 0.5, \cdots, x_4 = x_5 = 0.537274449$ so

$-1.315973778 < x < 0.537274449.$

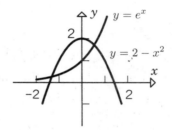

147. **(a)** $\dfrac{dS}{df} = \dfrac{(e^{bf} - 1)(3af^2) - af^3(be^{bf})}{(e^{bf} - 1)^2} = \dfrac{af^2(3e^{bf} - 3 - bfe^{bf})}{(e^{bf} - 1)^2}$, $dS/df = 0$ if $f = 0$ (where the minimum value occurs), and if $3e^{bf} - 3 - bfe^{bf} = 0$, or $3e^{-bf} + bf - 3 = 0$ (where the maximum value occurs).

(b) $f(x) = 3e^{-x} + x - 3, x_{n+1} = x_n - \dfrac{3e^{-x_n} + x_n - 3}{1 - 3e^{-x_n}}, x_1 = 3, \cdots, x_4 = x_5 = 2.821439372.$

(c) $f \approx 2.821439372/b = (2.821439372)T/(4.8043 \times 10^{-11}) \approx 5.87 \times 10^{10} T.$

149. $y = x^x, \ln y = x \ln x, y'/y = 1 + \ln x, y' = x^x(1 + \ln x); y' = 0$ when $\ln x = -1, x = e^{-1}$ at which there is a relative minimum and hence the absolute minimum for $x > 0$. The minimum value is $e^{-1/e}$.

151. $A = \int_0^b e^{-2x} dx = -\dfrac{1}{2}(e^{-2b} - 1) = \dfrac{1}{2} - \dfrac{1}{2}e^{-2b}.$ If $A = 1/4$ then $1/2 - e^{-2b}/2 = 1/4,$ $e^{-2b} = 1/2, b = (1/2)\ln 2.$ $\displaystyle\lim_{b \to +\infty} A = 1/2$

EXERCISE SET 7.3

1. (a) $+\infty$ (b) 0   **3.** (a) $+\infty$ (b) $+\infty$

5. (a) 1   (b) 1   **7.** (a) $+\infty$ (b) 0

9.

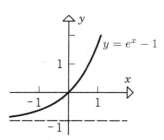

11.

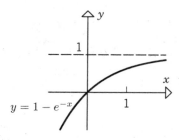

13. (a) yes, because $\lim\limits_{x \to 0} f(x) = f(0)$

     (b) no, because $\lim\limits_{x \to 0+} f'(x) = \lim\limits_{x \to 0+} e^x = 1$

         and $\lim\limits_{x \to 0-} f'(x) = \lim\limits_{x \to 0-} (-e^{-x}) = -1$ so

         $\lim\limits_{x \to 0+} f'(x) \neq \lim\limits_{x \to 0-} f'(x)$

     (c) $f(x) = \begin{cases} e^x, & x \geq 0 \\ e^{-x}, & x < 0 \end{cases}$

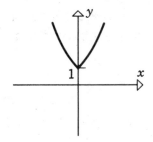

15. $e^x \cos x = e^x$ when $\cos x = 1$

       $x = 0, \pm 2\pi, \pm 4\pi, \ldots$

   $e^x \cos x = -e^x$ when $\cos x = -1$

       $x = \pm\pi, \pm 3\pi, \pm 5\pi, \ldots$

   $e^x \cos x = 0$ when $\cos x = 0$

       $x = \pm\pi/2, \pm 3\pi/2, \pm 5\pi/2, \ldots$

   $-1 \leq \cos x \leq 1$ thus

   $-e^x \leq e^x \cos x \leq e^x$ and so

   $\lim\limits_{x \to -\infty} e^x \cos x = 0$ by the Squeezing Theorem.

   $f(x) = e^x \cos x$, $f'(x) = e^x(\cos x - \sin x)$,

   $f'(x) = 0$ when $\sin x = \cos x$,

   $\tan x = 1$, $x = \pi/4 + n\pi$, $n = 0, \pm 1, \pm 2, \ldots$

   $f''(x) = -2e^x \sin x$, $f''(x) = 0$ when $x = 0, \pm\pi, \pm 2\pi, \ldots$

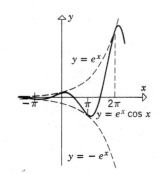

17. $\displaystyle\lim_{x\to-\infty} \frac{e^x + e^{-x}}{e^x - e^{-x}} = \lim_{x\to-\infty} \frac{e^{2x}+1}{e^{2x}-1} = -1$ **19.** 0

21. $\displaystyle\frac{d}{dx}[e^x]\Big|_{x=0} = \lim_{h\to0} \frac{e^{(0+h)}-e^0}{h} = \lim_{h\to0} \frac{e^h-1}{h} = 1$

23. $\displaystyle\lim_{x\to0} \frac{1-e^{-x}}{x} = \lim_{x\to0} -\frac{e^{-x}-e^0}{x-0} = -\frac{d}{dx}[e^{-x}]\Big|_{x=0} = e^{-x}\Big|_{x=0} = 1$

25. let $h = 1/x$ then $x = 1/h$ and $\displaystyle\lim_{x\to+\infty} x(e^{1/x}-1) = \lim_{h\to0^+} \frac{e^h-1}{h} = \frac{d}{dx}[e^x]\Big|_{x=0} = 1$

27. **(a)** $\displaystyle\lim_{x\to+\infty} xe^x = +\infty,\ \lim_{x\to-\infty} xe^x = 0.$

  (b) $y = xe^x$
  $y' = (x+1)e^x$
  $y'' = (x+2)e^x$

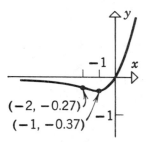

$(-2, -0.27)$
$(-1, -0.37)$

29. **(a)** $\displaystyle\lim_{x\to+\infty} \frac{x^2}{e^{2x}} = 0,\ \lim_{x\to-\infty} \frac{x^2}{e^{2x}} = +\infty.$

  (b) $y = x^2/e^{2x} = x^2 e^{-2x}$
  $y' = 2x(1-x)e^{-2x}$
  $y'' = 2(2x^2 - 4x + 1)e^{-2x}$
  $y'' = 0$ if $2x^2 - 4x + 1 = 0$
  $x = \dfrac{4 \pm \sqrt{16-8}}{2},$
  $= 1 \pm \sqrt{2}/2 \approx 0.29, 1.71$

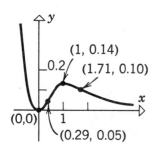

$(1, 0.14)$
$(1.71, 0.10)$
$(0,0)$
$(0.29, 0.05)$

31. **(a)**

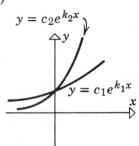

(b)

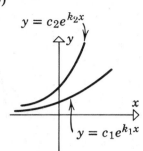

(c)

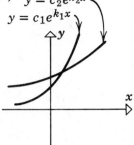

33. $m_{\text{line}} = \dfrac{1/e - 1}{1 - 0} = \dfrac{1 - e}{e}$, an equation

of the line is $y = \dfrac{1 - e}{e} x + 1$ so

$$A = \int_0^1 \left(\dfrac{1 - e}{e} x + 1 - e^{-x} \right) dx$$

$$= \dfrac{1 - e}{2e} x^2 + x + e^{-x} \Big]_0^1 = \dfrac{3 - e}{2e}$$

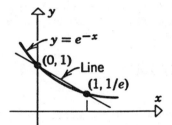

35. **(a)** If $\ln x < \sqrt{x}$, then $x < e^{\sqrt{x}}, 1/x > 1/e^{\sqrt{x}} = e^{-\sqrt{x}}$.

(b) $\displaystyle\lim_{x \to +\infty} e^{x - n\sqrt{x}} = \lim_{x \to +\infty} e^{\sqrt{x}(\sqrt{x} - n)} = +\infty$ so $\displaystyle\lim_{x \to +\infty} \dfrac{e^x}{x^n} = +\infty$ because $\dfrac{e^x}{x^n} > e^{x - n\sqrt{x}}$.

37. If $x > 1$, then $0 < \ln x < \sqrt{x}, 0 < \dfrac{\ln x}{x^n} < \dfrac{\sqrt{x}}{x^n} = \dfrac{1}{x^{(n - 1/2)}}$, but $\displaystyle\lim_{x \to +\infty} \dfrac{1}{x^{(n - 1/2)}} = 0$ so

$\displaystyle\lim_{x \to +\infty} \dfrac{\ln x}{x^n} = 0.$

39. $\displaystyle\lim_{x \to 0^+} x^n \ln x = \lim_{t \to +\infty} \dfrac{\ln(1/t)}{t^n} = \lim_{t \to +\infty} \dfrac{-\ln t}{t^n} = 0.$

41. $y = -\ln x$

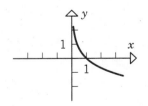

43. $y = \ln(x - 1)$

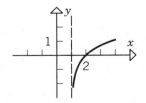

45. $y = \dfrac{\ln x}{x^2}$

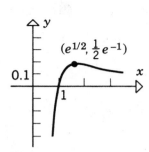

$y' = \dfrac{1 - 2\ln x}{x^3}$

$y'' = \dfrac{6\ln x - 5}{x^4}$

$y' = 0$ if $x = e^{1/2}$

$y'' = 0$ if $x = e^{5/6}$

$\lim\limits_{x \to +\infty} y = 0, \quad \lim\limits_{x \to 0^+} y = -\infty$

47. $(\ln x)/x \le 1/e$ so $e\ln x \le x, \ln(x^e) \le x, x^e \le e^x$ with equality only for $x = e$ because the maximum value of $(\ln x)/x$ occurs only at $x = e$.

EXERCISE SET 7.4

1. **(a)** $f(g(x)) = 4(x/4) = x, g(f(x)) = (4x)/4 = x, f$ and g are inverse functions

　　(b) $f(g(x)) = 3(3x - 1) + 1 = 9x - 2 \ne x$ so f and g are not inverse functions

　　(c) $f(g(x)) = \sqrt[3]{(x^3 + 2) - 2} = x, g(f(x)) = (x - 2) + 2 = x, f$ and g are inverse functions

　　(d) $f(g(x)) = (x^{1/4})^4 = x, g(f(x)) = (x^4)^{1/4} = |x| \ne x, f$ and g are not inverse functions

3. $f'(x) = 3; f$ is increasing on $(-\infty, +\infty)$ so f has an inverse.

5. $f(x) = (2 + x)(1 - x); f$ does not have an inverse because f is not one-to-one, for example $f(-2) = f(1) = 0$.

7. f does not have an inverse because f is not one-to-one, for example $f(0) = f(1) = -1$.

9. $f(x) = (x - 1)^3; f$ has an inverse because two different numbers cannot have the same cube so f is one-to-one.

11. $f'(x) = 10x^4 + 3x^2 + 3 \ge 3$ for $-\infty < x < +\infty; f$ is increasing on $(-\infty, +\infty)$ so f has an inverse.

13. $f'(x) = \cos x > 0$ for $-\pi/2 < x < \pi/2; f$ is increasing on $(-\pi/2, \pi/2)$ so f has an inverse.

15. $y = f^{-1}(x), x = f(y) = y^5, y = x^{1/5} = f^{-1}(x)$

17. $y = f^{-1}(x), x = f(y) = 7y - 6, y = \dfrac{1}{7}(x + 6) = f^{-1}(x)$

19. $y = f^{-1}(x)$, $x = f(y) = 3y^3 - 5$, $y = \sqrt[3]{(x+5)/3} = f^{-1}(x)$

21. $y = f^{-1}(x)$, $x = f(y) = \sqrt[3]{2y-1}$, $y = (x^3+1)/2 = f^{-1}(x)$

23. $y = f^{-1}(x)$, $x = f(y) = 3/y^2$, $y = -\sqrt{3/x} = f^{-1}(x)$

25. $y = f^{-1}(x)$, $x = f(y) = e^{1/y}$, $1/y = \ln x$, $y = \dfrac{1}{\ln x} = f^{-1}(x)$

27. $y = f^{-1}(x)$, $x = f(y) = 1 - \ln(3y)$, $\ln(3y) = 1 - x$, $y = \dfrac{1}{3}e^{1-x} = f^{-1}(x)$

29. $y = f^{-1}(x)$, $x = f(y) = \begin{cases} 5/2 - y, & y < 2 \\ 1/y, & y \geq 2 \end{cases}$, $y = f^{-1}(x) = \begin{cases} 5/2 - x, & x > 1/2 \\ 1/x, & 0 < x \leq 1/2 \end{cases}$

31. $y = f^{-1}(x)$, $x = f(y) = 5y^3 + y - 7$, $\dfrac{dx}{dy} = 15y^2 + 1$, $\dfrac{dy}{dx} = \dfrac{1}{15y^2 + 1}$;

  check: $1 = 15y^2 \dfrac{dy}{dx} + \dfrac{dy}{dx}$, $\dfrac{dy}{dx} = \dfrac{1}{15y^2 + 1}$.

33. $y = f^{-1}(x)$, $x = f(y) = \tan 2y$, $\dfrac{dx}{dy} = 2\sec^2 2y$, $\dfrac{dy}{dx} = \dfrac{1}{2\sec^2 2y}$;

  check: $1 = (2\sec^2 2y)\dfrac{dy}{dx}$, $\dfrac{dy}{dx} = \dfrac{1}{2\sec^2 2y}$.

35. $y = f^{-1}(x)$, $x = f(y) = 2y^5 + y^3 + 1$, $\dfrac{dx}{dy} = 10y^4 + 3y^2$, $\dfrac{dy}{dx} = \dfrac{1}{10y^4 + 3y^2}$;

  check: $1 = 10y^4 \dfrac{dy}{dx} + 3y^2 \dfrac{dy}{dx}$, $\dfrac{dy}{dx} = \dfrac{1}{10y^4 + 3y^2}$.

37. (a) $f(g(x)) = f(\sqrt{x})$
    $= (\sqrt{x})^2 = x, x > 1$;
    $g(f(x)) = g(x^2)$
    $= \sqrt{x^2} = x, x > 1$.

  (b)

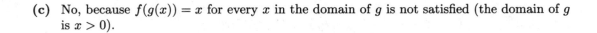

  (c) No, because $f(g(x)) = x$ for every x in the domain of g is not satisfied (the domain of g is $x > 0$).

39. $y = f^{-1}(x)$, $x = f(y) = (y+2)^4$ for $y \geq 0$, $y = f^{-1}(x) = x^{1/4} - 2$ for $x \geq 16$.

41. $y = f^{-1}(x)$, $x = f(y) = -\sqrt{3-2y}$ for $y \leq 3/2$, $y = f^{-1}(x) = (3-x^2)/2$ for $x \leq 0$.

43. $y = f^{-1}(x)$, $x = f(y) = y - 5y^2$ for $y \geq 1$, $5y^2 - y + x = 0$ for $y \geq 1$,
$y = f^{-1}(x) = (1 + \sqrt{1-20x})/10$ for $x \leq -4$.

45. **(a)** $f(f(x)) = \dfrac{3 - \dfrac{3-x}{1-x}}{1 - \dfrac{3-x}{1-x}} = \dfrac{3 - 3x - 3 + x}{1 - x - 3 + x} = x$ so $f = f^{-1}$

　　(b) symmetric about the line $y = x$

47. **(a)** $f(x) = x^3 - 3x^2 + 2x = x(x-1)(x-2)$ so $f(0) = f(1) = f(2) = 0$ thus f is not one-to-one.

　　(b) $f'(x) = 3x^2 - 6x + 2$, $f'(x) = 0$ when $x = \dfrac{6 \pm \sqrt{36-24}}{6} = 1 \pm \sqrt{3}/3$. $f'(x) > 0$ (f is increasing) if $x < 1 - \sqrt{3}/3$, $f'(x) < 0$ (f is decreasing) if $1 - \sqrt{3}/3 < x < 1 + \sqrt{3}/3$, so $f(x)$ takes on values less than $f(1 - \sqrt{3}/3)$ on both sides of $1 - \sqrt{3}/3$ thus $1 - \sqrt{3}/3$ is the largest value of k.

49. If $f^{-1}(x) = 1$, then $x = f(1) = 2(1)^3 + 5(1) + 3 = 10$.

51. $f'(x) = 3x^2 + 1$, $f'(2) = 13$ so $(f^{-1})'(10) = 1/13$.

53. $f'(x) = 2\cos 2x$, $f'(\pi/12) = \sqrt{3}$ so $(f^{-1})'(1/2) = 1/\sqrt{3}$.

55. **(a)** $f'(x) = \sqrt[3]{1+x^2} > 0$ on $(-\infty, +\infty)$ so f is one-to-one there because f is increasing.

　　(b) $f(1) = 0$, $f'(1) = \sqrt[3]{2}$ so $(f^{-1})'(0) = 1/\sqrt[3]{2}$.

57.

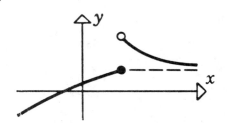

59. $F'(x) = 2f'(2g(x))g'(x)$ so $F'(3) = 2f'(2g(3))g'(3)$. By inspection $f(1) = 3$, so $g(3) = f^{-1}(3) = 1$ and $g'(3) = (f^{-1})'(3) = 1/f'(f^{-1}(3)) = 1/f'(1) = 1/7$ because $f'(x) = 4x^3 + 3x^2$. Thus $F'(3) = 2f'(2)(1/7) = 2(44)(1/7) = 88/7$.

EXERCISE SET 7.5

1. $3x + 2 > 0, x > -2/3$

3. $4 - x^2 > 0, x^2 < 4, -2 < x < 2$

5. $1 + \ln x \geq 0, \ln x \geq -1, x \geq e^{-1}$

7. **(a)** $x^2 > 0$; all $x \neq 0$

(b) $x > 0$

9. **(a)** $x^{-1}, x > 0$

(b) $x^2, x \neq 0$

    (c) $-x^2, -\infty < x < +\infty$

(d) $-x, -\infty < x < +\infty$

    (e) $x^3, x > 0$

(f) $\ln x + x, x > 0$

    (g) $x - \sqrt[3]{x}, -\infty < x < +\infty$

(h) $\dfrac{e^x}{x}, x > 0$

11. $f(\ln 2) = e^{\ln 2} + 3e^{-\ln 2} = 2 + 3e^{\ln(1/2)} = 2 + 3/2 = 7/2$

13. **(a)** $\pi^{-x} = e^{-x \ln \pi}$

(b) $x^{2x} = e^{2x \ln x}$

15. For any $\epsilon > 0, 0 < e^x < \epsilon$ if $\ln e^x < \ln \epsilon, x < \ln \epsilon$; choose $x_0 = \ln \epsilon$.

17. From 7.5.13, $\log_b x$ is $1/(\ln b)$ times $\ln x$. If $b > 1$, then $1/(\ln b) > 0$ so the graph of $\log_b x$ is a stretching or compression of the graph of $\ln x$ in the y-direction. If $0 < b < 1$, then $1/(\ln b) < 0$ so the graph of $\log_b x$ is a reflection of $\ln x$ about the x-axis followed by a stretching or compression of the graph in the y-direction. In both cases, the graph of b^x is the reflection of the graph of $\log_b x$ about the line $y = x$.

19. **(a)** The area under $1/t$ for $x \leq t \leq x + 1$ is less than the area of the rectangle with altitude $1/x$ and base 1, but greater than the area of the rectangle with altitude $1/(x + 1)$ and base 1.

    (b) $\displaystyle\int_x^{x+1} \frac{1}{t} dt = \ln t \Big]_x^{x+1} = \ln(x + 1) - \ln x = \ln(1 + 1/x)$, so
$1/(x + 1) < \ln(1 + 1/x) < 1/x$ for $x > 0$.

    (c) From part (b), $e^{1/(x+1)} < e^{\ln(1+1/x)} < e^{1/x}, e^{1/(x+1)} < 1 + 1/x < e^{1/x}$,
$e^{x/(x+1)} < (1 + 1/x)^x < e$; by the Squeezing Theorem, $\displaystyle\lim_{x \to +\infty} (1 + 1/x)^x = e$.

    (d) Use the inequality $e^{x/(x+1)} < (1 + 1/x)^x$ to get $e < (1 + 1/x)^{x+1}$ so
$(1 + 1/x)^x < e < (1 + 1/x)^{x+1}$.

EXERCISE SET 7.6

1.

	(a)	(b)	(c)	(d)	(e)	(f)
$\sinh x_0$	-2	$-3/4$	$-4/3$	$1/\sqrt{3}$	$8/15$	-1
$\cosh x_0$	$\sqrt{5}$	$5/4$	$5/3$	$2/\sqrt{3}$	$17/15$	$\sqrt{2}$
$\tanh x_0$	$-2/\sqrt{5}$	$-3/5$	$-4/5$	$1/2$	$8/17$	$-1/\sqrt{2}$
$\coth x_0$	$-\sqrt{5}/2$	$-5/3$	$-5/4$	2	$17/8$	$-\sqrt{2}$
$\text{sech}\, x_0$	$1/\sqrt{5}$	$4/5$	$3/5$	$\sqrt{3}/2$	$15/17$	$1/\sqrt{2}$
$\text{csch}\, x_0$	$-1/2$	$-4/3$	$-3/4$	$\sqrt{3}$	$15/8$	-1

(a) $\cosh^2 x_0 = 1 + \sinh^2 x_0 = 1 + (-2)^2 = 5,\ \cosh x_0 = \sqrt{5}$

(b) $\sinh^2 x_0 = \cosh^2 x_0 - 1 = \dfrac{25}{16} - 1 = \dfrac{9}{16},\ \sinh x_0 = -\dfrac{3}{4}$ (because $x_0 < 0$)

(c) $\text{sech}^2 x_0 = 1 - \tanh^2 x_0 = 1 - \left(-\dfrac{4}{5}\right)^2 = 1 - \dfrac{16}{25} = \dfrac{9}{25},\ \text{sech}\, x_0 = \dfrac{3}{5},$

$\cosh x_0 = \dfrac{1}{\text{sech}\, x_0} = \dfrac{5}{3},$ from $\dfrac{\sinh x_0}{\cosh x_0} = \tanh x_0$ we get $\sinh x_0 = \left(\dfrac{5}{3}\right)\left(-\dfrac{4}{5}\right) = -\dfrac{4}{3}$

(d) $\text{csch}^2 x_0 = \coth^2 x_0 - 1 = 4 - 1 = 3,\ \text{csch}\, x_0 = \sqrt{3},\ \sinh x_0 = \dfrac{1}{\text{csch}\, x_0} = \dfrac{1}{\sqrt{3}},$ from

$\dfrac{\cosh x_0}{\sinh x_0} = \coth x_0$ we get $\cosh x_0 = \left(\dfrac{1}{\sqrt{3}}\right)(2) = \dfrac{2}{\sqrt{3}}$

(e) $\cosh x_0 = \dfrac{1}{\text{sech}\, x_0} = \dfrac{17}{15},\ \sinh^2 x_0 = \cosh^2 x_0 - 1 = \dfrac{289}{225} - 1 = \dfrac{64}{255},\ \sinh x_0 = \dfrac{8}{15}$ (because $x_0 > 0$)

(f) $\sinh x_0 = \dfrac{1}{\text{csch}\, x_0} = -1,\ \cosh^2 x_0 = 1 + \sinh^2 x_0 = 2,\ \cosh x_0 = \sqrt{2}$

3. from (6b) and (1), $\cosh 2x = \cosh^2 x + \sinh^2 x$ and $\cosh^2 x = 1 + \sinh^2 x$ so $\cosh 2x = 2\sinh^2 x + 1$

5. $\cosh(-x) = \dfrac{1}{2}[e^{(-x)} + e^{-(-x)}] = \dfrac{1}{2}(e^{-x} + e^{x}) = \cosh x$

7. $\tanh(x+y) = \dfrac{\sinh(x+y)}{\cosh(x+y)} = \dfrac{\sinh x \cosh y + \cosh x \sinh y}{\cosh x \cosh y + \sinh x \sinh y}$

$= \dfrac{\dfrac{\sinh x \cosh y}{\cosh x \cosh y} + \dfrac{\cosh x \sinh y}{\cosh x \cosh y}}{\dfrac{\cosh x \cosh y}{\cosh x \cosh y} + \dfrac{\sinh x \sinh y}{\cosh x \cosh y}} = \dfrac{\tanh x + \tanh y}{1 + \tanh x \tanh y}$

9. $\tanh 2x = \dfrac{\sinh 2x}{\cosh 2x} = \dfrac{2 \sinh x \cosh x}{\cosh^2 x + \sinh^2 x}$ (from (6a) and (6b))

$\quad = \dfrac{2 \tanh x}{1 + \tanh^2 x}$ (after dividing numerator and denominator by $\cosh^2 x$)

11. from (7a) with x replaced by $\dfrac{x}{2}$: $\cosh x = 2 \sinh^2 \dfrac{x}{2} + 1$,

$\quad 2 \sinh^2 \dfrac{x}{2} = \cosh x - 1$, $\sinh^2 \dfrac{x}{2} = \dfrac{1}{2}(\cosh x - 1)$, $\sinh \dfrac{x}{2} = \pm\sqrt{\dfrac{1}{2}(\cosh x - 1)}$

13. add (4b) to (9b) then let $x = \dfrac{a+b}{2}$ and $y = \dfrac{a-b}{2}$.

15. **(a)** $\dfrac{d}{dx}(\sinh x) = \dfrac{d}{dx}\left[\dfrac{1}{2}(e^x - e^{-x})\right] = \dfrac{1}{2}(e^x + e^{-x}) = \cosh x$

$\quad$ **(b)** $\dfrac{d}{dx}(\coth x) = \dfrac{d}{dx}\left[\dfrac{e^x + e^{-x}}{e^x - e^{-x}}\right] = \dfrac{(e^x - e^{-x})(e^x - e^{-x}) - (e^x + e^{-x})(e^x + e^{-x})}{(e^x - e^{-x})^2}$

$\quad\quad = \dfrac{(e^{2x} - 2 + e^{-2x}) - (e^{2x} + 2 + e^{-2x})}{(e^x - e^{-x})^2} = -\dfrac{4}{(e^x - e^{-x})^2} = -\operatorname{csch}^2 x$

$\quad$ **(c)** $\dfrac{d}{dx}(\operatorname{sech} x) = \dfrac{d}{dx}\left[\dfrac{2}{e^x + e^{-x}}\right]$

$\quad\quad = \dfrac{d}{dx}[2(e^x + e^{-x})^{-1}] = -2(e^x + e^{-x})^{-2}(e^x - e^{-x})$

$\quad\quad = -\dfrac{2}{(e^x + e^{-x})}\dfrac{e^x - e^{-x}}{e^x + e^{-x}} = -\operatorname{sech} x \tanh x$

$\quad$ **(d)** proceed as in (c) using $\dfrac{2}{e^x - e^{-x}}$

17. $4 \cosh(4x - 8)$

19. $-\dfrac{1}{x} \operatorname{csch}^2(\ln x)$

21. $\dfrac{1}{x^2} \operatorname{csch}(1/x) \coth(1/x)$

23. $\dfrac{2 + 5 \cosh(5x) \sinh(5x)}{\sqrt{4x + \cosh^2(5x)}}$

25. $x^{5/2} \tanh(\sqrt{x}) \operatorname{sech}^2(\sqrt{x}) + 3x^2 \tanh^2(\sqrt{x})$

27. $\dfrac{1}{7} \sinh^7 x + C$

29. $\dfrac{2}{3}(\tanh x)^{3/2} + C$

31. $\ln(\cosh x) + C$

33. $-\dfrac{1}{3} \operatorname{sech}^3 x + C$

35. $2 \sinh(\sqrt{x}) + C$

37.   **(a)** $\dfrac{1}{2}(e^{\ln x} + e^{-\ln x}) = \dfrac{1}{2}\left(x + \dfrac{1}{x}\right) = \dfrac{x^2 + 1}{2x}$, $x > 0$

     (b) $\dfrac{1}{2}(e^{\ln x} - e^{-\ln x}) = \dfrac{1}{2}\left(x - \dfrac{1}{x}\right) = \dfrac{x^2 - 1}{2x}$, $x > 0$

     (c) $\dfrac{e^{2\ln x} - e^{-2\ln x}}{e^{2\ln x} + e^{-2\ln x}} = \dfrac{x^2 - 1/x^2}{x^2 + 1/x^2} = \dfrac{x^4 - 1}{x^4 + 1}$, $x > 0$

     (d) $\dfrac{1}{2}(e^{-\ln x} + e^{\ln x}) = \dfrac{1}{2}\left(\dfrac{1}{x} + x\right) = \dfrac{1 + x^2}{2x}$, $x > 0$

39.   positive on $(0, +\infty)$, negative on $(-\infty, 0)$, increasing on $(-\infty, +\infty)$ ($dy/dx = \operatorname{sech}^2 x > 0$) concave up on $(-\infty, 0)$, concave down on $(0, +\infty)$ ($d^2y/dx^2 = -2\operatorname{sech}^2 x \tanh x$)

41.   **(a)** $\cosh x = \sqrt{1 + \sinh^2 x} \geq \sqrt{1} = 1$

     (b) $\cosh x \geq 1$ (part (a)) so $0 < \dfrac{1}{\cosh x} \leq 1$, $0 < \operatorname{sech} x \leq 1$

43.   using $\sinh x + \cosh x = e^x$ (5a), $(\sinh x + \cosh x)^n = (e^x)^n = e^{nx} = \sinh nx + \cosh nx$

45.   **(a)** $\displaystyle \lim_{x \to +\infty} \dfrac{\cosh x}{e^x} = \lim_{x \to +\infty} \dfrac{e^x + e^{-x}}{2e^x} = \lim_{x \to +\infty} \dfrac{1}{2}(1 + e^{-2x}) = 1/2$

     (b) $\displaystyle \lim_{x \to +\infty} \dfrac{\sinh ax}{e^x} = \lim_{x \to +\infty} \dfrac{e^{ax} - e^{-ax}}{2e^x} = \lim_{x \to +\infty} \dfrac{1}{2}[e^{(a-1)x} - e^{-(a+1)x}]$

$$= \begin{cases} +\infty, & a > 1 \\ 1/2, & a = 1 \\ 0, & 0 < a < 1 \end{cases}$$

47. $A = \displaystyle \int_0^{\ln 3} \sinh 2x\, dx = \dfrac{1}{2}\cosh 2x \Big]_0^{\ln 3} = \dfrac{1}{2}[\cosh(2\ln 3) - 1]$,

     but $\cosh(2\ln 3) = \cosh(\ln 9) = \dfrac{1}{2}(e^{\ln 9} + e^{-\ln 9}) = \dfrac{1}{2}(9 + 1/9) = 41/9$ so $A = \dfrac{1}{2}[41/9 - 1] = 16/9$.

49. $V = \pi \displaystyle \int_0^5 (\cosh^2 2x - \sinh^2 2x)\, dx = \pi \int_0^5 dx = 5\pi$

51. $y' = \sinh x$, $1 + (y')^2 = 1 + \sinh^2 x = \cosh^2 x$

$$L = \int_0^{\ln 2} \cosh x\, dx = \sinh x \Big]_0^{\ln 2} = \sinh(\ln 2) = \dfrac{1}{2}(e^{\ln 2} - e^{-\ln 2}) = \dfrac{1}{2}\left(2 - \dfrac{1}{2}\right) = \dfrac{3}{4}$$

53. **(a)** $y' = \sinh(x/a), 1 + (y')^2 = 1 + \sinh^2(x/a) = \cosh^2(x/a)$

$$L = 2 \int_0^b \cosh(x/a)\, dx = 2a \sinh(x/a) \Big]_0^b = 2a \sinh(b/a)$$

(b) The highest point is at $x = b$, the lowest at $x = 0$,
so $S = a \cosh(b/a) - a \cosh(0) = a \cosh(b/a) - a$.

55. From part (b) of Exercise 53, $S = a \cosh(b/a) - a$ so $30 = a \cosh(200/a) - a$. Let $u = 200/a$,
then $a = 200/u$ so $30 = (200/u)[\cosh u - 1], \cosh u - 1 = 0.15u$. If $f(u) = \cosh u - 0.15u - 1$,
then $u_{n+1} = u_n - \dfrac{\cosh u_n - 0.15u_n - 1}{\sinh u_n - 0.15}$; $u_1 = 0.3, \cdots, u_4 = u_5 = 0.297792782 \approx 200/a$ so
$a \approx 671.6079505$. From part (a),
$L = 2a \sinh(b/a) \approx 2(671.6079505) \sinh(0.297792782) \approx 405.9\, \text{ft}$.

EXERCISE SET 7.7

1. $\dfrac{1}{y} dy = \dfrac{1}{x} dx, \ \ln|y| = \ln|x| + C_1, \ \ln\left|\dfrac{y}{x}\right| = C_1, \dfrac{y}{x} = \pm e^{C_1} = C, \ y = Cx$

3. $\dfrac{1}{1+y} dy = -\dfrac{x}{\sqrt{1+x^2}} dx, \ \ln|1+y| = -\sqrt{1+x^2} + C_1,$

$1 + y = \pm e^{-\sqrt{1+x^2}+C_1} = \pm e^{C_1} e^{-\sqrt{1+x^2}} = Ce^{-\sqrt{1+x^2}}, \ y = Ce^{-\sqrt{1+x^2}} - 1$

5. $e^y dy = \dfrac{\sin x}{\cos^2 x} dx = \sec x \tan x\, dx, \ e^y = \sec x + C, \ y = \ln(\sec x + C)$

7. $\mu = e^{\int 3dx} = e^{3x}, \ e^{3x} y = \int e^x\, dx = e^x + C, \ y = e^{-2x} + Ce^{-3x}$

9. $\mu = e^{\int dx} = e^x, \ e^x y = \int e^x \cos(e^x) dx = \sin(e^x) + C, \ y = e^{-x} \sin(e^x) + Ce^{-x}$

11. $y' + \dfrac{3}{x} y = -2x^3, \ \mu = e^{\int \frac{3}{x} dx} = e^{3 \ln x} = x^3,$

$x^3 y = \int -2x^6 dx = -\dfrac{2}{7} x^7 + C, \ y = -\dfrac{2}{7} x^4 + Cx^{-3}$

13. $\mu = e^{\int -x\, dx} = e^{-x^2/2}, \ e^{-x^2/2} y = \int xe^{-x^2/2} dx = -e^{-x^2/2} + C,$

$y = -1 + Ce^{x^2/2}, \ 3 = -1 + C, \ C = 4, \ y = -1 + 4e^{x^2/2}$

15. $\mu = e^{\int dt} = e^t$, $e^t y = \displaystyle\int 2e^t \, dt = 2e^t + C$, $y = 2 + Ce^{-t}$, $1 = 2 + C$, $C = -1$, $y = 2 - e^{-t}$

17. $y^2 t \dfrac{dy}{dt} = t - 1$, $y^2 \, dy = \left(1 - \dfrac{1}{t}\right) dt$, $\dfrac{1}{3} y^3 = t - \ln t + C$,

$9 = 1 + C$, $C = 8$, $\dfrac{1}{3} y^3 = t - \ln t + 8$, $y = \sqrt[3]{3t - 3\ln t + 24}$

19. $\dfrac{dy}{dx} = \dfrac{y^2}{3\sqrt{x}}$, $\dfrac{1}{y^2} \, dy = \dfrac{1}{3\sqrt{x}} \, dx$, $-\dfrac{1}{y} = \dfrac{2}{3}\sqrt{x} + C$; $y = -1$ when $x = 1$ so $1 = \dfrac{2}{3} + C$,

$C = \dfrac{1}{3}$, $-\dfrac{1}{y} = \dfrac{2}{3}\sqrt{x} + \dfrac{1}{3}$, $y = -\dfrac{3}{2\sqrt{x} + 1}$

21. $\dfrac{dy}{dx} = xe^y$, $e^{-y} \, dy = x \, dx$, $-e^{-y} = \dfrac{1}{2} x^2 + C$; $y = 0$ when $x = 2$ so $-1 = 2 + C$, $C = -3$,

$-e^{-y} = \dfrac{1}{2} x^2 - 3$, $y = -\ln\left(3 - \dfrac{1}{2} x^2\right)$

23. (a) $A(h) = \pi(1)^2 = \pi$, $\pi \dfrac{dh}{dt} = -0.025\sqrt{h}$, $\dfrac{\pi}{\sqrt{h}} dh = -0.025 dt$, $2\pi\sqrt{h} = -0.025t + C$; $h = 4$

when $t = 0$ so $4\pi = C$, $2\pi\sqrt{h} = -0.025t + 4\pi$, $\sqrt{h} = 2 - \dfrac{0.025}{2\pi} t$, $h \approx (2 - 0.003979t)^2$.

(b) $h = 0$ when $t \approx 2/0.003979 \approx 502.6\,\text{sec} \approx 8.4\,\text{min}$

25. $\dfrac{dv}{dt} = -0.04v^2$, $\dfrac{1}{v^2} dv = -0.04 dt$, $-\dfrac{1}{v} = -0.04t + C$; $v = 50$ when $t = 0$ so $-\dfrac{1}{50} = C$,

$-\dfrac{1}{v} = -0.04t - \dfrac{1}{50}$, $v = \dfrac{50}{2t + 1}$. But $v = \dfrac{dx}{dt}$ so $\dfrac{dx}{dt} = \dfrac{50}{2t + 1}$, $x = 25\ln(2t + 1) + C_1$; $x = 0$

when $t = 0$ so $C_1 = 0$, $x = 25\ln(2t + 1)$.

27. (a) $\dfrac{dv}{dt} = \dfrac{ck}{m_0 - kt} - g$, $v = -c\ln(m_0 - kt) - gt + C$; $v = 0$ when $t = 0$ so $0 = -c\ln m_0 + C$,

$C = c\ln m_0$, $v = c\ln m_0 - c\ln(m_0 - kt) - gt = c\ln\dfrac{m_0}{m_0 - kt} - gt$.

(b) $m_0 - kt = 0.2m_0$ when $t = 100$ so

$v = 2500\ln\dfrac{m_0}{0.2m_0} - 9.8(100) = 2500\ln 5 - 980 \approx 3044\,\text{m/sec}$.

29. (a) $v\,dv = -\dfrac{gR^2}{x^2} \, dx$, $\dfrac{1}{2} v^2 = \dfrac{gR^2}{x} + C$; $v = v_0$ when $x = R$ so $\dfrac{1}{2} v_0^2 = gR + C$,

$C = \dfrac{1}{2} v_0^2 - gR$, $\dfrac{1}{2} v^2 = \dfrac{gR^2}{x} + \dfrac{1}{2} v_0^2 - gR$, $v^2 = \dfrac{2gR^2}{x} + v_0^2 - 2gR$.

(b) From the result in part (a), $v^2 > 0$ for all $x \geq R$ if $v_0^2 - 2gR \geq 0$, $v_0 \geq \sqrt{2gR}$.

(c) $1 \text{ mi} = 5280 \text{ ft}$ so $32 \text{ ft/sec}^2 = 32/5280 \text{ mi/sec}^2$,
$$v_0 = \sqrt{2gR} = \sqrt{2(32/5280)(3960)} \approx 6.9 \text{ mi/sec}.$$

31. $\dfrac{dy}{dt} = $ rate in $-$ rate out, where y is the amount of salt at time t,

$$\frac{dy}{dt} = (4)(2) - \left(\frac{y}{50}\right)(2) = 8 - \frac{1}{25}y \text{ so } \frac{dy}{dt} + \frac{1}{25}y = 8 \text{ and } y(0) = 25.$$

$$\mu = e^{\int \frac{1}{25}dt} = e^{t/25}, \; e^{t/25}y = \int 8e^{t/25}dt = 200e^{t/25} + C,$$

$$y = 200 + Ce^{-t/25}, \; 25 = 200 + C, \; C = -175,$$

(a) $y = 200 - 175e^{-t/25}$

(b) when $t = 25$, $y = 200 - 175e^{-1} \approx 136$ lb

33. At time t there are $500 + (20 - 10)t = 500 + 10t$ gallons of brine in the tank so

$$\frac{dy}{dt} = 0 - \frac{y}{500 + 10t}(10) = -\frac{y}{50 + t}, \; \frac{dy}{dt} + \frac{1}{50 + t}y = 0 \text{ and } y(0) = 50,$$

$$\mu = e^{\int 1/(50+t)dt} = e^{\ln(50+t)} = 50 + t, \; (50+t)y = C, \; y = \frac{C}{50 + t},$$

$$50 = C/50, \; C = 2500, \; y = \frac{2500}{50 + t}$$

The tank reaches the point of overflowing when $500 + 10t = 1000$, $t = 50$ min so
$y = 2500/(50 + 50) = 25$ lb.

35. (a) From (24), $k = -\dfrac{1}{T}\ln 2 = -\dfrac{1}{140}\ln 2 \approx -0.005$ so $y = 10e^{-0.005t}$ (approximately).

(b) 10 weeks $= 70$ days so $y = 10e^{-0.35} \approx 7$ mg

37. $100e^{0.02t} = 5000$, $e^{0.02t} = 50$, $t = \dfrac{1}{0.02}\ln 50 \approx 196$ days

39. $y = y_0e^{kt}$, but $y = 0.6y_0$ when $t = 5$ so $0.6y_0 = y_0e^{5k}$, $e^{5k} = 0.6$
$k = \dfrac{1}{5}\ln 0.6$ so $T = -5\dfrac{\ln 2}{\ln 0.6} \approx 6.8$ years

41. (a) $y = 10,000e^{kt}$, but $y = 12,000$ when $t = 10$ so $10,000e^{10k} = 12,000$, $k = \dfrac{1}{10}\ln 1.2$.
When $t = 20$, $y = 10,000e^{20k} = 10,000e^{2\ln 1.2} = 10,000(1.44) = 14,400$.

(b) From (23), $T = 10\dfrac{\ln 2}{\ln 1.2} \approx 38$ years

43. $\dfrac{dT}{dt} = k(T - C),\ k < 0$

$\dfrac{dT}{dt} - kT = -kC,\ \mu = e^{\int -kdt} = e^{-kt},\ e^{-kt}T = \displaystyle\int -kCe^{-kt}dt = Ce^{-kt} + K,$

$T = C + Ke^{kt}$. But $T = T_0$ when $t = 0$ so $T_0 = C + K,\ K = T_0 - C,\ T = C + (T_0 - C)e^{kt}$

45. **(a)** In t years the interest will be compounded nt times at an interest rate of r/n each time. The value at the end of 1 interval is $P + (r/n)P = P(1 + r/n)$, at the end of 2 intervals it is $P(1 + r/n) + (r/n)P(1 + r/n) = P(1 + r/n)^2$, and continuing in this fashion the value at the end of nt intervals is $P(1 + r/n)^{nt}$.

(b) Let $x = r/n$, then $n = r/x$ and

$$\lim_{n \to +\infty} P(1 + r/n)^{nt} = \lim_{x \to 0+} P(1 + x)^{rt/x} = \lim_{x \to 0+} P[(1 + x)^{1/x}]^{rt} = Pe^{rt}.$$

(c) The rate of increase is $dA/dt = rPe^{rt} = rA$.

47. Let $y = y_0 e^{kt}$ with $y = y_1$ when $t = t_1$ and $y = y_1/2$ when $t = t_1 + T$ then $y_0 e^{kt_1} = y_1$ (i) and $y_0 e^{k(t_1+T)} = y_1/2$ (ii). Divide (i) by (ii) to get $e^{-kT} = 2,\ T = -\dfrac{1}{k}\ln 2$.

TECHNOLOGY EXERCISES 7

1. The area is given by $\displaystyle\int_1^k \ln x\, dx = k\ln k - k + 1 = 3$; solve for k to get $k = 4.319137$.

3. The volume is given by $2\pi \displaystyle\int_0^2 xe^{kx}dx = 2\pi\left[\dfrac{(2k-1)e^{2k}}{k^2} + \dfrac{1}{k^2}\right] = 1.5$; solve for k to get $k = -1.940953$.

5. $v \le 0$ for $0 \le t \le \ln 4$, $v \ge 0$ for $t \ge \ln 4$. The distance traveled for $0 \le t \le \ln 4$ is

$$\int_0^{\ln 4} (4e^{-t} - 1)dt = 3 - \ln 4 \approx 1.614\ \text{cm thus}$$

(a) if $t = a$ is the time to travel 1 cm, then $\displaystyle\int_0^a (4e^{-t} - 1)dt = -4e^{-a} - a + 4 = 1$ so $a = 0.45$ sec;

(b) if $t = b$ is the time to travel 4 cm, then

$$\int_0^{\ln 4} (4e^{-t} - 1)dt + \int_{\ln 4}^b (1 - 4e^{-t})dt = b + 4e^{-b} + 2 - 2\ln 4 = 4 \text{ so } b = 4.74 \text{ sec.}$$

7. Equate derivatives to get $1/(2\sqrt{x}) = 1/x$ which yields $x = 4$. At the point of tangency $\sqrt{x} + k = \ln x$ so $\sqrt{4} + k = \ln 4$, $k = \ln 4 - 2$.

9. Equate derivatives to get $kx^{k-1} = 1/x$, $x^k = 1/k$. At the point of tangency $x^k = \ln x$ so $1/k = \ln x$, $x = e^{1/k}$; since $x^k = 1/k$ it follows that $(e^{1/k})^k = 1/k$, $1/k = e$, $k = 1/e$.

11. $f'(x) = x^{\sin x} \left[\dfrac{\sin x}{x} + (\cos x) \ln x \right]$; if $0 < x < 6$, then $f'(x) = 0$ for $x = 0.352215$, 2.127616, 4.842558. A relative maximum of 1.898286 occurs at $x = 2.127616$, a relative minimum of 0.697684 at $x = 0.352215$, and a relative minimum of 0.209277 at $x = 4.842558$.

13. (a) $100,000/(\ln 100,000) \approx 8,686$ (b) $\displaystyle\int_{2}^{100,000} \dfrac{1}{\ln t}\,dt \approx 9,629$

15. (b) $r = 1$ when $t = 0.673080$ sec.   (c) $dr/dt = 4.48$ m/sec.

17. (a) $f'(x) = 0$ at $x = -0.914856$ where the minimum value of $f(x)$ occurs, so $k = -0.914856$.
    (b) $f(-0.914856) = -1.540463$; the domain of f^{-1} is $[-1.54046, +\infty)$ and the range is $[-0.91485, +\infty)$. Let $x = f^{-1}(5)$, then $f(x) = 5$ so $x^2 + 3\sin x = 5$; solve for the value of x in the range of f^{-1} to get $x = f^{-1}(5) = 1.425366$.

19. (a) $x_0 = 0$, $y_0 = 1$, $y_{n+1} = y_n + y_n(\cos x_n)(0.05)$; $y_6 \approx y(0.3) = 1.335729$.
    (b) Use separation of variables to obtain $(1/y)dy = (\cos x)dx$, $\ln y = \sin x + C$; $C = 0$ since $y(0) = 1$ so $\ln y = \sin x$, $y = e^{\sin x}$. If $x = 0.3$, then $y = e^{\sin 0.3} = 1.343825$. The percentage error is 0.6%.

CHAPTER 8

Inverse Trigonometric and Hyperbolic Functions

EXERCISE SET 8.1

1. (a) $-\pi/2$ (b) π (c) $-\pi/4$ (d) $\pi/4$ (e) 0 (f) $\pi/2$

3. $\theta = -\pi/3$; $\cos\theta = 1/2$, $\tan\theta = -\sqrt{3}$, $\cot\theta = -1/\sqrt{3}$, $\sec\theta = 2$, $\csc\theta = -2/\sqrt{3}$

5. $\tan\theta = 4/3$, $0 < \theta < \pi/2$; use the triangle shown to get $\sin\theta = 4/5$, $\cos\theta = 3/5$, $\cot\theta = 3/4$, $\sec\theta = 5/3$, $\csc\theta = 5/4$.

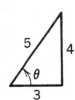

7. (a) $\pi/7$ (b) $\sin^{-1}(\sin\pi) = \sin^{-1}(\sin 0) = 0$
   (c) $\sin^{-1}(\sin(5\pi/7)) = \sin^{-1}(\sin(2\pi/7)) = 2\pi/7$
   (d) Note that $\pi/2 < 630 - 200\pi < \pi$ so
   $$\sin(630) = \sin(630 - 200\pi) = \sin(\pi - (630 - 200\pi)) = \sin(201\pi - 630)$$
   where $0 < 201\pi - 630 < \pi/2$; $\sin^{-1}(\sin 630) = \sin^{-1}(\sin(201\pi - 630)) = 201\pi - 630$.

9. (a) $0 \le x \le \pi$ (b) $-1 \le x \le 1$
   (c) $-\pi/2 < x < \pi/2$ (d) $-\infty < x < +\infty$
   (e) $0 < x \le \pi/2$ or $-\pi < x \le -\pi/2$ (f) $|x| \ge 1$

11. Let $\theta = \cos^{-1}(3/5)$,
    $$\sin 2\theta = 2\sin\theta\cos\theta$$
    $$= 2(4/5)(3/5) = 24/25$$

13. $\sin^{-1}(1) = \pi/2$

15. Let $\theta = \sec^{-1}(3/2)$,

$$\tan 2\theta = \frac{2\tan\theta}{1 - \tan^2\theta}$$

$$= \frac{2(\sqrt{5}/2)}{1 - 5/4} = -4\sqrt{5}$$

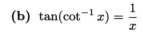

17. (a) $\tan^{-1}\dfrac{1}{2} + \tan^{-1}\dfrac{1}{3} = \tan^{-1}\dfrac{1/2 + 1/3}{1 - (1/2)(1/3)} = \tan^{-1}1 = \pi/4$

(b) $2\tan^{-1}\dfrac{1}{3} = \tan^{-1}\dfrac{1}{3} + \tan^{-1}\dfrac{1}{3} = \tan^{-1}\dfrac{1/3 + 1/3}{1 - (1/3)(1/3)} = \tan^{-1}\dfrac{3}{4}$,

$2\tan^{-1}\dfrac{1}{3} + \tan^{-1}\dfrac{1}{7} = \tan^{-1}\dfrac{3}{4} + \tan^{-1}\dfrac{1}{7} = \tan^{-1}\dfrac{3/4 + 1/7}{1 - (3/4)(1/7)} = \tan^{-1}1 = \pi/4$

19. (a) $\cos(\tan^{-1}x) = \dfrac{1}{\sqrt{1 + x^2}}$ **(b)** $\tan(\cot^{-1}x) = \dfrac{1}{x}$

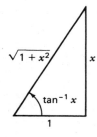   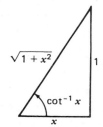

(c) $\sin(\sec^{-1}x) = \dfrac{\sqrt{x^2 - 1}}{x}$ **(d)** $\cot(\csc^{-1}x) = \sqrt{x^2 - 1}$

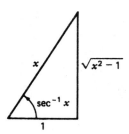   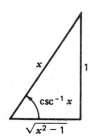

21. (a)

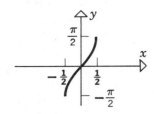

(b)

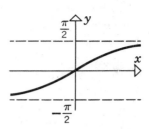

23. (a) let $\theta = \sin^{-1}(-x)$ then $\sin\theta = -x$, $-\pi/2 \le \theta \le \pi/2$. But $\sin(-\theta) = -\sin\theta$ and $-\pi/2 \le -\theta \le \pi/2$ so $\sin(-\theta) = -(-x) = x$, $-\theta = \sin^{-1} x$, $\theta = -\sin^{-1} x$.

(b) proof is similar to that in part (a).

25. (a) If $-1 \le x < 0$ then $0 < -x \le 1$ so $\sin^{-1}(-x) + \cos^{-1}(-x) = \pi/2$, but $\sin^{-1}(-x) = -\sin^{-1} x$ and $\cos^{-1}(-x) = \pi - \cos^{-1} x$ thus $-\sin^{-1} x + (\pi - \cos^{-1} x) = \pi/2$, $\sin^{-1} x + \cos^{-1} x = \pi/2$.

(b) If $x < 0$ then $-x > 0$ so $\sec(\tan^{-1}(-x)) = \sqrt{1 + (-x)^2} = \sqrt{1 + x^2}$, but $\tan^{-1}(-x) = -\tan^{-1} x$ and $\sec(-\tan^{-1} x) = \sec(\tan^{-1} x)$ thus $\sec(\tan^{-1} x) = \sqrt{1 + x^2}$.

27. (a) $55.0°$ **(b)** $33.6°$ **(c)** $25.8°$

29. $x = \pi + \tan^{-1} k$

31. $x = \pi - \sin^{-1}(0.37) \approx 2.7626$ **33.** $x = \tan^{-1}(3.16) - \pi \approx -1.8773$

35. $\theta = -\cos^{-1}(0.23) \approx -76.7°$ **37. (b)** $\theta = \sin^{-1}\frac{R}{R+h} = \sin^{-1}\frac{6378}{16,378} \approx 23°$

39. $\sin 2\theta = gR/v^2 = (9.8)(18)/(14)^2 = 0.9$, $2\theta = \sin^{-1}(0.9)$ or $2\theta = 180° - \sin^{-1}(0.9)$ so $\theta = \frac{1}{2}\sin^{-1}(0.9) \approx 32°$ or $\theta = 90° - \frac{1}{2}\sin^{-1}(0.9) \approx 58°$. The ball will have a lower parabolic trajectory for $\theta = 32°$ and hence will result in the shorter time of flight.

41. $y = 0$ when $x^2 = 6000v^2/g$, $x = 10v\sqrt{60/g} = 1000\sqrt{30}$ for $v = 400$ and $g = 32$; $\tan\theta = 3000/x = 3/\sqrt{30}$, $\theta = \tan^{-1}(3/\sqrt{30}) \approx 29°$.

43. (a) $\sin^{-1} x = \tan^{-1} \dfrac{x}{\sqrt{1-x^2}}$

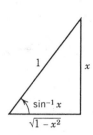

(b) $\sin^{-1} x + \cos^{-1} x = \pi/2$
$\cos^{-1} x = \pi/2 - \sin^{-1} x$
$= \pi/2 - \tan^{-1} \dfrac{x}{\sqrt{1-x^2}}$

EXERCISE SET 8.2

1. (a) $\dfrac{1}{\sqrt{1-x^2/9}}(1/3) = 1/\sqrt{9-x^2}$

(b) $-2/\sqrt{1-(2x+1)^2}$

3. (a) $\dfrac{1}{x^7\sqrt{x^{14}-1}}(7x^6) = \dfrac{7}{x\sqrt{x^{14}-1}}$

(b) $-1/\sqrt{e^{2x}-1}$

5. (a) $\dfrac{1}{\sqrt{1-1/x^2}}(-1/x^2) = -\dfrac{1}{|x|\sqrt{x^2-1}}$

(b) $\dfrac{\sin x}{\sqrt{1-\cos^2 x}} = \dfrac{\sin x}{|\sin x|} = \begin{cases} 1, & \sin x > 0 \\ -1, & \sin x < 0 \end{cases}$

7. (a) $\dfrac{e^x}{x\sqrt{x^2-1}} + e^x \sec^{-1} x$

(b) $\dfrac{3x^2(\sin^{-1}x)^2}{\sqrt{1-x^2}} + 2x(\sin^{-1}x)^3$

9. (a) $\dfrac{1}{1+(1-x)^2/(1+x)^2}\left[\dfrac{(1+x)(-1)-(1-x)(1)}{(1+x)^2}\right] = -\dfrac{2}{(1+x)^2+(1-x)^2} = -\dfrac{1}{x^2+1}$

(b) $10(1+x\csc^{-1}x)^9(-1/\sqrt{x^2-1}+\csc^{-1}x)$

11. (a) $\dfrac{1}{1+(1-x)/(1+x)}\dfrac{1}{2}\left(\dfrac{1-x}{1+x}\right)^{-1/2}\dfrac{(1+x)(-1)-(1-x)(1)}{(1+x)^2} = -\dfrac{1}{2\sqrt{1-x^2}}$

(b) $\dfrac{x+2x\ln x}{\sqrt{1-x^4\ln^2 x}}$

13. $\sin^{-1}(xy) = \cos^{-1}(x-y)$, $\dfrac{1}{\sqrt{1-x^2y^2}}(xy'+y) = -\dfrac{1}{\sqrt{1-(x-y)^2}}(1-y')$,

$y' = \dfrac{y\sqrt{1-(x-y)^2} + \sqrt{1-x^2y^2}}{\sqrt{1-x^2y^2} - x\sqrt{1-(x-y)^2}}$

15. $\tan^{-1}x\big]_{-1}^{1} = \tan^{-1}1 - \tan^{-1}(-1) = \pi/4 - (-\pi/4) = \pi/2$

17. $\sec^{-1}x\big]_{-\sqrt{2}}^{-2/\sqrt{3}} = \sec^{-1}(-2/\sqrt{3}) - \sec^{-1}(-\sqrt{2}) = 7\pi/6 - 5\pi/4 = -\pi/12$

19. $u = 4x$, $\dfrac{1}{4}\displaystyle\int \dfrac{1}{1+u^2}\,du = \dfrac{1}{4}\tan^{-1}4x + C$

21. $u = e^x$, $\displaystyle\int \dfrac{1}{1+u^2}\,du = \tan^{-1}(e^x) + C$

23. $u = \sqrt{x}$, $2\displaystyle\int_{1}^{\sqrt{3}} \dfrac{1}{u^2+1}\,du = 2\tan^{-1}u\bigg]_{1}^{\sqrt{3}} = 2(\tan^{-1}\sqrt{3} - \tan^{-1}1) = 2(\pi/3 - \pi/4) = \pi/6$

25. $u = \tan x$, $\displaystyle\int \dfrac{1}{\sqrt{1-u^2}}\,du = \sin^{-1}(\tan x) + C$

27. $u = \ln x$, $\displaystyle\int \dfrac{1}{\sqrt{1-u^2}}\,du = \sin^{-1}(\ln x) + C$

29. **(a)** $\sin^{-1}(x/3) + C$ **(b)** $(1/\sqrt{5})\tan^{-1}(x/\sqrt{5}) + C$
     (c) $(1/\sqrt{\pi})\sec^{-1}(x/\sqrt{\pi}) + C$

31. $u = \sqrt{3}x^2$, $\dfrac{1}{2\sqrt{3}}\displaystyle\int_{0}^{\sqrt{3}} \dfrac{1}{\sqrt{4-u^2}}\,du = \dfrac{1}{2\sqrt{3}}\sin^{-1}\dfrac{u}{2}\bigg]_{0}^{\sqrt{3}} = \dfrac{1}{2\sqrt{3}}\left(\dfrac{\pi}{3}\right) = \dfrac{\pi}{6\sqrt{3}}$

33. $u = 3x$, $\dfrac{1}{3}\displaystyle\int_{0}^{2\sqrt{3}} \dfrac{1}{4+u^2}\,du = \dfrac{1}{6}\tan^{-1}\dfrac{u}{2}\bigg]_{0}^{2\sqrt{3}} = \dfrac{1}{6}(\pi/3) = \pi/18$

35. $A = \displaystyle\int_{0}^{1/6} \dfrac{1}{\sqrt{1-9x^2}}\,dx = \dfrac{1}{3}\displaystyle\int_{0}^{1/2} \dfrac{1}{\sqrt{1-u^2}}\,du = \dfrac{1}{3}\sin^{-1}u\bigg]_{0}^{1/2} = \pi/18$

37. $V = \int_{-2}^{2} \pi \dfrac{1}{4+x^2} dx = \dfrac{\pi}{2} \tan^{-1}(x/2)\Big]_{-2}^{2} = \pi^2/4$

39. $A = \int_{0}^{\pi/2} (1 - \sin y) dy$

$\qquad = (y + \cos y)]_{0}^{\pi/2} = \pi/2 - 1$

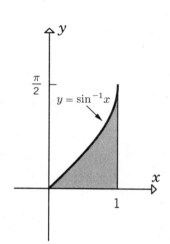

41. $\theta = \pi - (\alpha + \beta)$

$\qquad = \pi - \cot^{-1}(x-2) - \cot^{-1}\dfrac{5-x}{4},$

$\dfrac{d\theta}{dx} = \dfrac{1}{1+(x-2)^2} + \dfrac{-1/4}{1+(5-x)^2/16}$

$\qquad = -\dfrac{3(x^2 - 2x - 7)}{[1+(x-2)^2][16+(5-x)^2]}$

$d\theta/dx = 0$ when $x = \dfrac{2 \pm \sqrt{4+28}}{2} = 1 \pm 2\sqrt{2}$,

only $1 + 2\sqrt{2}$ is in $[2,5]$; $d\theta/dx > 0$ for x in $[2, 1+2\sqrt{2})$,

$d\theta/dx < 0$ for x in $(1+2\sqrt{2}, 5]$, θ is maximum when $x = 1 + 2\sqrt{2}$.

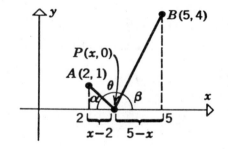

43. $\theta = \tan^{-1}(x/3)$

$\dfrac{d\theta}{dt} = \dfrac{3}{9+x^2}\dfrac{dx}{dt}, \dfrac{dx}{dt} = \dfrac{9+x^2}{3}\dfrac{d\theta}{dt}$

$\dfrac{dx}{dt}\Big|_{x=2} = \dfrac{9+4}{3}(4\pi) = 52\pi/3 \text{ mi/min}$

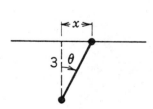

45. $\theta = \alpha - \beta$

$\quad = \cot^{-1}(x/12) - \cot^{-1}(x/2)$

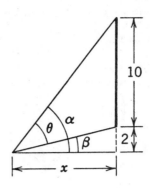

$\dfrac{d\theta}{dx} = -\dfrac{12}{144 + x^2} + \dfrac{2}{4 + x^2}$

$\phantom{\dfrac{d\theta}{dx}} = \dfrac{10(24 - x^2)}{(144 + x^2)(4 + x^2)}$

$d\theta/dx = 0$ when $x = \sqrt{24} = 2\sqrt{6}$, by the first derivative test θ is maximum there.

47. By the Mean-Value Theorem on the interval $[0, x]$,

$\dfrac{\tan^{-1} x - \tan^{-1} 0}{x - 0} = \dfrac{\tan^{-1} x}{x} = \dfrac{1}{1 + c^2}$ for c in $(0, x)$, but

$\dfrac{1}{1 + x^2} < \dfrac{1}{1 + c^2} < 1$ for c in $(0, x)$ so $\dfrac{1}{1 + x^2} < \dfrac{\tan^{-1} x}{x} < 1$, $\dfrac{x}{1 + x^2} < \tan^{-1} x < x$.

49. **(a)** $A = \displaystyle\int_0^{0.8} \dfrac{1}{\sqrt{1 - x^2}} dx = \sin^{-1} x \Big|_0^{0.8} = \sin^{-1}(0.8)$

 **(b)** The calculator was in degree mode instead of radian mode; the correct answer is 0.93.

EXERCISE SET 8.3

1. **(a)** let $y = \cosh^{-1} x$, then $x = \cosh y = \dfrac{1}{2}(e^y + e^{-y})$, $e^y - 2x + e^{-y} = 0$, $e^{2y} - 2xe^y + 1 = 0$,

$e^y = \dfrac{2x \pm \sqrt{4x^2 - 4}}{2} = x \pm \sqrt{x^2 - 1}$. To determine which sign to take, note that $y \geq 0$ so $e^{-y} \leq e^y$, $x = (e^y + e^{-y})/2 \leq (e^y + e^y)/2 = e^y$, hence $e^y \geq x$ thus $e^y = x + \sqrt{x^2 - 1}$, $y = \cosh^{-1} x = \ln(x + \sqrt{x^2 - 1})$.

 **(b)** $\dfrac{d}{dx}(\cosh^{-1} x) = \dfrac{1 + x/\sqrt{x^2 - 1}}{x + \sqrt{x^2 - 1}} = 1/\sqrt{x^2 - 1}$

3. **(a)** let $y = \operatorname{sech}^{-1} x$ then $x = \operatorname{sech} y = 1/\cosh y$, $\cosh y = 1/x$, $y = \cosh^{-1}(1/x)$; the proofs for the remaining two are similar.

 **(b)** $\dfrac{d}{dx}(\operatorname{sech}^{-1} x) = \dfrac{d}{dx}(\cosh^{-1}(1/x)) = \dfrac{(-1/x^2)}{\sqrt{1/x^2 - 1}} = -\dfrac{1}{x\sqrt{1 - x^2}}$; the remaining two are done similarly.

3. (c) $\operatorname{sech}^{-1}x = \cosh^{-1}(1/x) = \ln\left[\dfrac{1}{x} + \sqrt{\dfrac{1}{x^2} - 1}\right] = \ln\left[\dfrac{1 + \sqrt{1 - x^2}}{x}\right]$; the remaining two are done similarly.

5. (a) $\ln(3 + \sqrt{8})$ (b) $\ln(\sqrt{5} - 2)$

7. (a) $\dfrac{1}{\sqrt{1 + x^2/9}}\left(\dfrac{1}{3}\right) = 1/\sqrt{9 + x^2}$ (b) $2/\sqrt{(2x + 1)^2 - 1}$

9. (a) $-\dfrac{7x^6}{x^7\sqrt{1 - x^{14}}} = -\dfrac{7}{x\sqrt{1 - x^{14}}}$ (b) $-1/\sqrt{1 + e^{2x}}$

11. (a) $\dfrac{1}{\sqrt{1 + 1/x^2}}(-1/x^2) = -\dfrac{1}{|x|\sqrt{x^2 + 1}}$

      (b) $\dfrac{\sinh x}{\sqrt{\cosh^2 x - 1}} = \dfrac{\sinh x}{|\sinh x|} = \begin{cases} 1, & x > 0 \\ -1, & x < 0 \end{cases}$

13. (a) $-\dfrac{e^x}{x\sqrt{1 - x^2}} + e^x \operatorname{sech}^{-1}x$

      (b) $3x^2(\sinh^{-1}x)^2/\sqrt{1 + x^2} + 2x(\sinh^{-1}x)^3$

15. (a) $\dfrac{1}{1 - (1 - x)^2/(1 + x)^2}\left[-\dfrac{2}{(1 + x)^2}\right] = -1/(2x)$

      (b) $10(1 + x\operatorname{csch}^{-1}x)^9\left(-\dfrac{x}{|x|\sqrt{1 + x^2}} + \operatorname{csch}^{-1}x\right)$

17. $x = \sqrt{2}u$, $\displaystyle\int \dfrac{\sqrt{2}}{\sqrt{2u^2 - 2}}\,du = \int \dfrac{1}{\sqrt{u^2 - 1}}\,du = \cosh^{-1}(x/\sqrt{2}) + C$

19. $u = e^x$, $\displaystyle\int \dfrac{1}{u\sqrt{1 - u^2}}\,du = -\operatorname{sech}^{-1}(e^x) + C$

21. $u = x^3$, $\dfrac{1}{3}\displaystyle\int \dfrac{1}{u\sqrt{1 + u^2}}\,du = -\dfrac{1}{3}\operatorname{csch}^{-1}|x^3| + C$

23. $\coth^{-1}x\big]_2^3 = \coth^{-1}3 - \coth^{-1}2$

$\quad\quad\quad\quad = \dfrac{1}{2}\ln\dfrac{3 + 1}{3 - 1} - \dfrac{1}{2}\ln\dfrac{2 + 1}{2 - 1} = \dfrac{1}{2}\ln 2 - \dfrac{1}{2}\ln 3 \approx \dfrac{1}{2}(0.6931 - 1.0986) \approx -0.2028$

25. $\dfrac{d}{dx}(\text{sech}^{-1}|x|) = \dfrac{d}{dx}(\text{sech}^{-1}\sqrt{x^2}) = -\dfrac{1}{\sqrt{x^2}\sqrt{1-x^2}}\dfrac{x}{\sqrt{x^2}} = -\dfrac{1}{x\sqrt{1-x^2}}$

27. If $-1 < x < 1$ then $\dfrac{1+x}{1-x} > 0$, $\tanh^{-1}x = \dfrac{1}{2}\ln\dfrac{1+x}{1-x} = \dfrac{1}{2}\ln\left|\dfrac{1+x}{1-x}\right|$; if $|x| > 1$ then

$\dfrac{x+1}{x-1} > 0$, but $\dfrac{x+1}{x-1} = \left|\dfrac{1+x}{1-x}\right|$ so $\coth^{-1}x = \dfrac{1}{2}\ln\dfrac{x+1}{x-1} = \dfrac{1}{2}\ln\left|\dfrac{1+x}{1-x}\right|$,

$\displaystyle\int\dfrac{1}{1-x^2}dx = \dfrac{1}{2}\ln\left|\dfrac{1+x}{1-x}\right| + C$

29. Let $y = \sinh^{-1}x$ then $x = \sinh y$, $\dfrac{dy}{dx} = \dfrac{1}{dx/dy} = \dfrac{1}{\cosh y} = \dfrac{1}{\sqrt{1+\sinh^2 y}} = \dfrac{1}{\sqrt{1+x^2}}$.

31. Let $u = -x$, $\displaystyle\int\dfrac{1}{\sqrt{x^2-1}}dx = -\int\dfrac{1}{\sqrt{u^2-1}}du = -\cosh^{-1}u + C = -\cosh^{-1}(-x) + C$.

33. (a) $v\,dv = 16x\,dx$, $v^2/2 = 8x^2 + C$; $v = 0$ when $x = 1/4$ so $C = -1/2$, $v^2/2 = 8x^2 - 1/2$, $v^2 = 16x^2 - 1$, $v = \sqrt{16x^2 - 1}$.

(b) $\dfrac{dx}{dt} = \sqrt{16x^2 - 1}$, $\dfrac{1}{\sqrt{16x^2-1}}dx = dt$, $\dfrac{1}{4}\cosh^{-1}(4x) = t + C$; $x = 1/4$ when $t = 0$ so $C = \dfrac{1}{4}\cosh^{-1}(1) = 0$, $\dfrac{1}{4}\cosh^{-1}(4x) = t$, $x = \dfrac{1}{4}\cosh(4t)$. If $x = 2$, then $t = \dfrac{1}{4}\cosh^{-1}(8) \approx 0.7$ sec.

TECHNOLOGY EXERCISES 8

1. (a) Let $f(x) = 2\tan^{-1}x - \sin^{-1}x$, $f'(x) = 0$ for $x = 0.681250$; the maximum vertical separation is $f(0.681250) = 0.446593$.

   (b) Let $g(y) = \sin y - \tan(y/2)$, $g'(y) = 0$ for $y = 0.904557$; the maximum horizontal separation is $g(0.904557) = 0.300283$.

3. $f'(x) = ke^{-x}/(1+k^2x^2) - e^{-x}\tan^{-1}(kx)$ so $f'(0.8) = ke^{-0.8}/(1+0.64k^2) - e^{-0.8}\tan^{-1}(0.8k)$; solve $f'(0.8) = 0$ for k to get $k = 0.793387$.

5. The curves intersect at $x = a = 0$ and $x = b = 0.838422$ so the area is
$\displaystyle\int_a^b (\sin 2x - \sin^{-1}x)dx = 0.174192$.

7. The area of the regions are $\int_0^k 1/(1+3x^2)dx = (1/\sqrt{3})\tan^{-1}(\sqrt{3}k)$ and

$\int_0^k [1-1/(1+3x^2)]dx = k - (1\sqrt{3})\tan^{-1}(\sqrt{3}k)$. They are equal if

$(1/\sqrt{3})\tan^{-1}(\sqrt{3}k) = k - (1/\sqrt{3})\tan^{-1}(\sqrt{3}k)$; solve for k to get $k = 1.345874$.

9. (a) $\pi \int_0^1 (\sin^{-1} x)^2 dx = 1.468384$.

  (b) $2\pi \int_0^{\pi/2} y(1 - \sin y)dy = 1.468384$.

11. (a) $V_0 = \pi(18)^2(72)/231 = 317.26$ gallons.
   (c) At $h = 18$, $dV/dh = 11.2$ gallons/inch.
   (d) Solve $V = 70$ for h to get $h = 9.8$ inches.

CHAPTER 9
Techniques of Integration

EXERCISE SET 9.1

1. $3\displaystyle\int \frac{x}{4x-1}dx = \frac{3}{16}(4x + \ln|4x-1|) + C$ (#6; $a = -1$, $b = 4$)

3. $\dfrac{1}{5}\ln\left|\dfrac{x}{2x+5}\right| + C$ (#11; $a = 5$, $b = 2$)

5. $\dfrac{1}{30}(6x+6)(2x-3)^{3/2} + C = \dfrac{1}{5}(x+1)(2x-3)^{3/2} + C$ (#14; $a = -3$, $b = 2$)

7. $\dfrac{1}{2}\ln\left|\dfrac{\sqrt{4-3x}-2}{\sqrt{4-3x}+2}\right| + C$ (#20; $a = 4$, $b = -3$)

9. $\dfrac{1}{2\sqrt{5}}\ln\left|\dfrac{x+\sqrt{5}}{x-\sqrt{5}}\right| + C$ (#25; $a = \sqrt{5}$)

11. $\dfrac{x}{2}\sqrt{x^2-3} - \dfrac{3}{2}\ln\left|x+\sqrt{x^2-3}\right| + C$ (#28; $a^2 = 3$)

13. $\dfrac{x}{2}\sqrt{x^2+4} - 2\ln\left|x+\sqrt{x^2+4}\right| + C$ (#34; $a^2 = 4$)

15. $\dfrac{x}{2}\sqrt{9-x^2} + \dfrac{9}{2}\sin^{-1}\dfrac{x}{3} + C$ (#41; $a = 3$)

17. $\sqrt{3-x^2} - \sqrt{3}\ln\left|\dfrac{\sqrt{3}+\sqrt{3-x^2}}{x}\right| + C$ (#43; $a = \sqrt{3}$)

19. $-\dfrac{\sin 5x}{10} + \dfrac{\sin x}{2} + C$ (#80; $m = 3$, $n = 2$)

21. $\dfrac{x^4}{16}[4\ln x - 1] + C$ (#104; $n = 3$)

23. $\dfrac{e^{-2x}}{13}(-2\sin 3x - 3\cos 3x) + C$ (#106; $a = -2$, $b = 3$)

25. $\dfrac{1}{2}\displaystyle\int \dfrac{u}{(4-3u)^2}\,du = \dfrac{1}{18}\left[\dfrac{4}{4-3u} + \ln|4-3u|\right] + C \quad (\#8; \; a=4, b=-3)$

$\qquad\qquad = \dfrac{1}{18}\left[\dfrac{4}{4-3e^{2x}} + \ln|4-3e^{2x}|\right] + C$

27. $\dfrac{2}{3}\displaystyle\int \dfrac{1}{u^2+4}\,du = \dfrac{1}{3}\tan^{-1}\dfrac{u}{2} + C \quad (\#24; \; a=2) = \dfrac{1}{3}\tan^{-1}\dfrac{3\sqrt{x}}{2} + C$

29. $\dfrac{1}{3}\displaystyle\int \dfrac{1}{\sqrt{u^2-4}}\,du = \dfrac{1}{3}\ln\left|u+\sqrt{u^2-4}\right| + C \quad (\#27; \; a^2=4) = \dfrac{1}{3}\ln\left|3x+\sqrt{9x^2-4}\right| + C$

31. $\dfrac{1}{54}\displaystyle\int \dfrac{u^2}{\sqrt{5-u^2}}\,du = \dfrac{1}{54}\left[-\dfrac{u}{2}\sqrt{5-u^2} + \dfrac{5}{2}\sin^{-1}\dfrac{u}{\sqrt{5}}\right] + C \quad (\#45; \; a=\sqrt{5})$

$\qquad\qquad = \dfrac{1}{108}\left[-3x^2\sqrt{5-9x^4} + 5\sin^{-1}\dfrac{3x^2}{\sqrt{5}}\right] + C$

33. $\displaystyle\int \sin^2 u\,du = \dfrac{1}{2}u - \dfrac{1}{4}\sin 2u + C \quad (\#70) = \dfrac{1}{2}\ln x - \dfrac{1}{4}\sin(2\ln x) + C$

35. $\dfrac{1}{4}\displaystyle\int ue^u\,du = \dfrac{1}{4}e^u(u-1) + C \quad (\#98) = \dfrac{1}{4}e^{-2x}(-2x-1) + C = -\dfrac{1}{4}e^{-2x}(2x+1) + C$

37. $u=\cos 3x, \quad -\dfrac{1}{3}\displaystyle\int \dfrac{1}{u(u+1)^2}\,du = -\dfrac{1}{3}\left[\dfrac{1}{u+1} + \ln\left|\dfrac{u}{u+1}\right|\right] + C \quad (\#13; \; a=1, b=1)$

$\qquad\qquad = -\dfrac{1}{3}\left[\dfrac{1}{\cos 3x+1} + \ln\left|\dfrac{\cos 3x}{\cos 3x+1}\right|\right] + C$

39. $u=4x^2, \quad \dfrac{1}{8}\displaystyle\int \dfrac{1}{u^2-1}\,du = \dfrac{1}{16}\ln\left|\dfrac{u-1}{u+1}\right| + C \quad (\#26; \; a=1) = \dfrac{1}{16}\ln\left|\dfrac{4x^2-1}{4x^2+1}\right| + C$

41. $u=2e^x, \quad \dfrac{1}{2}\displaystyle\int \sqrt{3-u^2}\,du = \dfrac{1}{2}\left[\dfrac{u}{2}\sqrt{3-u^2} + \dfrac{3}{2}\sin^{-1}\dfrac{u}{\sqrt{3}}\right] + C \quad (\#41; \; a=\sqrt{3})$

$\qquad\qquad = \dfrac{1}{2}e^x\sqrt{3-4e^{2x}} + \dfrac{3}{4}\sin^{-1}\dfrac{2e^x}{\sqrt{3}} + C$

43. $u=3x,$

$\qquad \dfrac{1}{3}\displaystyle\int \sqrt{(5/3)u-u^2}\,du = \dfrac{1}{3}\left[\dfrac{u-5/6}{2}\sqrt{(5/3)u-u^2} + \dfrac{25}{72}\sin^{-1}\left(\dfrac{u-5/6}{5/6}\right)\right] + C \; (\#50; \; a=5/6)$

$\qquad\qquad = \dfrac{18x-5}{36}\sqrt{5x-9x^2} + \dfrac{25}{216}\sin^{-1}\dfrac{18x-5}{5} + C$

45. $u = 3x$, $\dfrac{1}{9}\displaystyle\int u \sin u\, du = \dfrac{1}{9}[\sin u - u \cos u] + C$ (#83) $= \dfrac{1}{9}[\sin 3x - 3x \cos 3x] + C$

47. $u = -\sqrt{x}$, $x = u^2$, $dx = 2u\, du$,

$2\displaystyle\int u e^u\, du = 2e^u(u - 1) + C$ (#98) $= 2e^{-\sqrt{x}}(-\sqrt{x} - 1) + C = -2e^{-\sqrt{x}}(\sqrt{x} + 1) + C$

49. $\displaystyle\int \dfrac{1}{(x+2)^2 - 9}\, dx$; $u = x + 2$,

$\displaystyle\int \dfrac{1}{u^2 - 9}\, du = \dfrac{1}{6}\ln\left|\dfrac{u-3}{u+3}\right| + C$ (#26; $a = 3$) $= \dfrac{1}{6}\ln\left|\dfrac{x-1}{x+5}\right| + C$

51. $\displaystyle\int \dfrac{x}{\sqrt{9 - (x-2)^2}}\, dx$; $u = x - 2$, $x = u + 2$,

$\displaystyle\int \dfrac{u+2}{\sqrt{9 - u^2}}\, du = \int \dfrac{u}{\sqrt{9 - u^2}}\, du + 2\int \dfrac{1}{\sqrt{9 - u^2}}\, du$

$\qquad\qquad = -\sqrt{9 - u^2} + 2\sin^{-1}\dfrac{u}{3} + C$ (#4 and #40)

$\qquad\qquad = -\sqrt{5 + 4x - x^2} + 2\sin^{-1}\dfrac{x-2}{3} + C$

53. $\displaystyle\int_2^x \dfrac{1}{t(4-t)}\, dt = \dfrac{1}{4}\ln\dfrac{t}{4-t}\Big]_2^x$ (#11; $a = 4, b = -1$)

$\qquad\qquad = \dfrac{1}{4}\left[\ln\dfrac{x}{4-x} - \ln 1\right] = \dfrac{1}{4}\ln\dfrac{x}{4-x}, \dfrac{1}{4}\ln\dfrac{x}{4-x} = 0.5, \ln\dfrac{x}{4-x} = 2,$

$\dfrac{x}{4-x} = e^2$, $x = 4e^2 - e^2 x$, $x(1 + e^2) = 4e^2$, $x = 4e^2/(1 + e^2) \approx 3.523188312$.

55. $A = \displaystyle\int_0^4 \sqrt{25 - x^2}\, dx = \left(\dfrac{1}{2}x\sqrt{25 - x^2} + \dfrac{25}{2}\sin^{-1}\dfrac{x}{5}\right)\Big]_0^4$ (#41; $a = 5$)

$\qquad\qquad = 6 + \dfrac{25}{2}\sin^{-1}\dfrac{4}{5} \approx 17.59119022$

57. $A = \displaystyle\int_0^1 \dfrac{1}{25 - 16x^2}\, dx$; $u = 4x$,

$A = \dfrac{1}{4}\displaystyle\int_0^4 \dfrac{1}{25 - u^2}\, du = \dfrac{1}{40}\ln\left|\dfrac{u+5}{u-5}\right|\Big]_0^4$ (#25; $a = 5$) $= \dfrac{1}{40}\ln 9 \approx 0.054930614$

59. $V = 2\pi\displaystyle\int_0^{\pi/2} x \cos x\, dx = 2\pi(\cos x + x \sin x)\big]_0^{\pi/2}$ (#84) $= \pi(\pi - 2) \approx 3.586419094$

61. $V = 2\pi \int_0^3 xe^{-x} dx; \; u = -x,$

$V = 2\pi \int_0^{-3} ue^u du = 2\pi e^u (u-1)]_0^{-3}$ (#98) $= 2\pi(1 - 4e^{-3}) \approx 5.031899801$

63. $L = \int_0^2 \sqrt{1 + 16x^2} \, dx; \; u = 4x,$

$L = \frac{1}{4} \int_0^8 \sqrt{1 + u^2} \, du = \frac{1}{4}\left(\frac{u}{2}\sqrt{1+u^2} + \frac{1}{2}\ln\left|u + \sqrt{1+u^2}\right|\right)\Big]_0^8$ (#28; $a^2 = 1$)

$= \sqrt{65} + \frac{1}{8}\ln(8 + \sqrt{65}) \approx 8.409316783$

65. $S = 2\pi \int_0^\pi (\sin x)\sqrt{1 + \cos^2 x} \, dx; \; u = \cos x,$

$S = -2\pi \int_1^{-1} \sqrt{1 + u^2} \, du = 4\pi \int_0^1 \sqrt{1 + u^2} \, du$

$= 4\pi \left(\frac{u}{2}\sqrt{1+u^2} + \frac{1}{2}\ln\left|u + \sqrt{1+u^2}\right|\right)\Big]_0^1$ (#28; $a^2 = 1$)

$= 2\pi[\sqrt{2} + \ln(1 + \sqrt{2})] \approx 14.42359944$

EXERCISE SET 9.2

1. $u = x, \; dv = e^{-x}dx, \; du = dx, \; v = -e^{-x}$

$\int xe^{-x} dx = -xe^{-x} + \int e^{-x} dx = -xe^{-x} - e^{-x} + C$

3. $u = x^2, \; dv = e^x dx, \; du = 2x \, dx, \; v = e^x; \; \int x^2 e^x dx = x^2 e^x - 2\int xe^x dx.$

For $\int xe^x dx$ use $u = x, \; dv = e^x dx, \; du = dx, \; v = e^x$ to get

$\int xe^x dx = xe^x - e^x + C_1$ so $\int x^2 e^x dx = x^2 e^x - 2xe^x + 2e^x + C$

5. $u = x, \; dv = \sin 2x \, dx, \; du = dx, \; v = -\frac{1}{2}\cos 2x$

$\int x \sin 2x \, dx = -\frac{1}{2}x\cos 2x + \frac{1}{2}\int \cos 2x \, dx = -\frac{1}{2}x\cos 2x + \frac{1}{4}\sin 2x + C$

7. $u = x^2$, $dv = \cos x\, dx$, $du = 2x\, dx$, $v = \sin x$; $\displaystyle\int x^2 \cos x\, dx = x^2 \sin x - 2\int x \sin x\, dx$

For $\displaystyle\int x \sin x\, dx$ use $u = x$, $dv = \sin x\, dx$ to get

$\displaystyle\int x \sin x\, dx = -x \cos x + \sin x + C_1$ so $\displaystyle\int x^2 \cos x\, dx = x^2 \sin x + 2x \cos x - 2 \sin x + C$

9. $u = \ln x$, $dv = \sqrt{x}\, dx$, $du = \dfrac{1}{x} dx$, $v = \dfrac{2}{3} x^{3/2}$

$\displaystyle\int \sqrt{x} \ln x\, dx = \frac{2}{3} x^{3/2} \ln x - \frac{2}{3}\int x^{1/2} dx = \frac{2}{3} x^{3/2} \ln x - \frac{4}{9} x^{3/2} + C$

11. $u = (\ln x)^2$, $dv = dx$, $du = 2\dfrac{\ln x}{x} dx$, $v = x$; $\displaystyle\int (\ln x)^2 dx = x(\ln x)^2 - 2\int \ln x\, dx.$

Use $u = \ln x$, $dv = dx$ to get

$\displaystyle\int \ln x\, dx = x \ln x - \int dx = x \ln x - x + C_1$ so $\displaystyle\int (\ln x)^2 dx = x(\ln x)^2 - 2x \ln x + 2x + C$

13. $u = \ln(2x + 3)$, $dv = dx$, $du = \dfrac{2}{2x + 3} dx$, $v = x$

$\displaystyle\int \ln(2x + 3) dx = x \ln(2x + 3) - \int \frac{2x}{2x + 3} dx$

but $\displaystyle\int \frac{2x}{2x + 3} dx = \int \left(1 - \frac{3}{2x + 3}\right) dx = x - \frac{3}{2} \ln(2x + 3) + C_1$ so

$\displaystyle\int \ln(2x + 3) dx = x \ln(2x + 3) - x + \frac{3}{2} \ln(2x + 3) + C$

15. $u = \sin^{-1} x$, $dv = dx$, $du = 1/\sqrt{1 - x^2}\, dx$, $v = x$

$\displaystyle\int \sin^{-1} x\, dx = x \sin^{-1} x - \int x/\sqrt{1 - x^2}\, dx = x \sin^{-1} x + \sqrt{1 - x^2} + C$

17. $u = \tan^{-1}(2x)$, $dv = dx$, $du = \dfrac{2}{1 + 4x^2} dx$, $v = x$

$\displaystyle\int \tan^{-1}(2x) dx = x \tan^{-1}(2x) - \int \frac{2x}{1 + 4x^2} dx = x \tan^{-1}(2x) - \frac{1}{4} \ln(1 + 4x^2) + C$

19. $u = e^x$, $dv = \sin x\, dx$, $du = e^x dx$, $v = -\cos x$; $\displaystyle\int e^x \sin x\, dx = -e^x \cos x + \int e^x \cos x\, dx.$

For $\displaystyle\int e^x \cos x\, dx$ use $u = e^x$, $dv = \cos x\, dx$ to get $\displaystyle\int e^x \cos x = e^x \sin x - \int e^x \sin x\, dx$ so

$$\int e^x \sin x\, dx = -e^x \cos x + e^x \sin x - \int e^x \sin x\, dx,$$

$$2\int e^x \sin x\, dx = e^x(\sin x - \cos x) + C_1, \quad \int e^x \sin x\, dx = \frac{1}{2}e^x(\sin x - \cos x) + C$$

21. $u = e^{ax}$, $dv = \sin bx\, dx$, $du = ae^{ax}\, dx$, $v = -\frac{1}{b}\cos bx$

$$\int e^{ax} \sin bx\, dx = -\frac{1}{b}e^{ax} \cos bx + \frac{a}{b}\int e^{ax} \cos bx\, dx. \text{ Use } u = e^{ax}, dv = \cos bx\, dx \text{ to get}$$

$$\int e^{ax} \cos bx\, dx = \frac{1}{b}e^{ax} \sin bx - \frac{a}{b}\int e^{ax} \sin bx\, dx \text{ so}$$

$$\int e^{ax} \sin bx\, dx = -\frac{1}{b}e^{ax} \cos bx + \frac{a}{b^2}e^{ax} \sin bx - \frac{a^2}{b^2}\int e^{ax} \sin bx\, dx,$$

$$\int e^{ax} \sin bx\, dx = \frac{e^{ax}}{a^2 + b^2}(a \sin bx - b \cos bx) + C$$

23. $u = \sin(\ln x)$, $dv = dx$, $du = \dfrac{\cos(\ln x)}{x}dx$, $v = x$

$$\int \sin(\ln x)dx = x \sin(\ln x) - \int \cos(\ln x)dx. \text{ Use } u = \cos(\ln x), dv = dx \text{ to get}$$

$$\int \cos(\ln x)dx = x \cos(\ln x) + \int \sin(\ln x)dx \text{ so}$$

$$\int \sin(\ln x)dx = x \sin(\ln x) - x \cos(\ln x) - \int \sin(\ln x)dx,$$

$$\int \sin(\ln x)dx = (x/2)[\sin(\ln x) - \cos(\ln x)] + C$$

25. $u = x$, $dv = \sec^2 x\, dx$, $du = dx$, $v = \tan x$

$$\int x \sec^2 x\, dx = x \tan x - \int \tan x\, dx = x \tan x - \int \frac{\sin x}{\cos x}dx = x \tan x + \ln|\cos x| + C$$

27. $u = x^2$, $dv = xe^{x^2}\, dx$, $du = 2x\, dx$, $v = \frac{1}{2}e^{x^2}$

$$\int x^3 e^{x^2}\, dx = \frac{1}{2}x^2 e^{x^2} - \int xe^{x^2}\, dx = \frac{1}{2}x^2 e^{x^2} - \frac{1}{2}e^{x^2} + C$$

29. $u = x$, $dv = e^{-5x}dx$, $du = dx$, $v = -\frac{1}{5}e^{-5x}$

$$\int_0^1 xe^{-5x}dx = -\frac{1}{5}xe^{-5x}\bigg]_0^1 + \frac{1}{5}\int_0^1 e^{-5x}dx$$

$$= -\frac{1}{5}e^{-5} - \frac{1}{25}e^{-5x}\bigg]_0^1 = -\frac{1}{5}e^{-5} - \frac{1}{25}(e^{-5} - 1) = (1 - 6e^{-5})/25$$

31. $u = \ln x$, $dv = x^2 dx$, $du = \dfrac{1}{x} dx$, $v = \dfrac{1}{3} x^3$

$$\int_1^e x^2 \ln x \, dx = \frac{1}{3} x^3 \ln x \Big]_1^e - \frac{1}{3} \int_1^e x^2 dx = \frac{1}{3} e^3 - \frac{1}{9} x^3 \Big]_1^e = \frac{1}{3} e^3 - \frac{1}{9}(e^3 - 1) = (2e^3 + 1)/9$$

33. $u = \ln(x+3)$, $dv = dx$, $du = \dfrac{1}{x+3} dx$, $v = x$

$$\int_{-2}^2 \ln(x+3) dx = x \ln(x+3)\Big]_{-2}^2 - \int_{-2}^2 \frac{x}{x+3} dx = 2\ln 5 + 2\ln 1 - \int_{-2}^2 \left[1 - \frac{3}{x+3}\right] dx$$

$$= 2\ln 5 - [x - 3\ln(x+3)]_{-2}^2 = 2\ln 5 - (2 - 3\ln 5) + (-2 - 3\ln 1) = 5\ln 5 - 4$$

35. $u = \sec^{-1}\sqrt{\theta}$, $dv = d\theta$, $du = \dfrac{1}{2\theta\sqrt{\theta-1}} d\theta$, $v = \theta$

$$\int_2^4 \sec^{-1}\sqrt{\theta} d\theta = \theta \sec^{-1}\sqrt{\theta}\Big]_2^4 - \frac{1}{2}\int_2^4 \frac{1}{\sqrt{\theta-1}} d\theta = 4\sec^{-1}2 - 2\sec^{-1}\sqrt{2} - \sqrt{\theta-1}\Big]_2^4$$

$$= 4\left(\frac{\pi}{3}\right) - 2\left(\frac{\pi}{4}\right) - \sqrt{3} + 1 = \frac{5\pi}{6} - \sqrt{3} + 1$$

37. $u = x$, $dv = \sin 4x \, dx$, $du = dx$, $v = -\dfrac{1}{4}\cos 4x$

$$\int_0^{\pi/2} x \sin 4x \, dx = -\frac{1}{4} x \cos 4x \Big]_0^{\pi/2} + \frac{1}{4}\int_0^{\pi/2} \cos 4x \, dx = -\pi/8 + \frac{1}{16}\sin 4x \Big]_0^{\pi/2} = -\pi/8$$

39. $u = \tan^{-1}\sqrt{x}$, $dv = \sqrt{x} dx$, $du = \dfrac{1}{2\sqrt{x}(1+x)} dx$, $v = \dfrac{2}{3} x^{3/2}$

$$\int_1^3 \sqrt{x} \tan^{-1}\sqrt{x} dx = \frac{2}{3} x^{3/2} \tan^{-1}\sqrt{x}\Big]_1^3 - \frac{1}{3}\int_1^3 \frac{x}{1+x} dx$$

$$= \frac{2}{3} x^{3/2} \tan^{-1}\sqrt{x}\Big]_1^3 - \frac{1}{3}\int_1^3 \left[1 - \frac{1}{1+x}\right] dx$$

$$= \left[\frac{2}{3} x^{3/2} \tan^{-1}\sqrt{x} - \frac{1}{3} x + \frac{1}{3}\ln|1+x|\right]_1^3 = (2\sqrt{3}\pi - \pi/2 - 2 + \ln 2)/3$$

41. $u = x^2$, $dv = \dfrac{x}{\sqrt{x^2+1}} dx$, $du = 2x \, dx$, $v = \sqrt{x^2+1}$

$$\int_0^1 \frac{x^3}{\sqrt{x^2+1}} dx = x^2\sqrt{x^2+1}\Big]_0^1 - 2\int_0^1 x(x^2+1)^{1/2} dx$$

$$= \sqrt{2} - \frac{2}{3}(x^2+1)^{3/2}\Big]_0^1 = \sqrt{2} - \frac{2}{3}[2\sqrt{2} - 1] = (2 - \sqrt{2})/3$$

43. (a) $A = \int_1^e \ln x \, dx = (x \ln x - x) \Big]_1^e = 1$

(b) $V = \pi \int_1^e (\ln x)^2 dx = \pi (x(\ln x)^2 - 2x \ln x + 2x) \Big]_1^e = \pi(e - 2)$

45. $V = 2\pi \int_0^\pi x \sin x \, dx = 2\pi (-x \cos x + \sin x) \Big]_0^\pi = 2\pi^2$

47. (a) $\int \sin^3 x \, dx = -\dfrac{1}{3} \sin^2 x \cos x + \dfrac{2}{3} \int \sin x \, dx = -\dfrac{1}{3} \sin^2 x \cos x - \dfrac{2}{3} \cos x + C$

(b) $\int \sin^4 x \, dx = -\dfrac{1}{4} \sin^3 x \cos x + \dfrac{3}{4} \int \sin^2 x \, dx,$

$\int \sin^2 x \, dx = -\dfrac{1}{2} \sin x \cos x + \dfrac{1}{2} x + C_1$ so

$\int_0^{\pi/4} \sin^4 x \, dx = -\dfrac{1}{4} \sin^3 x \cos x - \dfrac{3}{8} \sin x \cos x + \dfrac{3}{8} x \Big]_0^{\pi/4}$

$= -\dfrac{1}{4}(1/\sqrt{2})^3(1/\sqrt{2}) - \dfrac{3}{8}(1/\sqrt{2})(1/\sqrt{2}) + 3\pi/32 = 3\pi/32 - 1/4$

49. (a) $u = 5x,$

$\int \cos^3 5x \, dx = \dfrac{1}{5} \int \cos^3 u \, du = \dfrac{1}{5} \left[\dfrac{1}{3} \cos^2 u \sin u + \dfrac{2}{3} \int \cos u \, du \right]$

$= \dfrac{1}{15} \cos^2 u \sin u + \dfrac{2}{15} \sin u + C = \dfrac{1}{15} \cos^2 5x \sin 5x + \dfrac{2}{15} \sin 5x + C$

(b) $u = x^2,$

$\int x \cos^4(x^2) dx = \dfrac{1}{2} \int \cos^4 u \, du = \dfrac{1}{2} \left[\dfrac{1}{4} \cos^3 u \sin u + \dfrac{3}{4} \int \cos^2 u \, du \right]$

$= \dfrac{1}{8} \cos^3 u \sin u + \dfrac{3}{8} \left[\dfrac{1}{2} \cos u \sin u + \dfrac{1}{2} \int du \right]$

$= \dfrac{1}{8} \cos^3 u \sin u + \dfrac{3}{16} \cos u \sin u + \dfrac{3}{16} u + C$

$= \dfrac{1}{8} \cos^3(x^2) \sin(x^2) + \dfrac{3}{16} \cos(x^2) \sin(x^2) + \dfrac{3}{16} x^2 + C$

51. $u = \sin^{n-1} x,\ dv = \sin x \, dx,\ du = (n-1)\sin^{n-2} x \cos x \, dx,\ v = -\cos x$

$\int \sin^n x \, dx = -\sin^{n-1} x \cos x + (n-1) \int \sin^{n-2} x \cos^2 x \, dx$

$$= -\sin^{n-1} x \cos x + (n-1) \int \sin^{n-2} x (1 - \sin^2 x) dx$$

$$= -\sin^{n-1} x \cos x + (n-1) \int \sin^{n-2} x\, dx - (n-1) \int \sin^n x\, dx,$$

$$n \int \sin^n x\, dx = -\sin^{n-1} x \cos x + (n-1) \int \sin^{n-2} x\, dx,$$

$$\int \sin^n x\, dx = -\frac{1}{n} \sin^{n-1} x \cos x + \frac{n-1}{n} \int \sin^{n-2} x\, dx$$

53. **(a)** $u = x^n$, $dv = e^x dx$, $du = nx^{n-1} dx$, $v = e^x$; $\displaystyle\int x^n e^x dx = x^n e^x - n \int x^{n-1} e^x dx$

(b) $\displaystyle\int x^3 e^x dx = x^3 e^x - 3 \int x^2 e^x dx = x^3 e^x - 3 \left[x^2 e^x - 2 \int x e^x dx \right]$

$$= x^3 e^x - 3x^2 e^x + 6 \left[x e^x - \int e^x dx \right] = x^3 e^x - 3x^2 e^x + 6x e^x - 6e^x + C$$

55. $u = x$, $dv = f''(x) dx$, $du = dx$, $v = f'(x)$

$$\int_{-1}^1 x f''(x) dx = xf'(x)\big]_{-1}^1 - \int_{-1}^1 f'(x) dx$$

$$= f'(1) - f'(-1) - f(x)\big]_{-1}^1 = f'(1) - f'(-1) + f(-1) - f(1)$$

57. $du = -(1/x^2) dx$, $v = x$; $\displaystyle\int \frac{1}{x} dx = 1 + \int \frac{1}{x} dx$ so $0 = 1???$

The "obvious" cancellation of the indefinite integrals causes the problem. Instead, proceed as follows:

$$\int \frac{1}{x} dx - \int \frac{1}{x} dx = 1, \quad \int (0) dx = 1, \text{ but } \int (0) dx = C \text{ so } C = 1.$$

EXERCISE SET 9.3

1. $u = \cos x$, $\displaystyle -\int u^5 du = -\frac{1}{6} \cos^6 x + C$ 3. $u = \sin ax$, $\displaystyle \frac{1}{a} \int u\, du = \frac{1}{2a} \sin^2 ax + C$

5. $\displaystyle \int \sin^2 5\theta\, d\theta = \frac{1}{2} \int (1 - \cos 10\theta) d\theta = \frac{1}{2}\theta - \frac{1}{20} \sin 10\theta + C$

7. $\int \cos^4(x/4)dx = \frac{1}{4}\int [1+\cos(x/2)]^2 dx = \frac{1}{4}\int [1+2\cos(x/2)+\cos^2(x/2)]dx$

$$= \frac{1}{4}\int \left[1+2\cos(x/2)+\frac{1}{2}(1+\cos x)\right]dx = \frac{1}{4}\int \left[\frac{3}{2}+2\cos(x/2)+\frac{1}{2}\cos x\right]dx$$

$$= \frac{3}{8}x + \sin(x/2) + \frac{1}{8}\sin x + C$$

9. $\int \cos^5 \theta d\theta = \int (1-\sin^2\theta)^2 \cos\theta d\theta = \int (1-2\sin^2\theta + \sin^4\theta)\cos\theta d\theta$

$$= \sin\theta - \frac{2}{3}\sin^3\theta + \frac{1}{5}\sin^5\theta + C$$

11. $\int \sin^2 2t \cos^3 2t\, dt = \int \sin^2 2t(1-\sin^2 2t)\cos 2t\, dt = \int (\sin^2 2t - \sin^4 2t)\cos 2t\, dt$

$$= \frac{1}{6}\sin^3 2t - \frac{1}{10}\sin^5 2t + C$$

13. $\int \cos^4 x \sin^3 x\, dx = \int \cos^4 x(1-\cos^2 x)\sin x\, dx$

$$= \int (\cos^4 x - \cos^6 x)\sin x\, dx = -\frac{1}{5}\cos^5 x + \frac{1}{7}\cos^7 x + C$$

15. $\int \sin^5\theta \cos^4\theta d\theta = \int (1-\cos^2\theta)^2 \cos^4\theta \sin\theta d\theta$

$$= \int (\cos^4\theta - 2\cos^6\theta + \cos^8\theta)\sin\theta d\theta = -\frac{1}{5}\cos^5\theta + \frac{2}{7}\cos^7\theta - \frac{1}{9}\cos^9\theta + C$$

17. $\int \sin^2 x \cos^2 x\, dx = \frac{1}{4}\int (1-\cos 2x)(1+\cos 2x)dx = \frac{1}{4}\int (1-\cos^2 2x)dx = \frac{1}{4}\int \sin^2 2x\, dx$

$$= \frac{1}{8}\int (1-\cos 4x)dx = \frac{1}{8}x - \frac{1}{32}\sin 4x + C$$

19. $\int \sin x \cos 2x\, dx = \frac{1}{2}\int (\sin 3x - \sin x)dx = -\frac{1}{6}\cos 3x + \frac{1}{2}\cos x + C$

21. $\int \sin x \cos(x/2)dx = \frac{1}{2}\int [\sin(3x/2)+\sin(x/2)]dx = -\frac{1}{3}\cos(3x/2) - \cos(x/2) + C$

23. $u = \cos x, \; -\int u^{-8}du = 1/(7\cos^7 x) + C$

25. $\displaystyle\int_0^{\pi/4} \cos^3 x \, dx = \int_0^{\pi/4} (1 - \sin^2 x) \cos x \, dx$

$$= \sin x - \frac{1}{3}\sin^3 x \bigg]_0^{\pi/4} = (\sqrt{2}/2) - \frac{1}{3}(\sqrt{2}/2)^3 = 5\sqrt{2}/12$$

27. $\displaystyle\int_0^{\pi/3} \sin^4 3x \cos^3 3x \, dx = \int_0^{\pi/3} \sin^4 3x(1 - \sin^2 3x) \cos 3x \, dx = \frac{1}{15}\sin^5 3x - \frac{1}{21}\sin^7 3x \bigg]_0^{\pi/3} = 0$

29. $\displaystyle\int_0^{\pi/6} \sin 2x \cos 4x \, dx = \frac{1}{2}\int_0^{\pi/6}(\sin 6x - \sin 2x)dx = -\frac{1}{12}\cos 6x + \frac{1}{4}\cos 2x \bigg]_0^{\pi/6}$

$$= [(-1/12)(-1) + (1/4)(1/2)] - [-1/12 + 1/4] = 1/24$$

31. **(a)** $\displaystyle\int_0^{2\pi} \sin mx \cos nx \, dx = \frac{1}{2}\int_0^{2\pi}[\sin(m+n)x + \sin(m-n)x]dx$

$$= -\frac{\cos(m+n)x}{2(m+n)} - \frac{\cos(m-n)x}{2(m-n)}\bigg]_0^{2\pi},$$

$m + n$ and $m - n$ are integers so $\cos[(m+n)2\pi] = \cos[(m-n)2\pi] = 1$ thus

$$\int_0^{2\pi} \sin mx \cos nx \, dx = \left[\left(-\frac{1}{2(m+n)} - \frac{1}{2(m-n)}\right) + \left(\frac{1}{2(m+n)} + \frac{1}{2(m-n)}\right)\right] = 0$$

(b) $\displaystyle\int_0^{2\pi} \cos mx \cos nx \, dx = \frac{1}{2}\int_0^{2\pi}[\cos(m+n)x + \cos(m-n)x]dx$

$$= \frac{\sin(m+n)x}{2(m+n)} + \frac{\sin(m-n)x}{2(m-n)}\bigg]_0^{2\pi} = 0$$

(c) $\displaystyle\int_0^{2\pi} \sin mx \sin nx \, dx = \frac{1}{2}\int_0^{2\pi}[\cos(m-n)x - \cos(m+n)x]dx$

$$= \frac{\sin(m-n)x}{2(m-n)} + \frac{\sin(m+n)x}{2(m+n)}\bigg]_0^{2\pi} = 0$$

33. $\displaystyle V = \pi\int_0^{\pi/4}(\cos^2 x - \sin^2 x)dx = \pi\int_0^{\pi/4}\cos 2x \, dx = \frac{1}{2}\pi\sin 2x\bigg]_0^{\pi/4} = \pi/2$

35. **(a)** $\displaystyle\int_0^{\pi/2} \sin^3 x \, dx = \frac{2}{3}$
        (b) $\displaystyle\int_0^{\pi/2} \sin^4 x \, dx = \frac{1\cdot 3}{2\cdot 4}\cdot\frac{\pi}{2} = 3\pi/16$

    (c) $\displaystyle\int_0^{\pi/2} \sin^5 x \, dx = \frac{2\cdot 4}{3\cdot 5} = 8/15$
        (d) $\displaystyle\int_0^{\pi/2} \sin^6 x \, dx = \frac{1\cdot 3\cdot 5}{2\cdot 4\cdot 6}\cdot\frac{\pi}{2} = 5\pi/32$

EXERCISE SET 9.4

1. $\dfrac{1}{3}\tan(3x+1)+C$

3. $\dfrac{1}{2}\ln|\cos(e^{-2x})|+C$

5. $\dfrac{1}{2}\ln|\sec 2x + \tan 2x| + C$

7. $u = \tan x, \displaystyle\int u^2\,du = \dfrac{1}{3}\tan^3 x + C$

9. $\displaystyle\int \tan^3 4x(1+\tan^2 4x)\sec^2 4x\,dx = \int (\tan^3 4x + \tan^5 4x)\sec^2 4x\,dx$

$$= \dfrac{1}{16}\tan^4 4x + \dfrac{1}{24}\tan^6 4x + C$$

11. $\displaystyle\int \sec^4 x(\sec^2 x - 1)\sec x \tan x\,dx = \int (\sec^6 x - \sec^4 x)\sec x \tan x\,dx = \dfrac{1}{7}\sec^7 x - \dfrac{1}{5}\sec^5 x + C$

13. $\displaystyle\int (\sec^2 x - 1)^2 \sec x\,dx = \int (\sec^5 x - 2\sec^3 x + \sec x)\,dx$

$$= \int \sec^5 x\,dx - 2\int \sec^3 x\,dx + \int \sec x\,dx$$

$$= \dfrac{1}{4}\sec^3 x\tan x + \dfrac{3}{4}\int \sec^3 x\,dx - 2\int \sec^3 x\,dx + \ln|\sec x + \tan x|$$

$$= \dfrac{1}{4}\sec^3 x\tan x - \dfrac{5}{4}\left[\dfrac{1}{2}\sec x\tan x + \dfrac{1}{2}\ln|\sec x + \tan x|\right] + \ln|\sec x + \tan x| + C$$

$$= \dfrac{1}{4}\sec^3 x\tan x - \dfrac{5}{8}\sec x\tan x + \dfrac{3}{8}\ln|\sec x + \tan x| + C$$

15. $\displaystyle\int \sec^2 2t(\sec 2t \tan 2t)\,dt = \dfrac{1}{6}\sec^3 2t + C$

17. $\displaystyle\int \sec^4 x\,dx = \int (1+\tan^2 x)\sec^2 x\,dx = \int (\sec^2 x + \tan^2 x\sec^2 x)\,dx = \tan x + \dfrac{1}{3}\tan^3 x + C$

19. $u = \pi x$, use reduction formula (3) to get

$$\dfrac{1}{\pi}\int \sec^6 u\,du = \dfrac{1}{\pi}\left[\dfrac{1}{5}\sec^4 u \tan u + \dfrac{4}{5}\int \sec^4 u\,du\right]$$

$$= \dfrac{1}{5\pi}\sec^4 u \tan u + \dfrac{4}{5\pi}\left[\dfrac{1}{3}\sec^2 u \tan u + \dfrac{2}{3}\tan u\right] + C$$

$$= \dfrac{1}{5\pi}\sec^4 \pi x \tan \pi x + \dfrac{4}{15\pi}\sec^2 \pi x \tan \pi x + \dfrac{8}{15\pi}\tan \pi x + C$$

21. Use reduction formula (4) to get $\int \tan^4 x\, dx = \dfrac{1}{3}\tan^3 x - \tan x + x + C$

23. $u = \tan(x^2)$, $\dfrac{1}{2}\int u^2 du = \dfrac{1}{6}\tan^3(x^2) + C$

25. $\int (\csc^2 x - 1)\csc^2 x(\csc x \cot x)dx = \int (\csc^4 x - \csc^2 x)(\csc x \cot x)dx$

$$= -\dfrac{1}{5}\csc^5 x + \dfrac{1}{3}\csc^3 x + C$$

27. $\int (\csc^2 x - 1)\cot x\, dx = \int \csc x(\csc x \cot x)dt - \int \dfrac{\cos x}{\sin x}dx = -\dfrac{1}{2}\csc^2 x - \ln|\sin x| + C$

29. $\int \tan^{1/2} x(1 + \tan^2 x)\sec^2 x\, dx = \dfrac{2}{3}\tan^{3/2} x + \dfrac{2}{7}\tan^{7/2} x + C$

31. $\displaystyle\int_0^{\pi/6}(\sec^2 2x - 1)dx = \dfrac{1}{2}\tan 2x - x\Big]_0^{\pi/6} = \sqrt{3}/2 - \pi/6$

33. $u = x/2$,

$$2\int_0^{\pi/4}\tan^5 u\, du = \dfrac{1}{2}\tan^4 u - \tan^2 u - 2\ln|\cos u|\Big]_0^{\pi/4} = 1/2 - 1 - 2\ln(1/\sqrt{2}) = -1/2 + \ln 2$$

35. $y' = \tan x$, $1 + (y')^2 = 1 + \tan^2 x = \sec^2 x$,

$$L = \int_0^{\pi/4}\sqrt{\sec^2 x}\, dx = \int_0^{\pi/4}\sec x\, dx = \ln|\sec x + \tan x|\Big]_0^{\pi/4} = \ln(\sqrt{2} + 1)$$

37. **(a)** $\int \csc x\, dx = \int \sec(\pi/2 - x)dx = -\ln|\sec(\pi/2 - x) + \tan(\pi/2 - x)| + C$

$$= -\ln|\csc x + \cot x| + C$$

(b) $-\ln|\csc x + \cot x| = \ln\dfrac{1}{|\csc x + \cot x|} = \ln\dfrac{|\csc x - \cot x|}{|\csc^2 x - \cot^2 x|} = \ln|\csc x - \cot x|.$

$$-\ln|\csc x + \cot x| = -\ln\left|\dfrac{1}{\sin x} + \dfrac{\cos x}{\sin x}\right| = \ln\left|\dfrac{\sin x}{1 + \cos x}\right|$$

$$= \ln\left|\dfrac{2\sin(x/2)\cos(x/2)}{2\cos^2(x/2)}\right| = \ln|\tan(x/2)|$$

39. $a \sin x + b \cos x = \sqrt{a^2 + b^2} \left[\dfrac{a}{\sqrt{a^2 + b^2}} \sin x + \dfrac{b}{\sqrt{a^2 + b^2}} \cos x \right]$

$$= \sqrt{a^2 + b^2}(\sin x \cos \theta + \cos x \sin \theta)$$

where $\cos \theta = a/\sqrt{a^2 + b^2}$ and $\sin \theta = b/\sqrt{a^2 + b^2}$ so $a \sin x + b \cos x = \sqrt{a^2 + b^2} \sin(x + \theta)$

and $\displaystyle \int \dfrac{dx}{a \sin x + b \cos x} = \dfrac{1}{\sqrt{a^2 + b^2}} \int \csc(x + \theta) dx$

$$= -\dfrac{1}{\sqrt{a^2 + b^2}} \ln|\csc(x + \theta) + \cot(x + \theta)| + C$$

41. $\displaystyle \int \tan^m x \, dx = \int \tan^{m-2} x \tan^2 x \, dx = \int \tan^{m-2} x(\sec^2 x - 1)dx$

$$= \int \tan^{m-2} x \sec^2 x \, dx - \int \tan^{m-2} x \, dx$$

$$= \dfrac{\tan^{m-1} x}{m-1} - \int \tan^{m-2} x \, dx$$

EXERCISE SET 9.5

1. $x = 2 \sin \theta$, $dx = 2 \cos \theta \, d\theta$,

$$4 \int \cos^2 \theta \, d\theta = 2 \int (1 + \cos 2\theta)d\theta = 2\theta + \sin 2\theta + C$$

$$= 2\theta + 2 \sin \theta \cos \theta + C = 2 \sin^{-1}(x/2) + \dfrac{1}{2} x \sqrt{4 - x^2} + C$$

3. $x = 3 \sin \theta$, $dx = 3 \cos \theta \, d\theta$,

$$9 \int \sin^2 \theta \, d\theta = \dfrac{9}{2} \int (1 - \cos 2\theta)d\theta = \dfrac{9}{2}\theta - \dfrac{9}{4} \sin 2\theta + C = \dfrac{9}{2}\theta - \dfrac{9}{2} \sin \theta \cos \theta + C$$

$$= \dfrac{9}{2} \sin^{-1}(x/3) - \dfrac{1}{2} x \sqrt{9 - x^2} + C$$

5. $x = 2 \tan \theta$, $dx = 2 \sec^2 \theta \, d\theta$,

$$\dfrac{1}{8} \int \dfrac{1}{\sec^2 \theta} d\theta = \dfrac{1}{8} \int \cos^2 \theta \, d\theta = \dfrac{1}{16} \int (1 + \cos 2\theta)d\theta = \dfrac{1}{16}\theta + \dfrac{1}{32} \sin 2\theta + C$$

$$= \dfrac{1}{16}\theta + \dfrac{1}{16} \sin \theta \cos \theta + C = \dfrac{1}{16} \tan^{-1} \dfrac{x}{2} + \dfrac{x}{8(4 + x^2)} + C$$

7. $x = 3\sec\theta$, $dx = 3\sec\theta\tan\theta\,d\theta$,

$$3\int \tan^2\theta\,d\theta = 3\int(\sec^2\theta - 1)d\theta = 3\tan\theta - 3\theta + C = \sqrt{x^2 - 9} - 3\sec^{-1}\frac{x}{3} + C$$

9. $x = \sqrt{2}\sin\theta$, $dx = \sqrt{2}\cos\theta\,d\theta$,

$$2\sqrt{2}\int \sin^3\theta\,d\theta = 2\sqrt{2}\left(-\cos\theta + \frac{1}{3}\cos^3\theta\right) + C = -2\sqrt{2 - x^2} + \frac{1}{3}(2 - x^2)^{3/2} + C$$

11. $x = \sqrt{3}\tan\theta$, $dx = \sqrt{3}\sec^2\theta\,d\theta$,

$$\frac{1}{3}\int \frac{1}{\sec\theta}d\theta = \frac{1}{3}\int \cos\theta\,d\theta = \frac{1}{3}\sin\theta + C = \frac{x}{3\sqrt{3 + x^2}} + C$$

13. $x = \frac{3}{2}\sec\theta$, $dx = \frac{3}{2}\sec\theta\tan\theta\,d\theta$,

$$\frac{2}{9}\int \frac{1}{\sec\theta}d\theta = \frac{2}{9}\int \cos\theta\,d\theta = \frac{2}{9}\sin\theta + C = \frac{\sqrt{4x^2 - 9}}{9x} + C$$

15. $x = \sin\theta$, $dx = \cos\theta\,d\theta$, $\displaystyle\int \frac{1}{\cos^2\theta}d\theta = \int \sec^2\theta\,d\theta = \tan\theta + C = x/\sqrt{1 - x^2} + C$

17. $x = \sec\theta$, $dx = \sec\theta\tan\theta\,d\theta$

$$\int \sec\theta\,d\theta = \ln|\sec\theta + \tan\theta| + C = \ln\left|x + \sqrt{x^2 - 1}\right| + C$$

19. $x = \frac{3}{2}\sin\theta$, $dx = \frac{3}{2}\cos\theta\,d\theta$,

$$\frac{2}{9}\int \frac{1}{\sin^2\theta}d\theta = \frac{2}{9}\csc^2\theta\,d\theta = -\frac{2}{9}\cot\theta + C = -\frac{\sqrt{9 - 4x^2}}{9x} + C$$

21. $x = \frac{1}{3}\sec\theta$, $dx = \frac{1}{3}\sec\theta\tan\theta\,d\theta$,

$$\frac{1}{3}\int \frac{\sec\theta}{\tan^2\theta}d\theta = \frac{1}{3}\int \csc\theta\cot\theta\,d\theta = -\frac{1}{3}\csc\theta + C = -x/\sqrt{9x^2 - 1} + C$$

23. $e^x = \sin\theta$, $e^x dx = \cos\theta\,d\theta$,

$$\int \cos^2\theta\,d\theta = \frac{1}{2}\int(1 + \cos 2\theta)d\theta = \frac{1}{2}\theta + \frac{1}{4}\sin 2\theta + C = \frac{1}{2}\sin^{-1}(e^x) + \frac{1}{2}e^x\sqrt{1 - e^{2x}} + C$$

25. $x = 4\sin\theta$, $dx = 4\cos\theta\,d\theta$,

$$1024\int_0^{\pi/2} \sin^3\theta\cos^2\theta\,d\theta = 1024\left[-\frac{1}{3}\cos^3\theta + \frac{1}{5}\cos^5\theta\right]_0^{\pi/2} = 1024(1/3 - 1/5) = 2048/15$$

27. $x = \sec\theta$, $dx = \sec\theta\tan\theta\,d\theta$, $\displaystyle\int_{\pi/4}^{\pi/3}\frac{1}{\sec\theta}d\theta = \int_{\pi/4}^{\pi/3}\cos\theta\,d\theta = \sin\theta\Big]_{\pi/4}^{\pi/3} = (\sqrt{3} - \sqrt{2})/2$

29. $x = \sqrt{3}\tan\theta$, $dx = \sqrt{3}\sec^2\theta\,d\theta$,

$$\frac{1}{9}\int_{\pi/6}^{\pi/3}\frac{\sec\theta}{\tan^4\theta}d\theta = \frac{1}{9}\int_{\pi/6}^{\pi/3}\frac{\cos^3\theta}{\sin^4\theta}d\theta = \frac{1}{9}\int_{\pi/6}^{\pi/3}\frac{1-\sin^2\theta}{\sin^4\theta}\cos\theta\,d\theta$$

$$= \frac{1}{9}\int_{1/2}^{\sqrt{3}/2}\frac{1-u^2}{u^4}du \quad (u = \sin\theta)$$

$$= \frac{1}{9}\int_{1/2}^{\sqrt{3}/2}(u^{-4} - u^{-2})du = \frac{1}{9}\left[-\frac{1}{3u^3} + \frac{1}{u}\right]_{1/2}^{\sqrt{3}/2} = \frac{10\sqrt{3} + 18}{243}$$

31. $u = x^2 + 4$, $du = 2x\,dx$, $\dfrac{1}{2}\displaystyle\int\frac{1}{u}du = \frac{1}{2}\ln|u| + C = \frac{1}{2}\ln(x^2 + 4) + C$.

$x = 2\tan\theta$, $dx = 2\sec^2\theta\,d\theta$,

$$\int\tan\theta\,d\theta = \ln|\sec\theta| + C_1 = \ln\frac{\sqrt{x^2 + 4}}{2} + C_1 \doteq \ln(x^2+4)^{1/2} - \ln 2 + C_1$$

$$= \frac{1}{2}\ln(x^2 + 4) + C \text{ with } C = C_1 - \ln 2$$

33. $y' = \dfrac{1}{x}$, $1 + (y')^2 = 1 + \dfrac{1}{x^2} = \dfrac{x^2+1}{x^2}$,

$$L = \int_1^2\sqrt{\frac{x^2+1}{x^2}}dx = \int_1^2\frac{\sqrt{x^2+1}}{x}dx;\ x = \tan\theta,\ dx = \sec^2\theta\,d\theta,$$

$$L = \int_{\pi/4}^{\tan^{-1}2}\frac{\sec^3\theta}{\tan\theta}d\theta = \int_{\pi/4}^{\tan^{-1}2}\frac{\tan^2\theta + 1}{\tan\theta}\sec\theta\,d\theta = \int_{\pi/4}^{\tan^{-1}2}\left[\sec\theta\tan\theta + \frac{\sec\theta}{\tan\theta}\right]d\theta$$

$$= \int_{\pi/4}^{\tan^{-1}2}(\sec\theta\tan\theta + \csc\theta)d\theta = \sec\theta - \ln|\csc\theta + \cot\theta|\,\Big]_{\pi/4}^{\tan^{-1}2}$$

$$= \left[\sqrt{5} - \ln\left|\frac{\sqrt{5}}{2} + \frac{1}{2}\right|\right] - \left[\sqrt{2} - \ln|\sqrt{2} + 1|\right] = \sqrt{5} - \sqrt{2} + \ln\frac{2 + 2\sqrt{2}}{1 + \sqrt{5}}$$

35. $y' = 2x$, $1 + (y')^2 = 1 + 4x^2$,

$$S = 2\pi\int_0^1 x^2\sqrt{1 + 4x^2}dx;\ x = \frac{1}{2}\tan\theta,\ dx = \frac{1}{2}\sec^2\theta\,d\theta,$$

$$S = \frac{\pi}{4} \int_0^{\tan^{-1} 2} \tan^2 \theta \sec^3 \theta \, d\theta = \frac{\pi}{4} \int_0^{\tan^{-1} 2} (\sec^2 \theta - 1) \sec^3 \theta \, d\theta$$

$$= \frac{\pi}{4} \int_0^{\tan^{-1} 2} (\sec^5 \theta - \sec^3 \theta) d\theta$$

$$= \frac{\pi}{4} \left[\frac{1}{4} \sec^3 \theta \tan \theta - \frac{1}{8} \sec \theta \tan \theta - \frac{1}{8} \ln|\sec \theta + \tan \theta| \right]_0^{\tan^{-1} 2} = \frac{\pi}{32} [18\sqrt{5} - \ln(2 + \sqrt{5})]$$

37. **(a)** $x = 3 \sinh u$, $dx = 3 \cosh u \, du$, $\displaystyle\int du = u + C = \sinh^{-1}(x/3) + C$

(b) $x = 3 \tan \theta$, $dx = 3 \sec^2 \theta \, d\theta$,

$\displaystyle\int \sec \theta \, d\theta = \ln|\sec \theta + \tan \theta| + C = \ln(\sqrt{x^2 + 9}/3 + x/3) + C$

but $\sinh^{-1}(x/3) = \ln(x/3 + \sqrt{x^2/9 + 1}) = \ln(x/3 + \sqrt{x^2 + 9}/3)$ so the results agree.

39. $\displaystyle\int \frac{1}{(x-2)^2 + 9} dx = \frac{1}{3} \tan^{-1}\left(\frac{x-2}{3}\right) + C$

41. $\displaystyle\int \frac{1}{\sqrt{9 - (x-1)^2}} dx = \sin^{-1}\left(\frac{x-1}{3}\right) + C$

43. $\displaystyle\int \frac{1}{\sqrt{(x-3)^2 + 1}} dx = \sinh^{-1}(x-3) + C.$

Alternate solution: let $x - 3 = \tan \theta$,

$\displaystyle\int \sec \theta \, d\theta = \ln|\sec \theta + \tan \theta| + C = \ln(\sqrt{x^2 - 6x + 10} + x - 3) + C.$

45. $\displaystyle\int \sqrt{4 - (x+1)^2} dx$, let $x + 1 = 2 \sin \theta$,

$4 \displaystyle\int \cos^2 \theta \, d\theta = 2\theta + \sin 2\theta + C = 2\theta + 2 \sin \theta \cos \theta + C$

$= 2 \sin^{-1}\left(\frac{x+1}{2}\right) + \frac{1}{2}(x+1)\sqrt{3 - 2x - x^2} + C$

47. $\displaystyle\int \frac{1}{2(x+1)^2 + 5} dx = \frac{1}{2} \int \frac{1}{(x+1)^2 + 5/2} dx = \frac{1}{\sqrt{10}} \tan^{-1} \sqrt{2/5}(x+1) + C$

49. $\int \dfrac{2x+5}{(x+1)^2+4}\,dx$, let $u = x+1$,

$$\int \dfrac{2u+3}{u^2+4}\,du = \int \left[\dfrac{2u}{u^2+4} + \dfrac{3}{u^2+4}\right] du = \ln(u^2+4) + \dfrac{3}{2}\tan^{-1}(u/2) + C$$

$$= \ln(x^2+2x+5) + \dfrac{3}{2}\tan^{-1}\left(\dfrac{x+1}{2}\right) + C$$

51. $\int \dfrac{x+3}{\sqrt{(x+1)^2+1}}\,dx$, let $u = x+1$,

$$\int \dfrac{u+2}{\sqrt{u^2+1}}\,du = \int \left[u(u^2+1)^{-1/2} + \dfrac{2}{\sqrt{u^2+1}}\right] du = \sqrt{u^2+1} + 2\sinh^{-1}u + C$$

$$= \sqrt{x^2+2x+2} + 2\sinh^{-1}(x+1) + C$$

<u>Alternate solution</u>: let $x+1 = \tan\theta$,

$$\int (\tan\theta+2)\sec\theta\,d\theta = \int \sec\theta\tan\theta\,d\theta + 2\int \sec\theta\,d\theta = \sec\theta + 2\ln|\sec\theta+\tan\theta| + C$$

$$= \sqrt{x^2+2x+2} + 2\ln(\sqrt{x^2+2x+2}+x+1) + C.$$

53. $\int_0^1 \sqrt{4x-x^2}\,dx = \int_0^1 \sqrt{4-(x-2)^2}\,dx$, let $x-2 = 2\sin\theta$,

$$4\int_{-\pi/2}^{-\pi/6} \cos^2\theta\,d\theta = 2\theta + \sin 2\theta \Big]_{-\pi/2}^{-\pi/6} = \dfrac{2\pi}{3} - \dfrac{\sqrt{3}}{2}$$

EXERCISE SET 9.6

1. $\dfrac{A}{(x-2)} + \dfrac{B}{(x+5)}$

3. $\dfrac{2x-3}{x^2(x-1)} = \dfrac{A}{x} + \dfrac{B}{x^2} + \dfrac{C}{x-1}$

5. $\dfrac{A}{x} + \dfrac{B}{x^2} + \dfrac{C}{x^3} + \dfrac{Dx+E}{x^2+1}$

7. $\dfrac{Ax+B}{x^2+5} + \dfrac{Cx+D}{(x^2+5)^2}$

9. $\dfrac{1}{(x+4)(x-1)} = \dfrac{A}{x+4} + \dfrac{B}{x-1}$; $A = -\dfrac{1}{5}$, $B = \dfrac{1}{5}$

$$-\dfrac{1}{5}\int \dfrac{1}{x+4}\,dx + \dfrac{1}{5}\int \dfrac{1}{x-1}\,dx = -\dfrac{1}{5}\ln|x+4| + \dfrac{1}{5}\ln|x-1| + C = \dfrac{1}{5}\ln\left|\dfrac{x-1}{x+4}\right| + C$$

11. $\dfrac{x}{(x-2)(x-3)} = \dfrac{A}{x-2} + \dfrac{B}{x-3}$; $A = -2$, $B = 3$

$-2\displaystyle\int \frac{1}{x-2}\,dx + 3\int \frac{1}{x-3}\,dx = -2\ln|x-2| + 3\ln|x-3| + C$

13. $\dfrac{11x+17}{(2x-1)(x+4)} = \dfrac{A}{2x-1} + \dfrac{B}{x+4}$; $A = 5$, $B = 3$

$5\displaystyle\int \frac{1}{2x-1}\,dx + 3\int \frac{1}{x+4}\,dx = \frac{5}{2}\ln|2x-1| + 3\ln|x+4| + C$

15. $\dfrac{1}{(x-1)(x+2)(x-3)} = \dfrac{A}{x-1} + \dfrac{B}{x+2} + \dfrac{C}{x-3}$; $A = -\dfrac{1}{6}$, $B = \dfrac{1}{15}$, $C = \dfrac{1}{10}$

$-\dfrac{1}{6}\displaystyle\int \frac{1}{x-1}\,dx + \frac{1}{15}\int \frac{1}{x+2}\,dx + \frac{1}{10}\int \frac{1}{x-3}\,dx$

$= -\dfrac{1}{6}\ln|x-1| + \dfrac{1}{15}\ln|x+2| + \dfrac{1}{10}\ln|x-3| + C$

17. $\dfrac{2x^2 - 9x - 9}{x(x+3)(x-3)} = \dfrac{A}{x} + \dfrac{B}{x+3} + \dfrac{C}{x-3}$; $A = 1$, $B = 2$, $C = -1$

$\displaystyle\int \frac{1}{x}\,dx + 2\int \frac{1}{x+3}\,dx - \int \frac{1}{x-3}\,dx = \ln|x| + 2\ln|x+3| - \ln|x-3| + C = \ln\left|\frac{x(x+3)^2}{x-3}\right| + C$

19. $\dfrac{x^2+2}{x+2} = x - 2 + \dfrac{6}{x+2}$, $\displaystyle\int \left(x - 2 + \frac{6}{x+2}\right) dx = \frac{1}{2}x^2 - 2x + 6\ln|x+2| + C$

21. $\dfrac{3x^2 - 10}{x^2 - 4x + 4} = 3 + \dfrac{12x - 22}{x^2 - 4x + 4}$, $\dfrac{12x - 22}{(x-2)^2} = \dfrac{A}{x-2} + \dfrac{B}{(x-2)^2}$; $A = 12$, $B = 2$

$\displaystyle\int 3\,dx + 12\int \frac{1}{x-2}\,dx + 2\int \frac{1}{(x-2)^2}\,dx = 3x + 12\ln|x-2| - 2/(x-2) + C$

23. $\dfrac{x^3}{x^2 - 3x + 2} = x + 3 + \dfrac{7x - 6}{x^2 - 3x + 2}$, $\dfrac{7x - 6}{(x-1)(x-2)} = \dfrac{A}{x-1} + \dfrac{B}{x-2}$; $A = -1$, $B = 8$

$\displaystyle\int (x+3)\,dx - \int \frac{1}{x-1}\,dx + 8\int \frac{1}{x-2}\,dx = \frac{1}{2}x^2 + 3x - \ln|x-1| + 8\ln|x-2| + C$

25. $\dfrac{x^5 + 2x^2 + 1}{x^3 - x} = x^2 + 1 + \dfrac{2x^2 + x + 1}{x^3 - x}$,

$\dfrac{2x^2 + x + 1}{x(x+1)(x-1)} = \dfrac{A}{x} + \dfrac{B}{x+1} + \dfrac{C}{x-1}$; $A = -1$, $B = 1$, $C = 2$

$$\int (x^2 + 1)dx - \int \frac{1}{x}dx + \int \frac{1}{x+1}dx + 2\int \frac{1}{x-1}dx$$

$$= \frac{1}{3}x^3 + x - \ln|x| + \ln|x+1| + 2\ln|x-1| + C = \frac{1}{3}x^3 + x + \ln\left|\frac{(x+1)(x-1)^2}{x}\right| + C$$

27. $\dfrac{2x^2 + 3}{x(x-1)^2} = \dfrac{A}{x} + \dfrac{B}{x-1} + \dfrac{C}{(x-1)^2}; \; A = 3, \; B = -1, \; C = 5$

$$3\int \frac{1}{x}dx - \int \frac{1}{x-1}dx + 5\int \frac{1}{(x-1)^2}dx = 3\ln|x| - \ln|x-1| - 5/(x-1) + C$$

29. $\dfrac{x^2 + x - 16}{(x+1)(x-3)^2} = \dfrac{A}{x+1} + \dfrac{B}{x-3} + \dfrac{C}{(x-3)^2}; \; A = -1, \; B = 2, \; C = -1$

$$-\int \frac{1}{x+1}dx + 2\int \frac{1}{x-3}dx - \int \frac{1}{(x-3)^2}dx$$

$$= -\ln|x+1| + 2\ln|x-3| + \frac{1}{x-3} + C = \ln\frac{(x-3)^2}{|x+1|} + \frac{1}{x-3} + C$$

31. $\dfrac{x^2}{(x+2)^3} = \dfrac{A}{x+2} + \dfrac{B}{(x+2)^2} + \dfrac{C}{(x+2)^3}; \; A = 1, \; B = -4, \; C = 4$

$$\int \frac{1}{x+2}dx - 4\int \frac{1}{(x+2)^2}dx + 4\int \frac{1}{(x+2)^3}dx = \ln|x+2| + \frac{4}{x+2} - \frac{2}{(x+2)^2} + C$$

33. $\dfrac{2x^2 - 1}{(4x-1)(x^2+1)} = \dfrac{A}{4x-1} + \dfrac{Bx+C}{x^2+1}; \; A = -14/17, \; B = 12/17, \; C = 3/17$

$$\int \frac{2x^2 - 1}{4x^3 - x^2 + 4x - 1}dx = -\frac{7}{34}\ln|4x-1| + \frac{6}{17}\ln(x^2+1) + \frac{3}{17}\tan^{-1}x + C$$

35. $\dfrac{1}{(x+2)(x-2)(x^2+4)} = \dfrac{A}{x+2} + \dfrac{B}{x-2} + \dfrac{Cx+D}{x^2+4}; \; A = -1/32, \; B = 1/32, \; C = 0, \; D = -1/8$

$$\int \frac{dx}{x^4 - 16} = \frac{1}{32}\ln\left|\frac{x-2}{x+2}\right| - \frac{1}{16}\tan^{-1}(x/2) + C$$

37. $\dfrac{x^3 + 3x^2 + x + 9}{(x^2+1)(x^2+3)} = \dfrac{Ax+B}{x^2+1} + \dfrac{Cx+D}{x^2+3}; \; A = 0, \; B = 3, \; C = 1, \; D = 0$

$$\int \frac{x^3 + 3x^2 + x + 9}{(x^2+1)(x^2+3)}dx = 3\tan^{-1}x + \frac{1}{2}\ln(x^2+3) + C$$

39. $\dfrac{x^3 - 3x^2 + 2x - 3}{x^2 + 1} = x - 3 + \dfrac{x}{x^2 + 1}$,

$$\int \dfrac{x^3 - 3x^2 + 2x - 3}{x^2 + 1}\,dx = \dfrac{1}{2}x^2 - 3x + \dfrac{1}{2}\ln(x^2 + 1) + C$$

41. $\dfrac{x^2 + 1}{(x^2 + 2x + 3)^2} = \dfrac{Ax + B}{x^2 + 2x + 3} + \dfrac{Cx + D}{(x^2 + 2x + 3)^2}$; $A = 0$, $B = 1$, $C = D = -2$

$$\int \dfrac{x^2 + 1}{(x^2 + 2x + 3)^2}\,dx = \int \dfrac{1}{(x + 1)^2 + 2}\,dx - \int \dfrac{2x + 2}{(x^2 + 2x + 3)^2}\,dx$$

$$= \dfrac{1}{\sqrt{2}}\tan^{-1}\dfrac{x + 1}{\sqrt{2}} + 1/(x^2 + 2x + 3) + C$$

43. let $x = \sin\theta$ to get $\displaystyle\int \dfrac{1}{x^2 + 4x - 5}\,dx$,

$$\dfrac{1}{(x + 5)(x - 1)} = \dfrac{A}{x + 5} + \dfrac{B}{x - 1}; \; A = -1/6, \; B = 1/6$$

$$-\dfrac{1}{6}\int \dfrac{1}{x + 5}\,dx + \dfrac{1}{6}\int \dfrac{1}{x - 1}\,dx = \dfrac{1}{6}\ln\left|\dfrac{x - 1}{x + 5}\right| + C = \dfrac{1}{6}\ln\left|\dfrac{\sin\theta - 1}{\sin\theta + 5}\right| + C$$

45. let $u = e^x$ to get $\displaystyle\int \dfrac{dx}{1 + e^x} = \int \dfrac{e^x\,dx}{e^x(1 + e^x)} = \int \dfrac{du}{u(1 + u)}$,

$$\dfrac{1}{u(1 + u)} = \dfrac{A}{u} + \dfrac{B}{1 + u}; \; A = 1, \; B = -1$$

$$\int \dfrac{du}{u(1 + u)} = \ln u - \ln(1 + u) + C = \ln \dfrac{e^x}{1 + e^x} + C$$

47. (a) $x^4 + 1 = (x^4 + 2x^2 + 1) - 2x^2 = (x^2 + 1)^2 - 2x^2$

$$= [(x^2 + 1) + \sqrt{2}x][(x^2 + 1) - \sqrt{2}x]$$

$$= (x^2 + \sqrt{2}x + 1)(x^2 - \sqrt{2}x + 1); a = \sqrt{2}, b = -\sqrt{2}$$

(b) $\dfrac{x}{(x^2 + \sqrt{2}x + 1)(x^2 - \sqrt{2}x + 1)} = \dfrac{Ax + B}{x^2 + \sqrt{2}x + 1} + \dfrac{Cx + D}{x^2 - \sqrt{2}x + 1}$;

$$A = 0, \; B = -\dfrac{\sqrt{2}}{4}, \; C = 0, \; D = \dfrac{\sqrt{2}}{4} \text{ so}$$

$$\int_0^1 \frac{x}{x^4+1}dx = -\frac{\sqrt{2}}{4}\int_0^1 \frac{1}{x^2+\sqrt{2}x+1}dx + \frac{\sqrt{2}}{4}\int_0^1 \frac{1}{x^2-\sqrt{2}x+1}dx$$

$$= -\frac{\sqrt{2}}{4}\int_0^1 \frac{1}{(x+\sqrt{2}/2)^2+1/2}dx + \frac{\sqrt{2}}{4}\int_0^1 \frac{1}{(x-\sqrt{2}/2)^2+1/2}dx$$

$$= -\frac{\sqrt{2}}{4}\int_{\sqrt{2}/2}^{1+\sqrt{2}/2} \frac{1}{u^2+1/2}du + \frac{\sqrt{2}}{4}\int_{-\sqrt{2}/2}^{1-\sqrt{2}/2} \frac{1}{u^2+1/2}du$$

$$= -\frac{1}{2}\tan^{-1}\sqrt{2}u\Big]_{\sqrt{2}/2}^{1+\sqrt{2}/2} + \frac{1}{2}\tan^{-1}\sqrt{2}u\Big]_{-\sqrt{2}/2}^{1-\sqrt{2}/2}$$

$$= -\frac{1}{2}\tan^{-1}(\sqrt{2}+1) + \frac{1}{2}\left(\frac{\pi}{4}\right) + \frac{1}{2}\tan^{-1}(\sqrt{2}-1) - \frac{1}{2}\left(-\frac{\pi}{4}\right)$$

$$= \frac{\pi}{4} - \frac{1}{2}[\tan^{-1}(\sqrt{2}+1) - \tan^{-1}(\sqrt{2}-1)]$$

$$= \frac{\pi}{4} - \frac{1}{2}[\tan^{-1}(1+\sqrt{2}) + \tan^{-1}(1-\sqrt{2})]$$

$$= \frac{\pi}{4} - \frac{1}{2}\tan^{-1}\left[\frac{(1+\sqrt{2})+(1-\sqrt{2})}{1-(1+\sqrt{2})(1-\sqrt{2})}\right] \quad \text{(Exercise 16, 8.1)}$$

$$= \frac{\pi}{4} - \frac{1}{2}\tan^{-1}1 = \frac{\pi}{4} - \frac{1}{2}\left(\frac{\pi}{4}\right) = \frac{\pi}{8}$$

49. $V = \pi \int_0^2 \dfrac{x^4}{(9-x^2)^2}dx, \quad \dfrac{x^4}{x^4-18x^2+81} = 1 + \dfrac{18x^2-81}{x^4-18x^2+81},$

$$\frac{18x^2-81}{(9-x^2)^2} = \frac{18x^2-81}{(x+3)^2(x-3)^2} = \frac{A}{x+3} + \frac{B}{(x+3)^2} + \frac{C}{x-3} + \frac{D}{(x-3)^2};$$

$$A = -\frac{9}{4}, B = \frac{9}{4}, C = \frac{9}{4}, D = \frac{9}{4}$$

$$V = \pi\left[x - \frac{9}{4}\ln|x+3| - \frac{9/4}{x+3} + \frac{9}{4}\ln|x-3| - \frac{9/4}{x-3}\right]_0^2 = \pi\left(\frac{19}{5} - \frac{9}{4}\ln 5\right)$$

51. $\displaystyle\int \frac{1}{y^2-5y+6}dy = \int dx, \quad \frac{1}{(y-3)(y-2)} = \frac{1}{y-3} - \frac{1}{y-2},$

$$\ln|y-3| - \ln|y-2| = x + C_1, \ln\left|\frac{y-3}{y-2}\right| = x + C_1, \frac{y-3}{y-2} = \pm e^{C_1}e^x = Ce^x,$$

$$y - 3 = Ce^x y - 2Ce^x, y(1-Ce^x) = 3 - 2Ce^x, y = \frac{3-2Ce^x}{1-Ce^x}.$$

53. $\int \dfrac{1}{y^2+y}dy = \int \dfrac{1}{t(t-1)}dt$, $\dfrac{1}{y(y+1)} = \dfrac{1}{y} - \dfrac{1}{y+1}$ and $\dfrac{1}{t(t-1)} = -\dfrac{1}{t} + \dfrac{1}{t-1}$,

$\ln|y| - \ln|y+1| = -\ln|t| + \ln|t-1| + C_1$, $\ln\left|\dfrac{y}{y+1}\right| = \ln\left|\dfrac{t-1}{t}\right| + C_1$,

$\dfrac{y}{y+1} = C_2\dfrac{t-1}{t}$; solve for y to get $y = \dfrac{C_2(t-1)/t}{1 - C_2(t-1)/t} = \dfrac{1}{Ct/(t-1) - 1} = \dfrac{t-1}{Ct - t + 1}$.

55. (a) $\int \dfrac{1}{ay - by^2}dy = \int dt$, $\dfrac{1}{y(a-by)} = \dfrac{1/a}{y} + \dfrac{b/a}{a-by}$,

$\dfrac{1}{a}\ln|y| - \dfrac{1}{a}\ln|a-by| = t + C_1$, $\dfrac{1}{a}\ln\left|\dfrac{y}{a-by}\right| = t + C_1$,

$\dfrac{y}{a-by} = C_2 e^{at}$, $y = aC_2 e^{at} - bC_2 e^{at} y$, $y = \dfrac{aC_2 e^{at}}{1 + bC_2 e^{at}} = \dfrac{a}{Ce^{-at} + b}$.

  (b) $\lim\limits_{t \to +\infty} y = a/b$

57. (a) $x^3 - 6x^2 + 11x - 6 = (x-1)(x-2)(x-3)$
  (b) $x^3 - 3x^2 + x - 20 = (x-4)(x^2 + x + 5)$
  (c) $x^4 - 5x^3 + 7x^2 - 5x + 6 = (x-2)(x-3)(x^2 + 1)$

59. $x^4 - 3x^3 - 7x^2 + 27x - 18 = (x-1)(x-2)(x-3)(x+3)$,

$\dfrac{1}{(x-1)(x-2)(x-3)(x+3)} = \dfrac{A}{x-1} + \dfrac{B}{x-2} + \dfrac{C}{x-3} + \dfrac{D}{x+3}$;

$A = 1/8$, $B = -1/5$, $C = 1/12$, $D = -1/120$

$\int \dfrac{dx}{x^4 - 3x^3 - 7x^2 + 27x - 18} = \dfrac{1}{8}\ln|x-1| - \dfrac{1}{5}\ln|x-2| + \dfrac{1}{12}\ln|x-3| - \dfrac{1}{120}\ln|x+3| + C$

61. (a) $\sqrt{2}$ is a positive root of $x^2 - 2 = 0$. The only possible positive rational roots of $x^2 - 2 = 0$ are the integers 1 and 2 neither of which satisfies the equation so $\sqrt{2}$ is not rational.
  (b) $\sqrt{a}$ is a positive root of $x^2 - a = 0$. The only possible positive rational roots of $x^2 - a = 0$ are the positive integers that divide a. If one of these satisfies the equation then $\sqrt{a}$ is an integer, otherwise $\sqrt{a}$ cannot be rational.

EXERCISE SET 9.7

1. $u = \sqrt{x-2}$, $x = u^2 + 2$, $dx = 2u\,du$

$\int 2u^2(u^2 + 2)du = 2\int(u^4 + 2u^2)du = \dfrac{2}{5}u^5 + \dfrac{4}{3}u^3 + C = \dfrac{2}{5}(x-2)^{5/2} + \dfrac{4}{3}(x-2)^{3/2} + C$

3. $u = \sqrt{x-4}$, $x = u^2 + 4$, $dx = 2u\,du$

$$\int_0^2 \frac{2u^2}{u^2+4}\,du = 2\int_0^2 \left[1 - \frac{4}{u^2+4}\right]du = 2u - 4\tan^{-1}(u/2)\Big]_0^2 = 4 - \pi$$

5. $u = 3 + \sqrt{x}$, $x = (u-3)^2$, $dx = 2(u-3)du$

$$\int_3^5 \frac{2(u-3)}{u}\,du = 2\int_3^5 (1 - 3/u)du = 2u - 6\ln|u|\Big]_3^5 = 4 - 6\ln(5/3)$$

7. $u = \sqrt{x^3+1}$, $x^3 = u^2 - 1$, $3x^2 dx = 2u\,du$

$$\frac{2}{3}\int u^2(u^2-1)du = \frac{2}{3}\int (u^4 - u^2)du = \frac{2}{15}u^5 - \frac{2}{9}u^3 + C = \frac{2}{15}(x^3+1)^{5/2} - \frac{2}{9}(x^3+1)^{3/2} + C$$

9. $u = x^{1/6}$, $x = u^6$, $dx = 6u^5 du$

$$\int \frac{6u^5}{u^3+u^2}\,du = 6\int \frac{u^3}{u+1}\,du = 6\int\left[u^2 - u + 1 - \frac{1}{u+1}\right]du$$
$$= 2x^{1/2} - 3x^{1/3} + 6x^{1/6} - 6\ \ln(x^{1/6}+1) + C$$

11. $u = v^{1/4}$, $v = u^4$, $dv = 4u^3 du$; $4\int \frac{1}{u(1-u)}\,du = 4\int\left[\frac{1}{u} + \frac{1}{1-u}\right]du = 4\ln\frac{v^{1/4}}{|1-v^{1/4}|} + C$

13. $u = t^{1/6}$, $t = u^6$, $dt = 6u^5 du$

$$6\int \frac{u^3}{u-1}\,du = 6\int\left[u^2 + u + 1 + \frac{1}{u-1}\right]du = 2t^{1/2} + 3t^{1/3} + 6t^{1/6} + 6\ \ln|t^{1/6}-1| + C$$

15. $u = \sqrt{1+x^2}$, $x^2 = u^2 - 1$, $2x\,dx = 2u\,du$, $x\,dx = u\,du$

$$\int (u^2 - 1)du = \frac{1}{3}(1+x^2)^{3/2} - (1+x^2)^{1/2} + C$$

17. $z = \sqrt{x}$, $x = z^2$, $dx = 2z\,dz$, $2\int z\sin z\,dz$; use integration by parts:

$u = z$, $dv = \sin z\,dz$, $du = dz$, $v = -\cos z$

$$2\int z\sin z\,dz = 2\left(-z\cos z + \int \cos z\,dz\right) = -2z\cos z + 2\sin z + C$$

so $\int \sin\sqrt{x}dx = -2\sqrt{x}\cos\sqrt{x} + 2\sin\sqrt{x} + C$

19. $u = \sqrt{e^x + 1}, \; e^x = u^2 - 1, \; x = \ln(u^2 - 1), \; dx = \dfrac{2u}{u^2 - 1} du$

$$\int \frac{2}{u^2 - 1} du = \int \left[\frac{1}{u-1} - \frac{1}{u+1} \right] du = \ln|u-1| - \ln|u+1| + C = \ln \frac{\sqrt{e^x+1}-1}{\sqrt{e^x+1}+1} + C$$

21. $\displaystyle\int \frac{1}{1 + \dfrac{2u}{1+u^2} + \dfrac{1-u^2}{1+u^2}} \frac{2}{1+u^2} du = \int \frac{1}{u+1} du = \ln|\tan(x/2) + 1| + C$

23. $u = \tan(\theta/2), \; \displaystyle\int \frac{d\theta}{1 - \cos\theta} = \int \frac{1}{u^2} du = -\frac{1}{u} + C = -\cot(\theta/2) + C,$

$$\int_{\pi/2}^{\pi} \frac{d\theta}{1 - \cos\theta} = \left. -\cot(\theta/2) \right]_{\pi/2}^{\pi} = 1$$

25. $u = \tan(x/2), \; 2\displaystyle\int \frac{1-u^2}{(3u^2+1)(u^2+1)} du$

$$\frac{1-u^2}{(3u^2+1)(u^2+1)} = \frac{(0)u+2}{3u^2+1} + \frac{(0)u-1}{u^2+1} = \frac{2}{3u^2+1} - \frac{1}{u^2+1},$$

$$\int \frac{\cos x}{2 - \cos x} dx = \frac{4}{\sqrt{3}} \tan^{-1}[\sqrt{3}\tan(x/2)] - x + C$$

27. **(a)** $\displaystyle\int \sec x \, dx = \int \frac{1}{\cos x} dx = \int \frac{2}{1-u^2} du = \ln\left|\frac{1+u}{1-u}\right| + C = \ln\left|\frac{1+\tan(x/2)}{1-\tan(x/2)}\right| + C$

(b) $\dfrac{1+\tan(x/2)}{1-\tan(x/2)} = \dfrac{\tan(\pi/4)+\tan(x/2)}{1-\tan(\pi/4)\tan(x/2)} = \tan(\pi/4 + x/2)$ (trig identity)

(c) $\dfrac{1+\tan(x/2)}{1-\tan(x/2)} = \dfrac{\cos(x/2)+\sin(x/2)}{\cos(x/2)-\sin(x/2)} = \dfrac{[\cos(x/2)+\sin(x/2)]^2}{\cos^2(x/2)-\sin^2(x/2)}$

$$= \frac{\cos^2(x/2) + 2\sin(x/2)\cos(x/2) + \sin^2(x/2)}{\cos x}$$

$$= \frac{1+\sin x}{\cos x} = \sec x + \tan x$$

29. Let $u = \tanh(x/2)$ then $\cosh(x/2) = 1/\operatorname{sech}(x/2) = 1/\sqrt{1 - \tanh^2(x/2)} = 1/\sqrt{1 - u^2}$,

$\sinh(x/2) = \tanh(x/2)\cosh(x/2) = u/\sqrt{1 - u^2}$

so $\sinh x = 2\sinh(x/2)\cosh(x/2) = 2u/(1 - u^2)$,

$\cosh x = \cosh^2(x/2) + \sinh^2(x/2) = (1 + u^2)/(1 - u^2)$, $x = 2\tanh^{-1} u$, $dx = [1/(1 - u^2)]du$

$$\int \frac{dx}{2\cosh x + \sinh x} = \int \frac{1}{u^2 + u + 1} du$$

$$= \frac{2}{\sqrt{3}}\tan^{-1}\frac{2u + 1}{\sqrt{3}} + C = \frac{2}{\sqrt{3}}\tan^{-1}\frac{2\tanh(x/2) + 1}{\sqrt{3}} + C$$

31. $dx = -(1/u^2)du$, $-\displaystyle\int \frac{u}{\sqrt{3u^2 - 1}}du = -\frac{1}{3}\sqrt{3u^2 - 1} + C = -\frac{\sqrt{3 - x^2}}{3x} + C$

33. $dx = -(1/u^2)du$, $-\displaystyle\int u\sqrt{1 - 5u^2}du = \frac{1}{15}(1 - 5u^2)^{3/2} + C = \frac{(x^2 - 5)^{3/2}}{15x^3} + C$

EXERCISE SET 9.8

1. exact value $= 14/3 \approx 4.666666667$

   (a) 4.667600663, $|E_M| \approx 0.000933996$
   (b) 4.664795679, $|E_T| \approx 0.001870988$
   (c) 4.666651630, $|E_S| \approx 0.000015037$

3. exact value $= 2$

   (a) 2.008248408, $|E_M| \approx 0.008248408$
   (b) 1.983523538, $|E_T| \approx 0.016476462$
   (c) 2.000109517, $|E_S| \approx 0.000109517$

5. exact value $= e^{-1} - e^{-3} \approx 0.318092373$

   (a) 0.317562837, $|E_M| \approx 0.000529536$
   (b) 0.319151975, $\quad|E_T| \quad \approx \quad 0.001059602$
   (c) 0.318095187, $|E_S| \approx 0.000002814$

7. $f(x) = \sqrt{x + 1}$, $f''(x) = -\frac{1}{4}(x + 1)^{-3/2}$, $f^{(4)}(x) = -\frac{15}{16}(x + 1)^{-7/2}$; $K_2 = 1/4$, $K_4 = 15/16$

   (a) $|E_M| \le \dfrac{27}{2400}(1/4) \approx 0.002812500$ **(b)** $|E_T| \le \dfrac{27}{1200}(1/4) \approx 0.005625000$

   (c) $|E_S| \le \dfrac{243}{180 \times 10^4}(15/16) \approx 0.000126563$

9. $f(x) = \sin x$, $f''(x) = -\sin x$, $f^{(4)}(x) = \sin x$; $K_2 = K_4 = 1$

   (a) $|E_M| \le \dfrac{\pi^3}{2400}(1) \approx 0.012919282$ **(b)** $|E_T| \le \dfrac{\pi^3}{1200}(1) \approx 0.025838564$

   (c) $|E_S| \le \dfrac{\pi^5}{180 \times 10^4}(1) \approx 0.000170011$

11. $f(x) = e^{-x}$, $f''(x) = f^{(4)}(x) = e^{-x}$; $K_2 = K_4 = e^{-1}$

(a) $|E_M| \leq \dfrac{8}{2400}(e^{-1}) \approx 0.001226265$

(b) $|E_T| \leq \dfrac{8}{1200}(e^{-1}) \approx 0.002452530$

(c) $|E_S| \leq \dfrac{32}{180 \times 10^4}(e^{-1}) \approx 0.000006540$

13. (a) $n > \left[\dfrac{(27)(1/4)}{(24)(5 \times 10^{-4})}\right]^{1/2} \approx 23.7$; $n = 24$

(b) $n > \left[\dfrac{(27)(1/4)}{(12)(5 \times 10^{-4})}\right]^{1/2} \approx 33.5$; $n = 34$

(c) $n > \left[\dfrac{(243)(15/16)}{(180)(5 \times 10^{-4})}\right]^{1/4} \approx 7.1$; $n = 8$

15. (a) $n > \left[\dfrac{(\pi^3)(1)}{(24)(10^{-3})}\right]^{1/2} \approx 35.9$; $n = 36$

(b) $n > \left[\dfrac{(\pi^3)(1)}{(12)(10^{-3})}\right]^{1/2} \approx 50.8$; $n = 51$

(c) $n > \left[\dfrac{(\pi^5)(1)}{(180)(10^{-3})}\right]^{1/4} \approx 6.4$; $n = 8$

17. (a) $n > \left[\dfrac{(8)(e^{-1})}{(24)(10^{-6})}\right]^{1/2} \approx 350.2$; $n = 351$

(b) $n > \left[\dfrac{(8)(e^{-1})}{(12)(10^{-6})}\right]^{1/2} \approx 495.2$; $n = 496$

(c) $n > \left[\dfrac{(32)(e^{-1})}{(180)(10^{-6})}\right]^{1/4} \approx 15.99$; $n = 16$

19. (a) 0.747130878
(b) 0.746210796
(c) 0.746824948

21. (a) 2.129469966
(b) 2.130644002
(c) 2.129861595

23. (a) 0.809253858
(b) 0.795924733
(c) 0.805376152

25. (a) 3.142425985, $|E_M| \approx 0.000833331$
(b) 3.139925989, $|E_T| \approx 0.001666665$
(c) 3.141592614, $|E_S| \approx 0.000000040$

27. $S_{14} = 0.693147984$, $|E_S| \approx 0.000000803 = 8.03 \times 10^{-7}$; the method used in Example 5 results in a value of n which ensures that the magnitude of the error will be less than 10^{-6}, this is not necessarily the *smallest* value of n.

29. $f(x) = x \sin x$, $f''(x) = 2\cos x - x \sin x$, $|f''(x)| \leq 2|\cos x| + |x|\,|\sin x| \leq 2 + 2 = 4$ so $K_2 \leq 4$,

$n > \left[\dfrac{(8)(4)}{(24)(10^{-4})}\right]^{1/2} \approx 115.5$, $n = 116$.

31. $f(x) = \sqrt{x}$, $f''(x) = -\dfrac{1}{4x^{3/2}}$, $\lim\limits_{x \to 0^+} |f''(x)| = +\infty$

33. $L = \int_0^\pi \sqrt{1 + \cos^2 x}\, dx \approx 3.820$

35.

$t(\text{sec})$	0	0	10	15	20
$v\ (\text{mi/hr})$	0	40	60	73	84
$v\ (\text{ft/sec})$	0	58.67	88	107.07	123.2

$$\int_0^{20} v\, dt \approx \frac{20}{(3)(4)}[0 + 4(58.67) + 2(88) + 4(107.07) + 123.2] \approx 1604 \text{ ft}$$

37. $\int_0^{180} v\, dt \approx \frac{180}{(3)(6)}[0.00 + 4(0.03) + 2(0.08) + 4(0.16) + 2(0.27) + 4(0.42) + 0.65] = 37.9 \text{ mi}$

39. $V = \int_0^{16} \pi r^2 dy = \pi \int_0^{16} r^2 dy \approx \pi \frac{16}{(3)(4)}[(8.5)^2 + 4(11.5)^2 + 2(13.8)^2 + 4(15.4)^2 + (16.8)^2]$
$$\approx 9270 \text{ cm}^3 \approx 9.3 \text{ L}$$

41. right endpoint, trapezoidal, midpoint, left endpoint.

TECHNOLOGY EXERCISES 9

1. Let $u = x^4$ to get $\dfrac{1}{4} \displaystyle\int \dfrac{1}{\sqrt{1 - u^2}} du = \dfrac{1}{4} \sin^{-1} u + C = \dfrac{1}{4} \sin^{-1}(x^4) + C$

3. $\dfrac{\sqrt{1+x} + \sqrt{1-x}}{\sqrt{1+x} - \sqrt{1-x}} = \dfrac{(\sqrt{1+x} + \sqrt{1-x})^2}{2x} = \dfrac{1 + \sqrt{1 - x^2}}{x}$

$$\int \left[\frac{1}{x} + \frac{\sqrt{1 - x^2}}{x} \right] dx = \int \frac{1}{x} dx + \int \frac{\sqrt{1 - x^2}}{x} dx = \ln|x| + \int \frac{u^2}{u^2 - 1} du \quad (u = \sqrt{1 - x^2})$$

$$= \ln|x| + \int \left[1 + \frac{1}{u^2 - 1} \right] du = \ln|x| + u + \frac{1}{2} \ln \frac{1 - u}{1 + u} + C$$

$$= \ln|x| + \sqrt{1 - x^2} + \frac{1}{2} \ln \frac{1 - \sqrt{1 - x^2}}{1 + \sqrt{1 - x^2}} + C$$

$$= \sqrt{1 - x^2} + \frac{1}{2} \ln(x^2) + \frac{1}{2} \ln \frac{(1 - \sqrt{1 - x^2})^2}{x^2} + C$$

$$= \sqrt{1 - x^2} + \ln(1 - \sqrt{1 - x^2}) + C$$

5. $x - 1 = 1/u$, $x = 1 + 1/u$, $dx = -(1/u^2)du$

$$\int \frac{\sqrt{x+1}}{(x-1)^{5/2}} dx = -\int \sqrt{2u+1} du = -\frac{1}{3}(2u+1)^{3/2} + C = -\frac{1}{3}\left[\frac{x+1}{x-1}\right]^{3/2} + C$$

7. $\displaystyle\int \frac{1}{x^6(3+2x^{-5})} dx = -\frac{1}{10}\int \frac{1}{u} du \quad (u = 3+2x^{-5}) = -\frac{1}{10}\ln|u| + C = -\frac{1}{10}\ln|3+2x^{-5}| + C$

9. $\displaystyle\int \sqrt{x - \sqrt{x^2 - 4}} dx = \frac{1}{\sqrt{2}}\int (\sqrt{x+2} - \sqrt{x-2}) dx = \frac{\sqrt{2}}{3}[(x+2)^{3/2} - (x-2)^{3/2}] + C$

11. (a) With $u = \sqrt{x}$:

$$\int \frac{1}{\sqrt{x}\sqrt{2-x}} dx = 2\int \frac{1}{\sqrt{2-u^2}} du = 2\sin^{-1}(u/\sqrt{2}) + C = 2\sin^{-1}(\sqrt{x/2}) + C;$$

with $u = \sqrt{2-x}$:

$$\int \frac{1}{\sqrt{x}\sqrt{2-x}} dx = -2\int \frac{1}{\sqrt{2-u^2}} du = -2\sin^{-1}(u/\sqrt{2}) + C = -2\sin^{-1}(\sqrt{2-x}/\sqrt{2}) + C;$$

completing the square:

$$\int \frac{1}{\sqrt{1-(x-1)^2}} dx = \sin^{-1}(x-1) + C;$$

using a CAS: answers may vary.

(b) For the first three results in part (a) the antiderivatives differ by a constant, in particular
$2\sin^{-1}(\sqrt{x/2}) = \pi - 2\sin^{-1}(\sqrt{2-x}/\sqrt{2}) = \pi/2 + \sin^{-1}(x-1)$.

13. (a) The maximum value of $|f''(x)|$ is 3.844880.
    (b) $n = 18$
    (c) 0.904741

15. (a) The maximum value of $|f^{(4)}(x)|$ is 42.551816.
    (b) $n = 8$
    (c) 0.904524

17. $\displaystyle \frac{2}{x-2} - \frac{1}{(x-2)^2} + \frac{3x+1}{x^2 - x + 5}$

19. (a) $b = 1/2$ (b) $a = 0$ (c) $b = 1$

CHAPTER 10
Improper Integrals; L'Hôpital's Rule

EXERCISE SET 10.1

1. $\lim\limits_{\ell \to +\infty} (-e^{-x})\Big]_0^\ell = \lim\limits_{\ell \to +\infty} (-e^{-\ell} + 1) = 1$

3. $\lim\limits_{\ell \to +\infty} 2\sqrt{x}\Big]_1^\ell = \lim\limits_{\ell \to +\infty} 2(\sqrt{\ell} - 1) = +\infty$, divergent

5. $\lim\limits_{\ell \to +\infty} \ln\dfrac{x-1}{x+1}\Big]_4^\ell = \lim\limits_{\ell \to +\infty} \left(\ln\dfrac{\ell-1}{\ell+1} - \ln\dfrac{3}{5} \right) = -\ln\dfrac{3}{5} = \ln\dfrac{5}{3}$

7. $\lim\limits_{\ell \to +\infty} -\dfrac{1}{2\ln^2 x}\Big]_e^\ell = \lim\limits_{\ell \to +\infty} \left[-\dfrac{1}{2\ln^2 \ell} + \dfrac{1}{2} \right] = \dfrac{1}{2}$

9. $\lim\limits_{\ell \to +\infty} -\dfrac{1}{2(x^2+1)}\Big]_a^\ell = \lim\limits_{\ell \to +\infty} \left[-\dfrac{1}{2(\ell^2+1)} + \dfrac{1}{2(a^2+1)} \right] = \dfrac{1}{2(a^2+1)}$

11. $\lim\limits_{\ell \to -\infty} -\dfrac{1}{4(2x-1)^2}\Big]_\ell^0 = \lim\limits_{\ell \to -\infty} \dfrac{1}{4}[-1 + 1/(2\ell-1)^2] = -1/4$

13. $\lim\limits_{\ell \to -\infty} \dfrac{1}{3}e^{3x}\Big]_\ell^0 = \lim\limits_{\ell \to -\infty} \left[\dfrac{1}{3} - \dfrac{1}{3}e^{3\ell} \right] = \dfrac{1}{3}$

15. $\displaystyle\int_{-\infty}^{+\infty} x^3\,dx$ converges if $\displaystyle\int_{-\infty}^{0} x^3\,dx$ and $\displaystyle\int_0^{+\infty} x^3\,dx$ both converge; it diverges if either (or both)

    diverge. $\displaystyle\int_0^{+\infty} x^3\,dx = \lim\limits_{\ell \to +\infty} \dfrac{1}{4}x^4\Big]_0^\ell = \lim\limits_{\ell \to +\infty} \dfrac{1}{4}\ell^4 = +\infty$ so $\displaystyle\int_{-\infty}^{+\infty} x^3\,dx$ is divergent.

17. $\displaystyle\int_0^{+\infty} \dfrac{x}{(x^2+3)^2}\,dx = \lim\limits_{\ell \to +\infty} -\dfrac{1}{2(x^2+3)}\Big]_0^\ell = \lim\limits_{\ell \to +\infty} \dfrac{1}{2}[-1/(\ell^2+3) + 1/3] = \dfrac{1}{6}$,

    similarly $\displaystyle\int_{-\infty}^0 \dfrac{x}{(x^2+3)^2}\,dx = -1/6$ so $\displaystyle\int_{-\infty}^\infty \dfrac{x}{(x^2+3)^2}\,dx = 1/6 + (-1/6) = 0$

19. $\lim\limits_{\ell \to 3^+} -\dfrac{1}{x-3}\Big]_{\ell}^{4} = \lim\limits_{\ell \to 3^+} \left[-1 + \dfrac{1}{\ell - 3}\right] = +\infty$, divergent

21. $\lim\limits_{\ell \to \pi/2^-} -\ln(\cos x)\Big]_{0}^{\ell} = \lim\limits_{\ell \to \pi/2^-} -\ln(\cos \ell) = +\infty$, divergent

23. $\lim\limits_{\ell \to 1^-} \sin^{-1} x\Big]_{0}^{\ell} = \lim\limits_{\ell \to 1^-} \sin^{-1} \ell = \pi/2$

25. $\lim\limits_{\ell \to \pi/6^-} -\sqrt{1 - 2\sin x}\Big]_{0}^{\ell} = \lim\limits_{\ell \to \pi/6^-} (-\sqrt{1 - 2\sin \ell} + 1) = 1$

27. $\displaystyle\int_{0}^{2} \dfrac{dx}{x-2} = \lim\limits_{\ell \to 2^-} \ln|x - 2|\Big]_{0}^{\ell} = \lim\limits_{\ell \to 2^-} (\ln|\ell - 2| - \ln 2) = -\infty$, divergent

29. $\displaystyle\int_{0}^{8} x^{-1/3} dx = \lim\limits_{\ell \to 0^+} \dfrac{3}{2} x^{2/3}\Big]_{\ell}^{8} = \lim\limits_{\ell \to 0^+} \dfrac{3}{2}(4 - \ell^{2/3}) = 6,$

$\displaystyle\int_{-1}^{0} x^{-1/3} dx = \lim\limits_{\ell \to 0^-} \dfrac{3}{2} x^{2/3}\Big]_{-1}^{\ell} = \lim\limits_{\ell \to 0^-} \dfrac{3}{2}(\ell^{2/3} - 1) = -3/2$

so $\displaystyle\int_{-1}^{8} x^{-1/3} dx = 6 + (-3/2) = 9/2$

31. Define $\displaystyle\int_{0}^{+\infty} \dfrac{1}{x^2} dx = \int_{0}^{a} \dfrac{1}{x^2} dx + \int_{a}^{+\infty} \dfrac{1}{x^2} dx$ where $a > 0$; take $a = 1$ for convenience,

$\displaystyle\int_{0}^{1} \dfrac{1}{x^2} dx = \lim\limits_{\ell \to 0^+} (-1/x)\Big]_{\ell}^{1} = \lim\limits_{\ell \to 0^+} (1/\ell - 1) = +\infty$ so $\displaystyle\int_{0}^{+\infty} \dfrac{1}{x^2} dx$ is divergent.

33. $\displaystyle\int_{0}^{+\infty} \dfrac{e^{-\sqrt{x}}}{\sqrt{x}} dx = 2\int_{0}^{+\infty} e^{-u} du = 2\lim\limits_{\ell \to +\infty} (-e^{-u})\Big]_{0}^{\ell} = 2\lim\limits_{\ell \to +\infty} (1 - e^{-\ell}) = 2$

35. $\displaystyle\int_{0}^{+\infty} \dfrac{e^{-x}}{\sqrt{1 - e^{-x}}} dx = \int_{0}^{1} \dfrac{du}{\sqrt{u}} = \lim\limits_{\ell \to 0^+} 2\sqrt{u}\Big]_{\ell}^{1} = \lim\limits_{\ell \to 0^+} 2(1 - \sqrt{\ell}) = 2$

37. $\displaystyle\int_{0}^{+\infty} e^{-ax} dx = \lim\limits_{\ell \to +\infty} -\dfrac{1}{a} e^{-ax}\Big]_{0}^{\ell} = \lim\limits_{\ell \to +\infty} \left[-\dfrac{1}{a} e^{-a\ell} + \dfrac{1}{a}\right] = \dfrac{1}{a} = 5, \; a = \dfrac{1}{5}$

39. $u = \sqrt{x}, \; \displaystyle\int_{0}^{+\infty} \dfrac{e^{-x}}{\sqrt{x}} dx = 2\int_{0}^{+\infty} e^{-u^2} du = 2(\sqrt{\pi}/2) = \sqrt{\pi}$

41. $u = \sqrt{x}$, $\displaystyle\int_0^{+\infty} \frac{\sin x}{\sqrt{x}} dx = 2 \int_0^{+\infty} \sin(u^2) du = \sqrt{\pi/2}$

43. **(a)** $\displaystyle\int_0^{+\infty} \cos x\, dx = \lim_{\ell \to +\infty} \sin x \Big]_0^{\ell} = \lim_{\ell \to +\infty} \sin \ell$ which does not exist and does not become infinite.

(b) $u = \sqrt{x}$, $\displaystyle\int_0^{+\infty} \frac{\cos \sqrt{x}}{\sqrt{x}} dx = 2 \int_0^{+\infty} \cos u\, du$; $\displaystyle\int_0^{+\infty} \cos u\, du$ diverges

(c) $u = 1/x$, $\displaystyle\int_0^1 \frac{\cos(1/x)}{x^2} dx = -\int_{+\infty}^1 \cos u\, du = \int_1^{+\infty} \cos u\, du$ which diverges

45. $\displaystyle\lim_{\ell \to +\infty} \int_0^{\ell} e^{-x} \cos x\, dx = \lim_{\ell \to +\infty} \frac{1}{2} e^{-x} (\sin x - \cos x) \Big]_0^{\ell} = \lim_{\ell \to +\infty} \frac{1}{2} [e^{-\ell}(\sin \ell - \cos \ell) + 1]$

but both $e^{-\ell} \sin \ell$ and $e^{-\ell} \cos \ell \to 0$ as $\ell \to +\infty$ (by the Squeezing Theorem because

$-e^{-\ell} \le e^{-\ell} \sin \ell \le e^{-\ell}$ and $-e^{-\ell} \le e^{-\ell} \cos \ell \le e^{-\ell}$) so $\displaystyle\int_0^{+\infty} e^{-x} \cos x\, dx = 1/2$.

47. If $p = 1$, $\displaystyle\int_0^1 \frac{dx}{x} = \lim_{\ell \to 0^+} \ln x \Big]_{\ell}^1 = +\infty$;

if $p \ne 1$, $\displaystyle\int_0^1 \frac{dx}{x^p} = \lim_{\ell \to 0^+} \frac{x^{1-p}}{1-p} \Big]_{\ell}^1 = \lim_{\ell \to 0^+} [(1 - \ell^{1-p})/(1-p)] = \begin{cases} 1/(1-p), & p < 1 \\ +\infty, & p > 1 \end{cases}$

49. **(a)** $\displaystyle\int_2^{+\infty} \frac{x}{x^5 + 1} dx \le \int_2^{+\infty} \frac{dx}{x^4} = \lim_{\ell \to +\infty} -\frac{1}{3x^3} \Big]_2^{\ell} = 1/24$

(b) $\displaystyle\int_1^{+\infty} e^{-x^2} dx \le \int_1^{+\infty} x e^{-x^2} dx = \lim_{\ell \to +\infty} -\frac{1}{2} e^{-x^2} \Big]_1^{\ell} = \frac{1}{2} e^{-1}$

51. $A = \displaystyle\int_0^{+\infty} e^{-3x} dx = \lim_{\ell \to +\infty} -\frac{1}{3} e^{-3x} \Big]_0^{\ell} = \lim_{\ell \to +\infty} \frac{1}{3}(1 - e^{-3\ell}) = 1/3$.

53. **(a)** $V = \displaystyle\int_1^{+\infty} \frac{\pi}{x^2} dx = \lim_{\ell \to +\infty} -\pi/x \Big]_1^{\ell} = \pi$,

(b) $S = \displaystyle\int_1^{+\infty} 2\pi(1/x)\sqrt{1 + 1/x^4} dx$, but $\sqrt{1 + 1/x^4} \ge 1$ if $x \ge 1$ so

$(2\pi/x)\sqrt{1 + 1/x^4} \ge 2\pi/x$, $\displaystyle\lim_{\ell \to +\infty} \int_1^{\ell} (2\pi/x) dx = +\infty$; S is infinite.

55. $\displaystyle\int \frac{dx}{(r^2+x^2)^{3/2}} = \frac{1}{r^2}\int \cos\theta\, d\theta\,(x=r\tan\theta) = \frac{1}{r^2}\sin\theta + C = \frac{x}{r^2\sqrt{r^2+x^2}} + C$

$\displaystyle\text{so } u = \frac{2\pi NIr}{k}\lim_{\ell\to+\infty}\frac{x}{r^2\sqrt{r^2+x^2}}\Bigg]_a^\ell = \frac{2\pi NI}{kr}\lim_{\ell\to+\infty}(\ell/\sqrt{r^2+\ell^2} - a/\sqrt{r^2+a^2})$

$\displaystyle = \frac{2\pi NI}{kr}(1 - a/\sqrt{r^2+a^2})$

57. (a) $\displaystyle\int_{4000}^{4000+\ell} 9.6\times 10^{10}x^{-2}dx$

(b) $\displaystyle\int_{4000}^{+\infty} 9.6\times 10^{10}x^{-2}dx = \lim_{\ell\to+\infty} -9.6\times 10^{10}/x\Bigg]_{4000}^{\ell} = 2.4\times 10^7$

59. $\displaystyle \mathcal{L}\{1\} = \int_0^{+\infty} e^{-st}dt = \lim_{\ell\to+\infty} -\frac{1}{s}e^{-st}\Bigg]_0^\ell = \frac{1}{s}$

61. $\displaystyle \mathcal{L}\{\sin t\} = \int_0^{+\infty} e^{-st}\sin t\, dt = \lim_{\ell\to+\infty}\frac{e^{-st}}{s^2+1}(-s\sin t - \cos t)\Bigg]_0^\ell = \frac{1}{s^2+1}$

63. $\displaystyle 2\int_0^1 \cos(u^2)du \approx 1.809$

65. (a) $\displaystyle\int_0^4 \frac{1}{x^6+1}dx \approx 1.047;\ \pi/3 \approx 1.047$

(b) $\displaystyle\int_0^{+\infty}\frac{1}{x^6+1}dx = \int_0^4 \frac{1}{x^6+1}dx + \int_4^{+\infty}\frac{1}{x^6+1}dx$ so

$\displaystyle E = \int_4^{+\infty}\frac{1}{x^6+1}dx < \int_4^{+\infty}\frac{1}{x^6}dx = \frac{1}{5(4)^5} < 2\times 10^{-4}.$

EXERCISE SET 10.2

1. $\displaystyle\lim_{x\to 1}\frac{1/x}{1} = 1$ **3.** $\displaystyle\lim_{x\to 0}\frac{e^x}{\cos x} = 1$

5. $\displaystyle\lim_{\theta\to 0}\frac{\sec^2\theta}{1} = 1$ **7.** $\displaystyle\lim_{x\to 1}\frac{1/x}{\pi\sec^2\pi x} = 1/\pi$

9. $\displaystyle\lim_{x\to\pi^+}\frac{\cos x}{1} = -1$

11. $\displaystyle\lim_{x\to\pi/2^-} \frac{-\sin x}{-(1/2)(\pi/2-x)^{-1/2}} = \lim_{x\to\pi/2^-} 2(\sin x)\sqrt{\pi/2-x} = 0$

13. $\displaystyle\lim_{x\to0} \frac{e^x - e^{-x}}{2\sin 2x} = \lim_{x\to0} \frac{e^x + e^{-x}}{4\cos 2x} = 1/2$ **15.** $\displaystyle\lim_{x\to0} \frac{3\cos 3x}{2\cosh 2x} = 3/2$

17. $\displaystyle\lim_{x\to\pi/2} \frac{2\cos 2x}{8x} = -\frac{1}{2\pi}$

19. $\displaystyle\lim_{x\to0} \frac{1 - 1/(x+1)}{2\sin 2x} = \lim_{x\to0} \frac{x}{2(x+1)\sin 2x} = \lim_{x\to0} \frac{1}{4(x+1)\cos 2x + 2\sin 2x} = 1/4$

21. $\displaystyle\lim_{x\to0} \frac{-2x + 2\sin x}{4x^3} = \lim_{x\to0} \frac{-2 + 2\cos x}{12x^2} = \lim_{x\to0} -\frac{\sin x}{12x} = -1/12$

23. $\displaystyle\lim_{x\to0} \frac{x - \sin x}{x^3} = \lim_{x\to0} \frac{1 - \cos x}{3x^2} = \lim_{x\to0} \frac{\sin x}{6x} = 1/6$

25. $\displaystyle\lim_{x\to0} \frac{ae^{ax} - be^{bx}}{1} = a - b$

27. $\displaystyle\lim_{x\to0} \frac{1 - \sec^2 x}{\cos x - 1} = \lim_{x\to0} \frac{\cos^2 x - 1}{\cos^2 x(\cos x - 1)} = \lim_{x\to0} \frac{\cos x + 1}{\cos^2 x} = 2$

29. $\displaystyle\lim_{x\to+\infty} \frac{-\dfrac{1}{1+x^2}}{\dfrac{1}{1+1/x^2}\left(-\dfrac{2}{x^3}\right)} = \lim_{x\to+\infty} \frac{x}{2} = +\infty$

31. $\displaystyle\lim_{x\to0^+} \frac{-\sin x/\cos x}{-3\sin 3x/\cos 3x} = \lim_{x\to0^+} \frac{\tan x}{3\tan 3x} = \lim_{x\to0^+} \frac{\sec^2 x}{9\sec^2 3x} = 1/9$

33. **(a)** L'Hôpital's rule does not apply to the problem $\displaystyle\lim_{x\to1} \frac{3x^2 - 2x + 1}{3x^2 - 2x}$ because it is not a $\dfrac{0}{0}$ form

 (b) $\displaystyle\lim_{x\to1} \frac{3x^2 - 2x + 1}{3x^2 - 2x} = 2$

35. $\displaystyle\lim_{x\to0} \frac{k + \cos \ell x}{x^2}$ does not exist if $k \neq -1$ so suppose $k = -1$, then

 $\displaystyle\lim_{x\to0} \frac{-1 + \cos \ell x}{x^2} = \lim_{x\to0} \frac{-\ell \sin \ell x}{2x} = \lim_{x\to0} \frac{-\ell^2 \cos \ell x}{2} = -\ell^2/2 = -4$ if $\ell^2 = 8$, $\ell = \pm2\sqrt{2}$

37. $\lim\limits_{x\to 1}\dfrac{\ln x}{x^4-1}=\lim\limits_{x\to 1}\dfrac{1/x}{4x^3}=\dfrac{1}{4}; \lim\limits_{x\to 1}\sqrt{\dfrac{\ln x}{x^4-1}}=\sqrt{\lim\limits_{x\to 1}\dfrac{\ln x}{x^4-1}}=\dfrac{1}{2}$

39. $\lim\limits_{x\to+\infty}\dfrac{\ln\left(\dfrac{x+1}{x-1}\right)}{1/x}=\lim\limits_{x\to+\infty}\dfrac{-2/(x^2-1)}{-1/x^2}=\lim\limits_{x\to+\infty}\dfrac{2x^2}{x^2-1}=\lim\limits_{x\to+\infty}\dfrac{4x}{2x}=2$

41. $\lim\limits_{x\to 0+}\dfrac{\sin(1/x)}{(\sin x)/x}$, $\lim\limits_{x\to 0+}\dfrac{\sin x}{x}=1$ but $\lim\limits_{x\to 0+}\sin(1/x)$ does not exist because $\sin(1/x)$ oscillates between -1 and 1 as $x\to+\infty$, so $\lim\limits_{x\to 0+}\dfrac{x\sin(1/x)}{\sin x}$ does not exist.

43. $T(\theta)=\dfrac{1}{2}\sin\theta(1-\cos\theta),$

$S(\theta)=\dfrac{1}{2}(\theta-\sin\theta),$

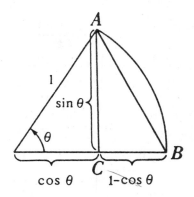

$$\lim_{\theta\to 0+}\frac{T(\theta)}{S(\theta)}=\lim_{\theta\to 0+}\frac{\sin\theta-\sin\theta\cos\theta}{\theta-\sin\theta}$$

$$=\lim_{\theta\to 0+}\frac{\cos\theta+\sin^2\theta-\cos^2\theta}{1-\cos\theta}$$

$$=\lim_{\theta\to 0+}\frac{\cos\theta-\cos 2\theta}{1-\cos\theta}$$

$$=\lim_{\theta\to 0+}\frac{-\sin\theta+2\sin 2\theta}{\sin\theta}$$

$$=\lim_{\theta\to 0+}\frac{-\cos\theta+4\cos 2\theta}{\cos\theta}=3$$

EXERCISE SET 10.3

1. $\lim\limits_{x\to+\infty}\dfrac{1/x}{1}=0$

3. $\lim\limits_{x\to 0+}\dfrac{-\csc^2 x}{1/x}=\lim\limits_{x\to 0+}\dfrac{-x}{\sin^2 x}=\lim\limits_{x\to 0+}\dfrac{-1}{2\sin x\cos x}=-\infty$

5. $\lim\limits_{x\to+\infty}\dfrac{1+\ln x}{1+1/x}=+\infty$

7. $\lim\limits_{x\to+\infty}\dfrac{100x^{99}}{e^x}=\lim\limits_{x\to+\infty}\dfrac{(100)(99)x^{98}}{e^x}=\cdots=\lim\limits_{x\to+\infty}\dfrac{(100)(99)(98)\cdots(1)}{e^x}=0$

9. $\displaystyle \lim_{x \to +\infty} xe^{-x} = \lim_{x \to +\infty} \frac{x}{e^x} = \lim_{x \to +\infty} \frac{1}{e^x} = 0$

11. $\displaystyle \lim_{x \to +\infty} x \sin(\pi/x) = \lim_{x \to +\infty} \frac{\sin(\pi/x)}{1/x} = \lim_{x \to +\infty} \frac{(-\pi/x^2)\cos(\pi/x)}{-1/x^2} = \lim_{x \to +\infty} \pi \cos(\pi/x) = \pi$

13. $\displaystyle \lim_{x \to +\infty} x(e^{\sin(2/x)} - 1) = \lim_{x \to +\infty} \frac{e^{\sin(2/x)} - 1}{1/x} = \lim_{x \to +\infty} \frac{(-2/x^2)e^{\sin(2/x)}}{-1/x^2} = \lim_{x \to +\infty} 2e^{\sin(2/x)} = 2$

15. $y = (1 - 3/x)^x$, $\displaystyle \lim_{x \to +\infty} \ln y = \lim_{x \to +\infty} \frac{\ln(1 - 3/x)}{1/x} = \lim_{x \to +\infty} \frac{-3}{1 - 3/x} = -3$, $\displaystyle \lim_{x \to +\infty} y = e^{-3}$

17. $y = (e^x + x)^{1/x}$, $\displaystyle \lim_{x \to 0} \ln y = \lim_{x \to 0} \frac{\ln(e^x + x)}{x} = \lim_{x \to 0} \frac{e^x + 1}{e^x + x} = 2$, $\displaystyle \lim_{x \to 0} y = e^2$

19. $y = (1 + 1/x^2)^x$

$\displaystyle \lim_{x \to +\infty} \ln y = \lim_{x \to +\infty} \frac{\ln(1 + 1/x^2)}{1/x} = \lim_{x \to +\infty} \frac{2x}{x^2 + 1} = \lim_{x \to +\infty} \frac{1}{x} = 0$, $\displaystyle \lim_{x \to +\infty} y = e^0 = 1$

21. $y = (1 + 1/x)^{x^2}$

$\displaystyle \lim_{x \to +\infty} \ln y = \lim_{x \to +\infty} \frac{\ln(1 + 1/x)}{1/x^2} = \lim_{x \to +\infty} \frac{x^2}{2(x + 1)} = \lim_{x \to +\infty} x = +\infty$, $\displaystyle \lim_{x \to +\infty} y = +\infty$

23. $y = (2 - x)^{\tan(\pi x/2)}$, $\displaystyle \lim_{x \to 1} \ln y = \lim_{x \to 1} \frac{\ln(2 - x)}{\cot(\pi x/2)} = \lim_{x \to 1} \frac{2\sin^2(\pi x/2)}{\pi(2 - x)} = 2/\pi$, $\displaystyle \lim_{x \to 1} y = e^{2/\pi}$

25. $y = x^{\sin x}$, $\displaystyle \lim_{x \to 0^+} \ln y = \lim_{x \to 0^+} \frac{\ln x}{\csc x} = \lim_{x \to 0^+} \frac{1/x}{-\csc x \cot x} = \lim_{x \to 0^+} \left(-\frac{\sin x}{x} \right) \tan x = 0$,

$\displaystyle \lim_{x \to 0^+} y = e^0 = 1$

27. $y = (\sin x)^{3/\ln x}$, $\displaystyle \lim_{x \to 0^+} \ln y = \lim_{x \to 0^+} \frac{3\ln \sin x}{\ln x} = \lim_{x \to 0^+} (3\cos x)\frac{x}{\sin x} = 3$, $\displaystyle \lim_{x \to 0^+} y = e^3$

29. $y = (\tan x)^{\cos x}$, $\displaystyle \lim_{x \to \pi/2^-} \ln y = \lim_{x \to \pi/2^-} \frac{\ln \tan x}{\sec x} = \lim_{x \to \pi/2^-} \frac{\sec^2 x/\tan x}{\sec x \tan x} = \lim_{x \to \pi/2^-} \frac{\cos x}{\sin^2 x} = 0$,

$\displaystyle \lim_{x \to \pi/2^-} y = e^0 = 1$

31. $y = (1 + x^2)^{1/\ln x}$

$\displaystyle \lim_{x \to +\infty} \ln y = \lim_{x \to +\infty} \frac{\ln(1 + x^2)}{\ln x} = \lim_{x \to +\infty} \frac{2x^2}{1 + x^2} = \lim_{x \to +\infty} 2 = 2$, $\displaystyle \lim_{x \to +\infty} y = e^2$

33. $\displaystyle\lim_{\theta\to 0}\left(\frac{1+\cos\theta}{1-\cos^2\theta}-\frac{2}{\sin^2\theta}\right)=\lim_{\theta\to 0}\frac{\cos\theta-1}{\sin^2\theta}=\lim_{\theta\to 0}\frac{-\sin\theta}{2\sin\theta\cos\theta}=-1/2$

35. $\displaystyle\lim_{x\to 0}\left(\frac{1}{\sin x}-\frac{1}{x}\right)=\lim_{x\to 0}\frac{x-\sin x}{x\sin x}=\lim_{x\to 0}\frac{1-\cos x}{x\cos x+\sin x}=\lim_{x\to 0}\frac{\sin x}{2\cos x-x\sin x}=0$

37. $\displaystyle\lim_{x\to+\infty}\frac{(x^2+x)-x^2}{\sqrt{x^2+x}+x}=\lim_{x\to+\infty}\frac{x}{\sqrt{x^2+x}+x}=\lim_{x\to+\infty}\frac{1}{\sqrt{1+1/x}+1}=1/2$

39. $\displaystyle\lim_{x\to+\infty}[x-\ln(x^2+1)]=\lim_{x\to+\infty}[\ln e^x-\ln(x^2+1)]=\lim_{x\to+\infty}\ln\frac{e^x}{x^2+1},$

$\displaystyle\lim_{x\to+\infty}\frac{e^x}{x^2+1}=\lim_{x\to+\infty}\frac{e^x}{2x}=\lim_{x\to+\infty}\frac{e^x}{2}=+\infty$ so $\displaystyle\lim_{x\to+\infty}[x-\ln(x^2+1)]=+\infty$

41. $\displaystyle\lim_{x\to 0^+}\frac{\cot x}{\cot 2x}=\lim_{x\to 0^+}\frac{\tan 2x}{\tan x}=\lim_{x\to 0^+}\frac{2\sec^2 2x}{\sec^2 x}=2$

43. In each case, apply L'Hôpital's rule n times.

    (a) $\displaystyle\lim_{x\to+\infty}\frac{x^n}{e^x}=\cdots=\lim_{x\to+\infty}\frac{n(n-1)(n-2)\cdots(1)}{e^x}=0$

    (b) $\displaystyle\lim_{x\to+\infty}\frac{e^x}{x^n}=\cdots=\lim_{x\to+\infty}\frac{e^x}{n(n-1)(n-2)\cdots(1)}=+\infty$

45. **(a)** 0   **(b)** $+\infty$ **(c)** 0   **(d)** $-\infty$
       (e) $+\infty$ **(f)** $+\infty$ **(g)** $-\infty$ **(h)** $-\infty$

47. Let $y=x^x$, $\displaystyle\lim_{x\to+\infty}y=+\infty$ and

$\displaystyle\lim_{x\to 0^+}y=1$. Use logarithmic differ-

entiation to get $dy/dx=x^x(1+\ln x)$

so $dy/dx=0$ when $x=e^{-1}$, $dy/dx<0$

if $x<e^{-1}$, $dy/dx>0$ if $x>e^{-1}$,

and $dy/dx\to-\infty$ as $x\to 0^+$. Also,

$d^2y/dx^2=x^x[1/x+(1+\ln x)^2]>0$

for $x>0$ so the curve is always

concave up.

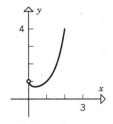

49. $\int \ln x \, dx = x \ln x - x + C$,

$$\int_0^1 \ln x \, dx = \lim_{\ell \to 0^+} \int_\ell^1 \ln x \, dx = \lim_{\ell \to 0^+} (x \ln x - x)\Big]_\ell^1 = \lim_{\ell \to 0^+} (-1 - \ell \ln \ell + \ell),$$

but $\displaystyle \lim_{\ell \to 0^+} \ell \ln \ell = \lim_{\ell \to 0^+} \frac{\ln \ell}{1/\ell} = \lim_{\ell \to 0^+} (-\ell) = 0$ so $\displaystyle \int_0^1 \ln x \, dx = -1$

51. $\int x e^{-3x} \, dx = -\frac{1}{3} x e^{-3x} - \frac{1}{9} e^{-3x} + C$,

$$\int_0^{+\infty} x e^{-3x} \, dx = \lim_{\ell \to +\infty} \int_0^\ell x e^{-3x} \, dx = \lim_{\ell \to +\infty} \left(-\frac{1}{3} x e^{-3x} - \frac{1}{9} e^{-3x} \right)\Big]_0^\ell$$

$$= \lim_{\ell \to +\infty} \left(-\frac{1}{3} \ell e^{-3\ell} - \frac{1}{9} e^{-3\ell} + \frac{1}{9} \right),$$

but $\displaystyle \lim_{\ell \to +\infty} \ell e^{-3\ell} = \lim_{\ell \to +\infty} \frac{\ell}{e^{3\ell}} = \lim_{\ell \to +\infty} \frac{1}{3e^{3\ell}} = 0$ so $\displaystyle \int_0^{+\infty} x e^{-3x} \, dx = 1/9$

53. $V = 2\pi \displaystyle\int_0^{+\infty} x e^{-x} \, dx = 2\pi \lim_{\ell \to +\infty} -e^{-x}(x+1)\Big]_0^\ell = 2\pi \lim_{\ell \to +\infty} [1 - e^{-\ell}(\ell+1)]$,

but $\displaystyle \lim_{\ell \to +\infty} e^{-\ell}(\ell+1) = \lim_{\ell \to +\infty} \frac{\ell+1}{e^\ell} = \lim_{\ell \to +\infty} \frac{1}{e^\ell} = 0$ so $V = 2\pi$.

55. (a) $\displaystyle \int_0^\ell \sqrt{1+t^3} \, dt \geq \int_0^\ell t^{3/2} \, dt = \frac{2}{5} t^{5/2}\Big]_0^\ell = \frac{2}{5} \ell^{5/2}$,

$$\lim_{\ell \to +\infty} \int_0^\ell t^{3/2} \, dt = \lim_{\ell \to +\infty} \frac{2}{5} \ell^{5/2} = +\infty \text{ so } \int_0^{+\infty} \sqrt{1+t^3} \, dt = +\infty$$

(b) $\displaystyle \lim_{x \to +\infty} \frac{2\sqrt{1+8x^3}}{(5/2)x^{3/2}} = \lim_{x \to +\infty} \frac{4}{5} \sqrt{1/x^3 + 8} = 8\sqrt{2}/5$

57. $\displaystyle \lim_{x \to +\infty} \frac{1 + 2\cos 2x}{1}$ does not exist, nor is it $\pm\infty$; $\displaystyle \lim_{x \to +\infty} \frac{x + \sin 2x}{x} = \lim_{x \to +\infty} \left(1 + \frac{\sin 2x}{x} \right) = 1$.

59. $\displaystyle \lim_{x \to +\infty} (2 + x\cos x + \sin x)$ does not exist, nor is it $\pm\infty$; $\displaystyle \lim_{x \to +\infty} \frac{x(2 + \sin x)}{x + 1} = \lim_{x \to +\infty} \frac{2 + \sin x}{1 + 1/x}$, which does not exist because $\sin x$ oscillates between -1 and 1 as $x \to +\infty$.

61. (a) $t = 1/x, x = 1/t$,

$$\lim_{x \to +\infty} x\left(k^{1/x} - 1\right) = \lim_{t \to 0^+} \frac{k^t - 1}{t} = \lim_{t \to 0^+} = \frac{e^{t \ln k} - 1}{t} = \lim_{t \to 0^+} (\ln k)e^{t \ln k} = \ln k$$

(b) $1024 = 2^{10}$ so enter the value of k, press the square root key ten times, subtract 1, and multiply the result by 1024: with $k = 0.3$ we get -1.203265293 ($\ln 0.3 \approx -1.203972804$), with $k = 2$ we get 0.693381829 ($\ln 2 \approx 0.693147181$)

63. $u = t/\sqrt{x}, \; x = t^2/u^2$,

$$\lim_{x \to +\infty} xf(t/\sqrt{x}) = \lim_{u \to 0^+} t^2 \frac{f(u)}{u^2} = t^2 \lim_{u \to 0^+} \frac{f'(u)}{2u} = t^2 \lim_{u \to 0^+} \frac{f''(u)}{2} = \frac{1}{2}at^2$$

65. $\displaystyle \mathcal{L}\{t\} = \int_0^{+\infty} te^{-st}\,dt = \left[-\frac{1}{s}te^{-st} - \frac{1}{s^2}e^{-st}\right]_0^{+\infty}$, but

$$\lim_{t \to +\infty} te^{-st} = 0 \text{ and } \lim_{t \to +\infty} e^{-st} = 0 \text{ so } \mathcal{L}\{t\} = 1/s^2.$$

67. (a) $y = x^{\frac{\ln a}{1 + \ln x}}$,

$$\lim_{x \to 0^+} \ln y = \lim_{x \to 0^+} \frac{(\ln a)\ln x}{1 + \ln x} = \lim_{x \to 0^+} \frac{(\ln a)/x}{1/x} = \lim_{x \to 0^+} \ln a = \ln a, \; \lim_{x \to 0^+} y = e^{\ln a} = a$$

(b) same as part (a) with $x \to +\infty$

(c) $y = (x+1)^{\frac{\ln a}{x}}$, $\displaystyle \lim_{x \to 0} \ln y = \lim_{x \to 0} \frac{(\ln a)\ln(x+1)}{x} = \lim_{x \to 0} \frac{\ln a}{x+1} = \ln a, \; \lim_{x \to 0} y = e^{\ln a} = a$

69. (a) $t = -\ln x, x = e^{-t}, dx = -e^{-t}dt$,

$$\int_0^1 (\ln x)^n\,dx = -\int_{+\infty}^0 (-t)^n e^{-t}\,dt = (-1)^n \int_0^{+\infty} t^n e^{-t}\,dt = (-1)^n \Gamma(n+1).$$

(b) $t = x^n, x = t^{1/n}, dx = (1/n)t^{1/n-1}dt$,

$$\int_0^{+\infty} e^{-x^n}\,dx = (1/n)\int_0^{+\infty} t^{1/n-1}e^{-t}\,dt = (1/n)\Gamma(1/n) = \Gamma(1/n + 1).$$

TECHNOLOGY EXERCISES 10

1.   0.504067   **3.**   2.804364

9. $\displaystyle \int_0^1 \sqrt{\frac{1+x}{1-x}}\,dx = 2\int_0^1 \sqrt{2 - u^2}\,du = 2.570796$

11. $1 + \left(\dfrac{dy}{dx}\right)^2 = 1 + \dfrac{16x^2}{9 - 4x^2} = \dfrac{9 + 12x^2}{9 - 4x^2}$; the arc length is $\displaystyle\int_0^{3/2} \sqrt{\dfrac{9 + 12x^2}{9 - 4x^2}}\, dx = 3.633168.$

13. **(c)** $\displaystyle\int_{\theta_0}^{\pi} \sqrt{\dfrac{1 - \cos\theta}{\cos\theta_0 - \cos\theta}}\, d\theta = \int_{\theta_0}^{\pi} \sqrt{\dfrac{2\sin^2(\theta/2)}{2\cos^2(\theta_0/2) - 2\cos^2(\theta/2)}}\, d\theta$

$$= \int_{\theta_0}^{\pi} \frac{\sin(\theta/2)}{\cos(\theta_0/2)\sqrt{1 - \cos^2(\theta/2)/\cos^2(\theta_0/2)}}\, d\theta$$

$$= -2 \int_1^0 \frac{1}{\sqrt{1 - u^2}}\, du$$

$$= 2 \int_0^1 \frac{1}{\sqrt{1 - u^2}}\, du = 2\sin^{-1}(1) = \pi$$

15. **(a)** Use logarithmic differentiation to obtain

$$f'(x) = \frac{1}{x^{12}(1.0001)^{1/x}}(\ln 1.0001 - 10x),\ f'(x) = 0 \text{ when } x = (\ln 1.0001)/10 \approx 10^{-5};$$

the maximum value is approximately $10^{50}/(1.0001)^{10^5} \approx 4.5 \times 10^{45}$.

(b) Let $x = 1/t$, then $\displaystyle\lim_{x \to 0^+} f(x) = \lim_{t \to +\infty} \frac{t^{10}}{(1.0001)^t} = \lim_{t \to +\infty} \frac{t^{10}}{e^{t \ln 1.0001}}$;

apply L'Hôpital's Rule ten times to obtain $\displaystyle\lim_{x \to 0^+} f(x) = \lim_{t \to +\infty} \frac{(10)(9)(8)\cdots(1)}{(\ln 1.0001)^{10} e^{t \ln 1.0001}} = 0.$

In a similar manner it is possible to show that $\displaystyle\lim_{x \to 0^+} f'(x) = 0.$

CHAPTER 11
Infinite Series

EXERCISE SET 11.1

1. $1/3, 2/4, 3/5, 4/6, 5/7, \ldots$; $\lim\limits_{n \to +\infty} \dfrac{n}{n+2} = 1$, converges

3. $2, 2, 2, 2, 2, \ldots$; $\lim\limits_{x \to +\infty} 2 = 2$, converges

5. $\dfrac{\ln 1}{1}, \dfrac{\ln 2}{2}, \dfrac{\ln 3}{3}, \dfrac{\ln 4}{4}, \dfrac{\ln 5}{5}, \ldots$; $\lim\limits_{x \to +\infty} \dfrac{\ln n}{n} = \lim\limits_{x \to +\infty} \dfrac{1}{n} = 0$, converges

7. $0, 2, 0, 2, 0, \ldots$; diverges

9. $-1, 16/9, -54/28, 128/65, -250/126, \ldots$; diverges because odd-numbered terms approach -2, even-numbered terms approach 2.

11. $6/2, 12/8, 20/18, 30/32, 42/50, \ldots$; $\lim\limits_{n \to +\infty} \dfrac{1}{2}(1+1/n)(1+2/n) = 1/2$, converges

13. $\cos(3), \cos(3/2), \cos(1), \cos(3/4), \cos(3/5), \ldots$; $\lim\limits_{n \to +\infty} \cos(3/n) = 1$, converges

15. $e^{-1}, 4e^{-2}, 9e^{-3}, 16e^{-4}, 25e^{-5}, \ldots$; $\lim\limits_{x \to +\infty} x^2 e^{-x} = \lim\limits_{x \to +\infty} \dfrac{x^2}{e^x} = 0$, so $\lim\limits_{n \to +\infty} n^2 e^{-n} = 0$, converges

17. $2, (5/3)^2, (6/4)^3, (7/5)^4, (8/6)^5, \ldots$; let $y = \left[\dfrac{x+3}{x+1}\right]^x$, converges because
$$\lim\limits_{x \to +\infty} \ln y = \lim\limits_{x \to +\infty} \dfrac{\ln \dfrac{x+3}{x+1}}{1/x} = \lim\limits_{x \to +\infty} \dfrac{2x^2}{(x+1)(x+3)} = 2, \text{ so } \lim\limits_{n \to +\infty} \left[\dfrac{n+3}{n+1}\right]^n = e^2$$

19. $\left\{\dfrac{2n-1}{2n}\right\}_{n=1}^{+\infty}$; $\lim\limits_{n \to +\infty} \dfrac{2n-1}{2n} = 1$, converges

21. $\{1/3^n\}_{n=1}^{+\infty}$; $\lim_{n \to +\infty} 1/3^n = 0$, converges

23. $\left\{\dfrac{1}{n} - \dfrac{1}{n+1}\right\}_{n=1}^{+\infty}$; $\lim\limits_{n \to +\infty} \left(\dfrac{1}{n} - \dfrac{1}{n+1}\right) = 0$, converges

233

25. $\{\sqrt{n+1} - \sqrt{n+2}\}_{n=1}^{+\infty}$; converges because

$$\lim_{n \to +\infty} (\sqrt{n+1} - \sqrt{n+2}) = \lim_{n \to +\infty} \frac{(n+1) - (n+2)}{\sqrt{n+1} + \sqrt{n+2}} = \lim_{n \to +\infty} \frac{-1}{\sqrt{n+1} + \sqrt{n+2}} = 0$$

27. **(a)** $\sqrt{6}, \sqrt{6+\sqrt{6}}, \sqrt{6+\sqrt{6+\sqrt{6}}}$

(b) $\lim_{n \to +\infty} a_{n+1} = \lim_{n \to +\infty} \sqrt{6 + a_n}$, $L = \sqrt{6 + L}$, $L^2 - L - 6 = 0$, $(L-3)(L+2) = 0$, $L = -2$
(reject, because the terms in the sequence are positive) or $L = 3$; $\lim_{n \to +\infty} a_n = 3$.

29. **(a)** $1, 1, 2, 3, 5, 8, 13, 21$

(b) $a_{n+2}/a_{n+1} = a_n/a_{n+1} + 1 = 1/(a_{n+1}/a_n) + 1$,
$\lim_{n \to +\infty} (a_{n+2}/a_{n+1}) = \lim_{n \to +\infty} [1/(a_{n+1}/a_n) + 1]$, with $L = \lim_{n \to +\infty} (a_{n+1}/a_n)$, $L = 1/L + 1$,
$L^2 - L - 1 = 0$, $L = (1 \pm \sqrt{5})/2$ so $L = (1 + \sqrt{5})/2$ because the limit cannot be negative.

31. **(a)** $1, \dfrac{1}{4} + \dfrac{2}{4}, \dfrac{1}{9} + \dfrac{2}{9} + \dfrac{3}{9}, \dfrac{1}{16} + \dfrac{2}{16} + \dfrac{3}{16} + \dfrac{4}{16}, \cdots = 1, 3/4, 2/3, 5/8, \cdots$

(b) $a_n = \dfrac{1}{n^2}(1 + 2 + \cdots + n) = \dfrac{1}{n^2}\dfrac{1}{2}n(n+1) = \dfrac{1}{2}\dfrac{n+1}{n}$, $\lim_{n \to +\infty} a_n = 1/2$

33. $\left| \dfrac{1}{n} - 0 \right| = \dfrac{1}{n} < \epsilon$ if $n > 1/\epsilon$

(a) $1/\epsilon = 1/0.5 = 2$, $N = 3$ **(b)** $1/\epsilon = 1/0.1 = 10$, $N = 11$
(c) $1/\epsilon = 1/0.001 = 1000$, $N = 1001$

35. **(a)** $\left| \dfrac{1}{n} - 0 \right| = \dfrac{1}{n} < \epsilon$ if $n > 1/\epsilon$, choose any $N > 1/\epsilon$.

(b) $\left| \dfrac{n}{n+1} - 1 \right| = \dfrac{1}{n+1} < \epsilon$ if $n > 1/\epsilon - 1$, choose any $N > 1/\epsilon - 1$.

37. **(a)** Let s_n denote the length of each
of the n sides of the polygon, then
$$\sin(\pi/n) = \frac{s_n/2}{r}$$
$$s_n = 2r \sin(\pi/n)$$
$$p_n = n\, s_n = 2r\, n \sin(\pi/n)$$

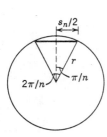

(b) $\lim_{n \to +\infty} 2r\, n \sin(\pi/n) = \lim_{n \to +\infty} \dfrac{2r \sin(\pi/n)}{1/n}$

$= \lim_{n \to +\infty} 2\pi r \cos(\pi/n) = 2\pi r$

39. Let $y = (2^x + 3^x)^{1/x}$,

$$\lim_{x \to +\infty} \ln y = \lim_{x \to +\infty} \frac{\ln(2^x + 3^x)}{x} = \lim_{x \to +\infty} \frac{2^x \ln 2 + 3^x \ln 3}{2^x + 3^x} = \lim_{x \to +\infty} \frac{(2/3)^x \ln 2 + \ln 3}{(2/3)^x + 1} = \ln 3$$

so $\displaystyle\lim_{x \to +\infty} (2^n + 3^n)^{1/n} = e^{\ln 3} = 3$

EXERCISE SET 11.2

1. $a_{n+1} - a_n = \dfrac{1}{n+1} - \dfrac{1}{n} = -\dfrac{1}{n(n+1)} < 0$ for $n \geq 1$, so decreasing.

3. $a_{n+1} - a_n = \dfrac{n+1}{2n+3} - \dfrac{n}{2n+1} = \dfrac{1}{(2n+1)(2n+3)} > 0$ for $n \geq 1$, so increasing.

5. $a_{n+1} - a_n = (n + 1 - 2^{n+1}) - (n - 2^n) = 1 - 2^n < 0$ for $n \geq 1$, so decreasing.

7. $\dfrac{a_{n+1}}{a_n} = \dfrac{(n+1)/(2n+3)}{n/(2n+1)} = \dfrac{(n+1)(2n+1)}{n(2n+3)} = \dfrac{2n^2 + 3n + 1}{2n^2 + 3n} > 1$ for $n \geq 1$, so increasing.

9. $\dfrac{a_{n+1}}{a_n} = \dfrac{(n+1)e^{-(n+1)}}{ne^{-n}} = (1 + 1/n)e^{-1} < 1$ for $n \geq 1$, so decreasing.

9. $\dfrac{a_{n+1}}{a_n} = \dfrac{(n+1)^{n+1}}{(n+1)!} \cdot \dfrac{n!}{n^n} = \dfrac{(n+1)^n}{n^n} = (1 + 1/n)^n > 1$ for $n \geq 1$, so increasing.

13. $f(x) = x/(2x+1)$, $f'(x) = 1/(2x+1)^2 > 0$ for $x \geq 1$, so increasing.

15. $f(x) = 1/(x + \ln x)$, $f'(x) = -\dfrac{1 + 1/x}{(x + \ln x)^2} < 0$ for $x \geq 1$, so decreasing.

17. $f(x) = \dfrac{\ln(x+2)}{x+2}$, $f'(x) = \dfrac{1 - \ln(x+2)}{(x+2)^2} < 0$ for $x \geq 1$, so decreasing.

19. $f(x) = 2x^2 - 7x$, $f'(x) = 4x - 7 > 0$ for $x \geq 2$, so eventually increasing.

21. $f(x) = \dfrac{x}{x^2 + 10}$, $f'(x) = \dfrac{10 - x^2}{(x^2 + 10)^2} < 0$ for $x \geq 4$, so eventually decreasing.

23. $\dfrac{a_{n+1}}{a_n} = \dfrac{(n+1)!}{3^{n+1}} \cdot \dfrac{3^n}{n!} = \dfrac{n+1}{3} > 1$ for $n \geq 3$, so eventually increasing.

25. If $x = 0$ the result follows immediately. Let $a_n = |x|^n/n!$, $x \neq 0$. Then
$a_{n+1}/a_n = |x|/(n+1) < 1$ for $n > |x|$, so a_n is eventually decreasing; 0 is a lower bound for a_n, so $a_n \to L$ as $n \to +\infty$. Use $a_{n+1}/a_n = |x|/(n+1)$ to obtain

$$a_{n+1} = \frac{|x|}{n+1} a_n$$

thus $L = 0 \cdot L = 0$. But $-|x^n| \le x^n \le |x^n|$ so $-|x|^n/n! \le x^n/n! \le |x|^n/n!$ and hence, by the Squeezing Theorem, $x^n/n! \to 0$ as $n \to +\infty$ because $|x|^n/n! \to 0$ as $n \to +\infty$.

27. (a) $\sqrt{2}, \sqrt{2 + \sqrt{2}}, \sqrt{2 + \sqrt{2 + \sqrt{2}}}$

(b) $a_1 = \sqrt{2} < 2$ so $a_2 = \sqrt{2 + a_1} < \sqrt{2 + 2} = 2$, $a_3 = \sqrt{2 + a_2} < \sqrt{2 + 2} = 2$, and so on indefinitely.

(c) $a_{n+1}^2 - a_n^2 = (2 + a_n) - a_n^2 = 2 + a_n - a_n^2 = (2 - a_n)(1 + a_n)$

(d) $a_n > 0$ and, from part (a), $a_n < 2$ so $2 - a_n > 0$ and $1 + a_n > 0$ thus, from part (c), $a_{n+1}^2 - a_n^2 > 0$, $a_{n+1} - a_n > 0$, $a_{n+1} > a_n$; $\{a_n\}$ is an increasing sequence.

(e) The sequence is increasing and has 2 as an upper bound so it must converge to a limit L, $\lim\limits_{n \to +\infty} a_{n+1} = \lim\limits_{n \to +\infty} \sqrt{2 + a_n}$, $L = \sqrt{2 + L}$, $L^2 - L - 2 = 0$, $(L - 2)(L + 1) = 0$ thus $\lim\limits_{n \to +\infty} a_n = 2$.

29. (a) If $\{a_n\}_{n=1}^{+\infty}$ is nonincreasing then $a_n \ge a_{n+1}$ for all n, thus $-a_n \le -a_{n+1}$ so $\{-a_n\}_{n=1}^{+\infty}$ is nondecreasing.

(b) From part (a), $\{-a_n\}_{n=1}^{+\infty}$ is nondecreasing so Theorem 11.2.2 applies. Suppose there is a finite constant M_1 such that $-a_n \le M_1$ and $\lim\limits_{n \to +\infty} (-a_n) = L_1 \le M_1$, then $a_n \ge -M_1$, and $\lim\limits_{n \to +\infty} a_n = -L_1 \ge -M_1$; part (a) of Theorem 11.2.3 follows by letting $M = -M_1$ and $L = -L_1$. Suppose no such constant exists, then $\lim\limits_{n \to +\infty} (-a_n) = +\infty$ so $\lim\limits_{n \to +\infty} a_n = -\infty$.

31. From part (b) of Exercise 30, $\left[\dfrac{n^n}{e^{n-1}} \right]^{1/n} < \sqrt[n]{n!} < \left[\dfrac{(n+1)^{n+1}}{e^n} \right]^{1/n}$,

$\dfrac{n}{e^{1-1/n}} < \sqrt[n]{n!} < \dfrac{(n+1)^{1+1/n}}{e}$, $\dfrac{1}{e^{1-1/n}} < \dfrac{\sqrt[n]{n!}}{n} < \dfrac{(1 + 1/n)(n+1)^{1/n}}{e}$,

but $\dfrac{1}{e^{1-1/n}} \to \dfrac{1}{e}$ and $\dfrac{(1 + 1/n)(n+1)^{1/n}}{e} \to \dfrac{1}{e}$ as $n \to +\infty$ so $\lim\limits_{n \to +\infty} \dfrac{\sqrt[n]{n!}}{n} = \dfrac{1}{e}$

33. (a) $\dfrac{a_{n+1}}{a_n} = \dfrac{(n+1)^{n+1}}{(n+1)! e^{n+1}} \cdot \dfrac{n! e^n}{n^n} = \dfrac{(1 + 1/n)^n}{e} < 1$ for $n \ge 1$ because $(1 + 1/n)^n < e$ for $n \ge 1$.

(b) Yes, because it is decreasing and bounded below by 0.

EXERCISE SET 11.3

1. **(a)** $s_1 = 2$, $s_2 = 12/5$, $s_3 = 62/25$, $s_4 = 312/125$

$s_n = \dfrac{2 - 2(1/5)^n}{1 - 1/5} = \dfrac{5}{2} - \dfrac{5}{2}(1/5)^n$, $\lim\limits_{n \to +\infty} s_n = 5/2$, converges

(b) $\dfrac{1}{(k+1)(k+2)} = \dfrac{1}{k+1} - \dfrac{1}{k+2}$, $s_1 = 1/6$, $s_2 = 1/4$, $s_3 = 3/10$, $s_4 = 1/3$ $s_n = \dfrac{1}{2} - \dfrac{1}{n+2}$,

$\lim\limits_{n \to +\infty} s_n = 1/2$, converges

(c) $s_1 = 1/4$, $s_2 = 3/4$, $s_3 = 7/4$, $s_4 = 15/4$

$s_n = \dfrac{(1/4) - (1/4)2^n}{1 - 2} = -\dfrac{1}{4} + \dfrac{1}{4}(2^n)$, $\lim\limits_{n \to +\infty} s_n = +\infty$, diverges

3. geometric, $a = 1$, $r = -3/4$, sum $= \dfrac{1}{1 - (-3/4)} = 4/7$

5. geometric, $a = 7$, $r = -1/6$, sum $= \dfrac{7}{1 + 1/6} = 6$

7. geometric, $r = -3/2$, diverges

9. $s_n = \sum\limits_{k=1}^{n} \left(\dfrac{1}{k+2} - \dfrac{1}{k+3} \right) = \dfrac{1}{3} - \dfrac{1}{n+3}$, $\lim\limits_{n \to +\infty} s_n = 1/3$

11. $s_n = \sum\limits_{k=1}^{n} \left(\dfrac{1/3}{3k-1} - \dfrac{1/3}{3k+2} \right) = \dfrac{1}{6} - \dfrac{1/3}{3n+2}$, $\lim\limits_{n \to +\infty} s_n = 1/6$

13. $\sum\limits_{k=1}^{\infty} \dfrac{4^{k+2}}{7^{k-1}} = \sum\limits_{k=1}^{\infty} 64 \left(\dfrac{4}{7} \right)^{k-1}$; geometric, $a = 64$, $r = 4/7$, sum $= \dfrac{64}{1 - 4/7} = 448/3$

15. geometric, $a = -1/2$, $r = -1/2$, sum $= \dfrac{-1/2}{1 + 1/2} = -1/3$

17. $0.4444 \cdots = 0.4 + 0.04 + 0.004 + \cdots = \dfrac{0.4}{1 - 0.1} = 4/9$

19. $5.373737 \cdots = 5 + 0.37 + 0.0037 + 0.000037 + \cdots = 5 + \dfrac{0.37}{1 - 0.01} = 5 + 37/99 = 532/99$

21. $0.782178217821\cdots = 0.7821 + 0.00007821 + 0.000000007821 + \cdots$

$$= \frac{0.7821}{1 - 0.0001} = 7821/9999 = 869/1111$$

23. $s_n = \ln\dfrac{1}{2} + \ln\dfrac{2}{3} + \ln\dfrac{3}{4} + \cdots + \ln\dfrac{n}{n+1} = \ln\left(\dfrac{1}{2}\cdot\dfrac{2}{3}\cdot\dfrac{3}{4}\cdots\dfrac{n}{n+1}\right) = \ln\dfrac{1}{n+1} = -\ln(n+1),$

$\displaystyle\lim_{n\to+\infty} s_n = -\infty$, series diverges.

25. $\ln 1 - 1/k^2 = \ln\dfrac{k^2-1}{k^2} = \ln\dfrac{(k-1)(k+1)}{k^2} = \ln\dfrac{k-1}{k} + \ln\dfrac{k+1}{k} = \ln\dfrac{k-1}{k} - \ln\dfrac{k}{k+1},$

$$s_n = \sum_{k=2}^{n+1}\left[\ln\frac{k-1}{k} - \ln\frac{k}{k+1}\right]$$

$$= \left(\ln\frac{1}{2} - \ln\frac{2}{3}\right) + \left(\ln\frac{2}{3} - \ln\frac{3}{4}\right) + \left(\ln\frac{3}{4} - \ln\frac{4}{5}\right) + \cdots + \left(\ln\frac{n}{n+1} - \ln\frac{n+1}{n+2}\right)$$

$$= \ln\frac{1}{2} - \ln\frac{n+1}{n+2}, \lim_{n\to+\infty} s_n = \ln\frac{1}{2} = -\ln 2$$

27. $s_n = (1 - 1/3) + (1/2 - 1/4) + (1/3 - 1/5) + (1/4 - 1/6) + \cdots + [1/n - 1/(n+2)]$

$$= (1 + 1/2 + 1/3 + \cdots + 1/n) - (1/3 + 1/4 + 1/5 + \cdots + 1/(n+2))$$

$$= 3/2 - 1/(n+1) - 1/(n+2), \lim_{n\to+\infty} s_n = 3/2$$

29. $s_n = \displaystyle\sum_{k=1}^{n}\frac{1}{(2k-1)(2k+1)} = \sum_{k=1}^{n}\left[\frac{1/2}{2k-1} - \frac{1/2}{2k+1}\right] = \frac{1}{2}\left[\sum_{k=1}^{n}\frac{1}{2k-1} - \sum_{k=1}^{n}\frac{1}{2k+1}\right]$

$$= \frac{1}{2}\left[\sum_{k=1}^{n}\frac{1}{2k-1} - \sum_{k=2}^{n+1}\frac{1}{2k-1}\right] = \frac{1}{2}\left[1 - \frac{1}{2n+1}\right]; \lim_{n\to+\infty} s_n = \frac{1}{2}$$

31.   **(a)** $\displaystyle\sum_{k=0}^{\infty}(-1)^k x^k = 1 - x + x^2 - x^3 + \cdots = \frac{1}{1-(-x)} = \frac{1}{1+x}$ if $|-x| < 1$, $|x| < 1$, $-1 < x < 1$.

     (b) $\displaystyle\sum_{k=0}^{\infty}(x-3)^k = 1 + (x-3) + (x-3)^2 + \cdots = \frac{1}{1-(x-3)} = \frac{1}{4-x}$ if $|x-3| < 1$, $2 < x < 4$.

     (c) $\displaystyle\sum_{k=0}^{\infty}(-1)^k x^{2k} = 1 - x^2 + x^4 - x^6 + \cdots = \frac{1}{1-(-x^2)} = \frac{1}{1+x^2}$ if $|-x^2| < 1$, $|x| < 1$, $-1 < x < 1$.

33. Geometric series, $a = 1/x^2$, $r = 2/x$. Converges for $|2/x| < 1$, $|x| > 2$;

$$S = \frac{1/x^2}{1 - 2/x} = \frac{1}{x^2 - 2x}.$$

35. Geometric series, $a = \sin x$, $r = -\frac{1}{2}\sin x$. Converges for $|-\frac{1}{2}\sin x| < 1$, $|\sin x| < 2$, so converges for all values of x. $S = \dfrac{\sin x}{1 + \dfrac{1}{2}\sin x} = \dfrac{2\sin x}{2 + \sin x}$.

37. $a_2 = \dfrac{1}{2}a_1 + \dfrac{1}{2}$, $a_3 = \dfrac{1}{2}a_2 + \dfrac{1}{2} = \dfrac{1}{2^2}a_1 + \dfrac{1}{2^2} + \dfrac{1}{2}$, $a_4 = \dfrac{1}{2}a_3 + \dfrac{1}{2} = \dfrac{1}{2^3}a_1 + \dfrac{1}{2^3} + \dfrac{1}{2^2} + \dfrac{1}{2}$,

$a_5 = \dfrac{1}{2}a_4 + \dfrac{1}{2} = \dfrac{1}{2^4}a_1 + \dfrac{1}{2^4} + \dfrac{1}{2^3} + \dfrac{1}{2^2} + \dfrac{1}{2}, \cdots, a_n = \dfrac{1}{2^{n-1}}a_1 + \dfrac{1}{2^{n-1}} + \dfrac{1}{2^{n-2}} + \cdots + \dfrac{1}{2}$,

$$\lim_{n \to +\infty} a_n = \lim_{n \to +\infty} \frac{a_1}{2^{n-1}} + \sum_{n=1}^{\infty}\left(\frac{1}{2}\right)^n = 0 + \frac{1/2}{1 - 1/2} = 1$$

39. By inspection, $\dfrac{\theta}{2} - \dfrac{\theta}{4} + \dfrac{\theta}{8} - \dfrac{\theta}{16} + \cdots = \dfrac{\theta/2}{1 - (-1/2)} = \theta/3$

41. The series converges to $1/(1 - x)$ only if $-1 < x < 1$.

EXERCISE SET 11.4

1. $\displaystyle\sum_{k=1}^{\infty} \frac{1}{2^k} = \frac{1/2}{1 - 1/2} = 1$; $\displaystyle\sum_{k=1}^{\infty} \frac{1}{4^k} = \frac{1/4}{1 - 1/4} = 1/3$; $\displaystyle\sum_{k=1}^{\infty}\left(\frac{1}{2^k} + \frac{1}{4^k}\right) = 1 + 1/3 = 4/3$

3. $\displaystyle\sum_{k=2}^{\infty} \frac{1}{k^2 - 1} = \sum_{k=2}^{\infty}\left[\frac{1/2}{k-1} - \frac{1/2}{k+1}\right]$,

$$s_n = \frac{1}{2}\left[\left(1 - \frac{1}{3}\right) + \left(\frac{1}{2} - \frac{1}{4}\right) + \left(\frac{1}{3} - \frac{1}{5}\right) + \cdots + \left(\frac{1}{n} - \frac{1}{n+2}\right)\right]$$

$$= \frac{1}{2}\left[\left(1 + \frac{1}{2} + \frac{1}{3} + \cdots + \frac{1}{n}\right) - \left(\frac{1}{3} + \frac{1}{4} + \frac{1}{5} + \cdots + \frac{1}{n+2}\right)\right]$$

$$= \frac{3}{4} - \frac{1}{2}\left(\frac{1}{n+1} + \frac{1}{n+2}\right),$$

$$\sum_{k=2}^{\infty} \frac{1}{k^2 - 1} = \lim_{n \to +\infty} s_n = 3/4; \quad \sum_{k=2}^{\infty} \frac{7}{10^{k-1}} = \frac{7/10}{1 - 1/10} = 7/9;$$

so $\displaystyle\sum_{k=2}^{\infty}\left[\frac{1}{k^2 - 1} - \frac{7}{10^{k-1}}\right] = 3/4 - 7/9 = -1/36$

5. **(a)** $p = 3$, converges

  (c) $p = 1$, diverges

  (e) $p = 4/3$, converges

  (g) $p = 5/3$, converges

(b) $p = 1/2$, diverges

(d) $p = 2/3$, diverges

(f) $p = 1/4$, diverges

(h) $p = \pi$, converges

7. **(a)** $\displaystyle\lim_{k \to +\infty} \frac{k^2 + k + 3}{2k^2 + 1} = \frac{1}{2}$

(b) $\displaystyle\lim_{k \to +\infty} \left(1 + \frac{1}{k}\right)^k = e$

9. $\displaystyle\sum_{k=1}^{\infty} \frac{1}{k+6} = \sum_{k=7}^{\infty} \frac{1}{k}$, diverges because the harmonic series diverges.

11. $\displaystyle\int_{1}^{+\infty} \frac{1}{5x+2} = \lim_{\ell \to +\infty} \frac{1}{5} \ln(5x+2) \bigg]_{1}^{\ell} = +\infty$, the series diverges by the integral test.

13. $\displaystyle\int_{1}^{+\infty} \frac{1}{1+9x^2} dx = \lim_{\ell \to +\infty} \frac{1}{3} \tan^{-1} 3x \bigg]_{1}^{\ell} = \frac{1}{3}\left(\pi/2 - \tan^{-1} 3\right)$, the series converges by the integral test.

15. $\displaystyle\sum_{k=1}^{\infty} \frac{1}{\sqrt{k+5}} = \sum_{k=6}^{\infty} \frac{1}{\sqrt{k}}$, diverges because the p-series with $p = 1/2 \leq 1$ diverges.

17. $\displaystyle\int_{1}^{+\infty} (2x-1)^{-1/3} dx = \lim_{\ell \to +\infty} \frac{3}{4}(2x-1)^{2/3} \bigg]_{1}^{\ell} = +\infty$, the series diverges by the integral test.

19. $\displaystyle\lim_{k \to +\infty} \frac{k}{\ln(k+1)} = \lim_{k \to +\infty} \frac{1}{1/(k+1)} = +\infty$, the series diverges because $\displaystyle\lim_{k \to +\infty} u_k \neq 0$.

21. $\displaystyle\int_{1}^{+\infty} \frac{[\ln(x+1)]^{-2}}{x+1} dx = \lim_{\ell \to +\infty} -1/\ln(x+1) \bigg]_{1}^{\ell} = 1/\ln 2$, the series converges by the integral test.

23. $\displaystyle\lim_{k \to +\infty} (1 + 1/k)^{-k} = 1/e \neq 0$, the series diverges.

25. $\displaystyle\int_{1}^{+\infty} \frac{\tan^{-1} x}{1+x^2} dx = \lim_{\ell \to +\infty} \frac{1}{2}\left(\tan^{-1} x\right)^2 \bigg]_{1}^{\ell} = 3\pi^2/32$, the series converges by the integral test.

27. $\displaystyle\sum_{k=5}^{\infty} 7k^{-p} = \sum_{k=5}^{\infty} 7\left(\frac{1}{k^p}\right)$, converges because a p-series with $p > 1$ converges.

29. $\lim\limits_{k \to +\infty} k^2 \sin^2(1/k) = 1 \ne 0$, the series diverges.

31. Use the integral test with $\displaystyle\int_2^{+\infty} \frac{dx}{x(\ln x)^p}$ to get $\lim\limits_{\ell \to +\infty} \ln(\ln x)\Big]_2^\ell = +\infty$ if $p = 1$,

$$\lim\limits_{\ell \to +\infty} \frac{(\ln x)^{1-p}}{1-p}\Big]_2^\ell = \begin{cases} +\infty & \text{if } p < 1 \\ \dfrac{1}{(p-1)(\ln 2)^{p-1}} & \text{if } p > 1 \end{cases}$$

33. Suppose $\Sigma(u_k + v_k)$ converges then so does $\Sigma[(u_k + v_k) - u_k]$, but $\Sigma[(u_k + v_k) - u_k] = \Sigma v_k$ so Σv_k converges which contradicts the assumption that Σv_k diverges. Suppose $\Sigma(u_k - v_k)$ converges then so does $\Sigma[u_k - (u_k - v_k)] = \Sigma v_k$ which leads to the same contradiction as before.

35. **(a)** diverges because $\displaystyle\sum_{k=1}^{\infty}(2/3)^{k-1}$ converges and $\displaystyle\sum_{k=1}^{\infty} 1/k$ diverges.

(b) diverges because $\displaystyle\sum_{k=1}^{\infty} \frac{k^2}{1+k^2}$ diverges and $\displaystyle\sum_{k=1}^{\infty} \frac{1}{k(k+1)}$ converges.

(c) diverges because $\displaystyle\sum_{k=1}^{\infty} 1/(3k+2)$ diverges and $\displaystyle\sum_{k=1}^{\infty} 1/k^{3/2}$ converges.

(d) converges because both $\displaystyle\sum_{k=2}^{\infty} \frac{1}{k(\ln k)^2}$ and $\displaystyle\sum_{k=2}^{\infty} 1/k^2$ converge.

37. **(a)** $s_{10} \approx 1.1975$; let $f(x) = \dfrac{1}{x^3}$, then $\displaystyle\int_{10}^{+\infty} \frac{1}{x^3}\, dx = \frac{1}{200} = 0.0050$ and

$$\int_{11}^{+\infty} \frac{1}{x^3}\, dx = \frac{1}{242} \approx 0.0041 \text{ so } 1.2016 < S < 1.2026$$

(b) $\displaystyle\int_n^{+\infty} \frac{1}{x^3}\, dx = \frac{1}{2n^2} < 10^{-3},\ n > \sqrt{500} \approx 22.4;\ n = 23.$

39. **(a)** Let $F(x) = \dfrac{1}{x}$, then $\displaystyle\int_1^n \frac{1}{x}\, dx = \ln n$ and $\displaystyle\int_1^{n+1} \frac{1}{x}\, dx = \ln(n+1)$, $u_1 = 1$ so
$\ln(n+1) < s_n < 1 + \ln n$; $\ln(1,000,001) < s_{1,000,000} < 1 + \ln(1,000,000)$,
$13 < s_{1,000,000} < 15.$

(b) $s_n > \ln(n+1) \ge 100,\ n \ge e^{100} - 1 \approx 2.688 \times 10^{43};\ n = 2.69 \times 10^{43}$

41. **(a)** From (2) with $s_n = \sum_{k=1}^{n} u_k$, $\int_{1}^{n+1} f(x)dx - \int_{1}^{n} f(x)dx < \sum_{k=1}^{n} u_k - \int_{1}^{n} f(x)dx < u_1$ but

$$\int_{1}^{n+1} f(x)dx - \int_{1}^{n} f(x)dx = \int_{n}^{n+1} f(x)dx > 0 \text{ because } f(x) > 0 \text{ so } 0 < a_n < u_1.$$

(b) $a_{n+1} - a_n = \left[\sum_{k=1}^{n+1} u_k - \int_{1}^{n+1} f(x)dx\right] - \left[\sum_{k=1}^{n} u_k - \int_{1}^{n} f(x)dx\right]$

$$= \left[\sum_{k=1}^{n+1} u_k - \sum_{k=1}^{n} u_k\right] - \left[\int_{1}^{n+1} f(x)dx - \int_{1}^{n} f(x)dx\right]$$

$$= u_{n+1} - \int_{n}^{n+1} f(x)dx$$

Since $f(x)$ is decreasing and positive on $[n, n+1]$ it follows that $f(x) > f(n+1) = u_{n+1}$ on $[n, n+1)$ so $\int_{n}^{n+1} f(x)dx > \int_{n}^{n+1} u_{n+1}dx = u_{n+1}$ thus $a_{n+1} - a_n < 0$.

(c) From part (a), 0 is a lower bound for a_n so from Theorem 11.2.3 the sequence converges.

EXERCISE SET 11.5

1. $\rho = \lim_{k \to +\infty} \dfrac{3^{k+1}/(k+1)!}{3^k/k!} = \lim_{k \to +\infty} \dfrac{3}{k+1} = 0$, the series converges.

3. $\rho = \lim_{k \to +\infty} \dfrac{k}{k+1} = 1$, the result is inconclusive.

5. $\rho = \lim_{k \to +\infty} \dfrac{(k+1)!/(k+1)^3}{k!/k^3} = \lim_{k \to +\infty} \dfrac{k^3}{(k+1)^2} = +\infty$, the series diverges.

7. $\rho = \lim_{k \to +\infty} \dfrac{3k+2}{2k-1} = 3/2$, the series diverges.

9. $\rho = \lim_{k \to +\infty} \dfrac{k^{1/k}}{5} = 1/5$, the series converges.

11. ratio test, $\rho = \lim_{k \to +\infty} \dfrac{2k^3}{(k+1)^3} = 2$, diverges

13. ratio test, $\rho = \lim_{k \to +\infty} 7/(k+1) = 0$, converges

15. ratio test, $\rho = \lim\limits_{k \to +\infty} \dfrac{(k+1)^2}{5k^2} = 1/5$, converges

17. ratio test, $\rho = \lim\limits_{k \to +\infty} e^{-1}(k+1)^{50}/k^{50} = e^{-1} < 1$, converges

19. root test, $\rho = \lim\limits_{k \to +\infty} k^{1/k}(2/3) = 2/3$, converges

21. diverges (integral test)

23. root test, $\rho = \lim\limits_{k \to +\infty} 4/(7k - 1) = 0$, converges

25. ratio test, $\rho = \lim\limits_{k \to +\infty} \dfrac{(k+1)^2}{(2k+2)(2k+1)} = 1/4$, converges

27. integral test, $\displaystyle\int_1^{+\infty} \dfrac{dx}{1 + \sqrt{x}} = \lim\limits_{\ell \to +\infty} 2[\sqrt{x} - \ln(1 + \sqrt{x})]_1^{\ell} = +\infty$, diverges

29. ratio test, $\rho = \lim\limits_{k \to +\infty} \dfrac{\ln(k+1)}{e \ln k} = \lim\limits_{k \to +\infty} \dfrac{k}{e(k+1)} = 1/e < 1$, converges

31. ratio test, $\rho = \lim\limits_{k \to +\infty} \dfrac{k+5}{4(k+1)} = 1/4$, converges

33. $u_k = \dfrac{k!}{1 \cdot 3 \cdot 5 \cdots (2k-1)}$, by the ratio test $\rho = \lim\limits_{k \to +\infty} \dfrac{k+1}{2k+1} = 1/2$; converges

35. $u_k = \dfrac{(k+1)!}{1 \cdot 4 \cdot 7 \cdots (3k-2)}$, by the ratio test $\rho = \lim\limits_{k \to +\infty} \dfrac{k+2}{3k+1} = \dfrac{1}{3}$; converges

37. **(a)** $\lim\limits_{x \to +\infty} \ln y = \lim\limits_{x \to +\infty} \dfrac{\ln(\ln x)}{x} = \lim\limits_{x \to +\infty} \dfrac{1}{x \ln x} = 0$ so $\lim\limits_{x \to +\infty} y = e^0 = 1 = \lim\limits_{k \to +\infty} (\ln k)^{1/k}$

(b) $\rho = \lim\limits_{k \to +\infty} \dfrac{1}{3}(\ln k)^{1/k} = 1/3$

(c) $\rho = \lim\limits_{k \to +\infty} \dfrac{\ln(k+1)}{3 \ln k} = \lim\limits_{k \to +\infty} \dfrac{k}{3(k+1)} = 1/3$

39. **(a)** $s_5 \approx 1.71667$, $r_k = \dfrac{1}{k+1}$ which is decreasing for $k \geq 6$ and $r_6 < 1$; $u_6 = 1/6!$, $r_6 = 1/7$,

$S - s_5 < \dfrac{1/6!}{1 - 1/7} < 0.00163$.

(b) $u_{n+1} = \dfrac{1}{(n+1)!}$, $r_{n+1} = \dfrac{1}{n+2}$, $\dfrac{u_{n+1}}{1 - r_{n+1}} = \dfrac{n+2}{(n+1)(n+1)!} \leq 10^{-5}$ if $n = 8$.

41. **(a)** $s_7 \approx 0.69226$, $r_k = \dfrac{k}{2(k+1)}$ which is increasing for $k \geq 8$ and $\rho = \lim\limits_{k \to +\infty} r_k = 1/2$,

$u_8 = 1/[(8)(2^8)]$, $S - s_7 < \dfrac{1/2^{11}}{1 - 1/2} < 0.00098$.

(b) $u_{n+1} = \dfrac{1}{(n+1)2^{n+1}}$, $\dfrac{u_{n+1}}{1 - \rho} = \dfrac{1}{(n+1)2^n} \leq 10^{-5}$ if $n = 13$.

EXERCISE SET 11.6

1. compare with the convergent series $\displaystyle\sum_{k=1}^{\infty} 1/k^5$, $\rho = \lim\limits_{k \to +\infty} \dfrac{4k^7 - 2k^6 + 6k^5}{8k^7 + k - 8} = 1/2$, converges

3. compare with the convergent series $\displaystyle\sum_{k=1}^{\infty} 5/3^k$, $\rho = \lim\limits_{k \to +\infty} \dfrac{3^k}{3^k + 1} = 1$, converges

5. compare with the divergent series $\displaystyle\sum_{k=1}^{\infty} \dfrac{1}{k^{2/3}}$,

$\rho = \lim\limits_{k \to +\infty} \dfrac{k^{2/3}}{(8k^2 - 3k)^{1/3}} = \lim\limits_{k \to +\infty} \dfrac{1}{(8 - 3/k)^{1/3}} = 1/2$, diverges

7. $\dfrac{1}{3^k + 5} < \dfrac{1}{3^k}$, $\displaystyle\sum_{k=1}^{\infty} \dfrac{1}{3^k}$ converges

9. $\dfrac{1}{5k^2 - k} \leq \dfrac{1}{5k^2 - k^2} = \dfrac{1}{4k^2}$, $\displaystyle\sum_{k=1}^{\infty} \dfrac{1}{4k^2}$ converges

11. $\dfrac{2^k - 1}{3^k + 2k} < \dfrac{2^k}{3^k} = (2/3)^k$, $\displaystyle\sum_{k=1}^{\infty} (2/3)^k$ converges

13. $\dfrac{3}{k - 1/4} > \dfrac{3}{k}$, $\displaystyle\sum_{k=1}^{\infty} 3/k$ diverges

15. $\dfrac{9}{\sqrt{k} + 1} \geq \dfrac{9}{\sqrt{k} + \sqrt{k}} = \dfrac{9}{2\sqrt{k}}$, $\displaystyle\sum_{k=1}^{\infty} \dfrac{9}{2\sqrt{k}}$ diverges

17. $\dfrac{k^{4/3}}{8k^2 + 5k + 1} \geq \dfrac{k^{4/3}}{8k^2 + 5k^2 + k^2} = \dfrac{1}{14k^{2/3}}$, $\displaystyle\sum_{k=1}^{\infty} \dfrac{1}{14k^{2/3}}$ diverges

19. $\dfrac{1}{k^3 + 2k + 1} < \dfrac{1}{k^3}$, $\displaystyle\sum_{k=1}^{\infty} 1/k^3$ converges so $\displaystyle\sum_{k=1}^{\infty} \dfrac{1}{k^3 + 2k + 1}$ converges by the comparison test

21. $\dfrac{1}{9k - 2} > \dfrac{1}{9k}$, $\displaystyle\sum_{k=1}^{\infty} \dfrac{1}{9k}$ diverges so $\displaystyle\sum_{k=1}^{\infty} \dfrac{1}{9k - 2}$ diverges by the comparison test

23. limit comparison test, compare with the convergent series $\displaystyle\sum_{k=1}^{\infty} 1/k^{5/2}$,

$\rho = \displaystyle\lim_{k \to +\infty} \dfrac{k^3}{k^3 + 1} = 1$, converges

25. limit comparison test, compare with the divergent series $\displaystyle\sum_{k=1}^{\infty} 1/k$,

$\rho = \displaystyle\lim_{k \to +\infty} \dfrac{k}{\sqrt{k^2 + k}} = 1$, diverges

27. limit comparison test, compare with the convergent series $\displaystyle\sum_{k=1}^{\infty} 1/k^{5/2}$,

$\rho = \displaystyle\lim_{k \to +\infty} \dfrac{k^3 + 2k^{5/2}}{k^3 + 3k^2 + 3k} = 1$, converges

29. diverges because $\displaystyle\lim_{k \to +\infty} \dfrac{1}{4 + 2^{-k}} = 1/4 \neq 0$

31. $\dfrac{\tan^{-1} k}{k^2} < \dfrac{\pi/2}{k^2}$, $\displaystyle\sum_{k=1}^{\infty} \dfrac{\pi/2}{k^2}$ converges so $\displaystyle\sum_{k=1}^{\infty} \dfrac{\tan^{-1} k}{k^2}$ converges

33. $\displaystyle\sum_{k=1}^{\infty} \dfrac{\ln k}{k\sqrt{k}} = \sum_{k=2}^{\infty} \dfrac{\ln k}{k\sqrt{k}}$ because $\ln 1 = 0$,

$\displaystyle\int_{2}^{+\infty} \dfrac{\ln x}{x^{3/2}} dx = \lim_{\ell \to +\infty} \left[-\dfrac{2\ln x}{x^{1/2}} - \dfrac{4}{x^{1/2}} \right]_{2}^{\ell} = \sqrt{2}(\ln 2 + 2)$ so $\displaystyle\sum_{k=2}^{\infty} \dfrac{\ln k}{k^{3/2}}$ converges.

35. $\rho = \lim\limits_{k \to +\infty} \dfrac{1 - \cos(1/k)}{1/k^2}$, but

$\lim\limits_{x \to +\infty} \dfrac{1 - \cos(1/x)}{1/x^2} = \lim\limits_{x \to +\infty} \dfrac{(-1/x^2)\sin(1/x)}{-2/x^3} = \lim\limits_{x \to +\infty} \dfrac{\sin(1/x)}{2(1/x)} = 1/2$, so $\rho = 1/2$;

the series converges.

37. $\dfrac{\ln k}{k^2} < \dfrac{\sqrt{k}}{k^2} = \dfrac{1}{k^{3/2}}$, $\sum\limits_{k=1}^{\infty} \dfrac{1}{k^{3/2}}$ converges so $\sum\limits_{k=1}^{\infty} \dfrac{\ln k}{k^2}$ converges

39. limit comparison test, compare with $\sum\limits_{k=1}^{\infty} 1/k^p$,

$\rho = \lim\limits_{k \to +\infty} \dfrac{k^p}{(a + bk)^p} = \lim\limits_{k \to +\infty} \dfrac{1}{(a/k + b)^p} = 1/b^p$, converges for $p > 1$.

41. Compare with the convergent series $\sum\limits_{k=1}^{\infty} 1/k!$, $\rho = \lim\limits_{k \to +\infty} \dfrac{(k+1)^2 k!}{(k+2)!} = \lim\limits_{k \to +\infty} \dfrac{k+1}{k+2} = 1$, converges

43. $k! = k(k-1)(k-2)\cdots(2)(1) \geq 2 \cdot 2 \cdot 2 \cdots 2 \cdot 1 = 2^{k-1}$, $1/k! \leq 1/2^{k-1}$, $\sum\limits_{k=1}^{\infty} 1/2^{k-1}$ converges

so $\sum\limits_{k=1}^{\infty} 1/k!$ converges

EXERCISE SET 11.7

1. converges

3. diverges because $\lim\limits_{k \to +\infty} a_k = \lim\limits_{k \to +\infty} \dfrac{k+1}{3k+1} = 1/3 \neq 0$

5. converges

7. $\rho = \lim\limits_{k \to +\infty} \dfrac{(3/5)^{k+1}}{(3/5)^k} = 3/5$, converges absolutely

9. $\rho = \lim\limits_{k \to +\infty} \dfrac{3k^2}{(k+1)^2} = 3$, diverges

11. $\rho = \lim\limits_{k \to +\infty} \dfrac{(k+1)^3}{ek^3} = 1/e$, converges absolutely

13. conditionally convergent, $\sum\limits_{k=1}^{\infty} \dfrac{(-1)^{k+1}}{3k}$ converges by the alternating series test but $\sum\limits_{k=1}^{\infty} \dfrac{1}{3k}$ diverges

15. divergent, $\lim\limits_{k \to +\infty} a_k \neq 0$

17. $\sum\limits_{k=1}^{\infty} \dfrac{\cos k\pi}{k} = \sum\limits_{k=1}^{\infty} \dfrac{(-1)^k}{k}$ is conditionally convergent, $\sum\limits_{k=1}^{\infty} \dfrac{(-1)^k}{k}$ converges by the alternating series test but $\sum\limits_{k=1}^{\infty} 1/k$ diverges.

19. absolutely convergent, $\sum\limits_{k=1}^{\infty} \left[\dfrac{k+2}{3k-1} \right]^k$ converges by the root test.

21. conditionally convergent, $\sum\limits_{k=1}^{\infty} (-1)^{k+1} \dfrac{k+2}{k(k+3)}$ converges by the alternating series test but

$\sum\limits_{k=1}^{\infty} \dfrac{k+2}{k(k+3)}$ diverges (limit comparison test with $\sum 1/k$)

23. $\sum\limits_{k=1}^{\infty} \sin(k\pi/2) = 1+0-1+0+1+0-1+0+\cdots$, divergent ($\lim\limits_{k\to+\infty} \sin(k\pi/2)$ does not exist)

25. conditionally convergent, $\sum\limits_{k=2}^{\infty} \dfrac{(-1)^k}{k \ln k}$ converges by the alternating series test but $\sum\limits_{k=2}^{\infty} \dfrac{1}{k \ln k}$ diverges (integral test)

27. absolutely convergent, $\sum\limits_{k=2}^{\infty} (1/\ln k)^k$ converges by the root test

29. conditionally convergent, let $f(x) = \dfrac{x^2+1}{x^3+2}$ then $f'(x) = \dfrac{x(4-3x-x^3)}{(x^3+2)^2} \leq 0$ for $x \geq 2$

so $\{a_k\}_{k=2}^{+\infty} = \left\{ \dfrac{k^2+1}{k^3+2} \right\}_{k=2}^{+\infty}$ is nonincreasing, $\lim\limits_{k\to+\infty} a_k = 0$; the series converges by the alternating series test but $\sum\limits_{k=2}^{\infty} \dfrac{k^2+1}{k^3+2}$ diverges (limit comparison test with $\sum 1/k$)

31. $|\text{error}| < a_8 = 1/8 = 0.125$ **33.** $|\text{error}| < a_{100} = 1/\sqrt{100} = 0.1$

35. $|\text{error}| < 0.0001$ if $a_{n+1} \le 0.0001$, $1/(n+1) \le 0.0001$, $n+1 \ge 10,000$, $n \ge 9,999$; $n = 9,999$

37. $|\text{error}| < 0.005$ if $a_{n+1} \le 0.005$, $1/\sqrt{n+1} \le 0.005$, $\sqrt{n+1} \ge 200$, $n+1 \ge 40,000$, $n \ge 39,999$; $n = 39,999$

39. $a_k = \dfrac{3}{2^{k+1}}$, $|\text{error}| < a_{11} = \dfrac{3}{2^{12}} < 0.00074$; $s_{10} \approx 0.4995$; $S = \dfrac{3/4}{1-(-1/2)} = 0.5$.

41. $a_k = \dfrac{1}{(2k-1)!}$, $a_{n+1} = \dfrac{1}{(2n+1)!} \le 10^{-4}$, $(2n+1)! \ge 10,000$, $2n+1 \ge 8$, $n \ge 3.5$; $n = 4$.
$s_4 \approx 0.84147$, $\sin(1) \approx 0.841470985$.

43. $a_k = \dfrac{1}{k2^k}$, $a_{n+1} = \dfrac{1}{(n+1)2^{n+1}} \le 10^{-4}$, $(n+1)2^{n+1} \ge 10,000$, $n+1 \ge 10$, $n \ge 9$; $n = 9$.
$s_9 \approx 0.40553$, $\ln\dfrac{3}{2} \approx 0.405465108$.

45. **(a)** $a_k = \dfrac{1}{k^2}$, $a_{n+1} = \dfrac{1}{(n+1)^2} \le 5 \times 10^{-3}$, $(n+1)^2 \ge 200$, $n \ge 14$; $n = 14$

     (b) $s_{10} \approx 0.817962176$, $\pi^2/12 \approx 0.822467033$, $|\text{error}| \approx 0.004504858$

47. Suppose $\Sigma|a_k|$ converges, then $\lim\limits_{k \to +\infty} |a_k| = 0$ so $|a_k| < 1$ for $k \ge K$ and thus $|a_k|^2 < |a_k|$, $a_k^2 < |a_k|$ hence Σa_k^2 converges by the comparison test.

49. Suppose $a_1 - a_2 + a_3 - a_4 + \cdots + (-1)^{k+1}a_k + \cdots$ satisfies (a) and (b) of Theorem 11.7.1, then the series converges and by Theorem 11.4.3b so does $- \left(a_1 - a_2 + a_3 - a_4 + \cdots + (-1)^{k+1}a_k + \cdots\right)$ which equals $-a_1 + a_2 - a_3 + a_4 - \cdots + (-1)^k a_k + \cdots$

51. $\left(1 - \dfrac{1}{2} - \dfrac{1}{4}\right) + \left(\dfrac{1}{3} - \dfrac{1}{6} - \dfrac{1}{8}\right) + \left(\dfrac{1}{5} - \dfrac{1}{10} - \dfrac{1}{12}\right) + \cdots$

$= \left(\dfrac{1}{2} - \dfrac{1}{4}\right) + \left(\dfrac{1}{6} - \dfrac{1}{8}\right) + \left(\dfrac{1}{10} - \dfrac{1}{12}\right) + \cdots = \dfrac{1}{2}\left(1 - \dfrac{1}{2} + \dfrac{1}{3} - \dfrac{1}{4} + \dfrac{1}{5} - \dfrac{1}{6} + \cdots\right) = S/2$

53. $1 + \dfrac{1}{3^2} + \dfrac{1}{5^2} + \cdots = \left[1 + \dfrac{1}{2^2} + \dfrac{1}{3^2} + \cdots\right] - \left[\dfrac{1}{2^2} + \dfrac{1}{4^2} + \dfrac{1}{6^2} + \cdots\right]$

$= \dfrac{\pi^2}{6} - \dfrac{1}{2^2}\left[1 + \dfrac{1}{2^2} + \dfrac{1}{3^2} + \cdots\right] = \dfrac{\pi^2}{6} - \dfrac{1}{4}\dfrac{\pi^2}{6} = \dfrac{\pi^2}{8}$

55. $1 - \dfrac{1}{2^2} + \dfrac{1}{3^2} - \cdots = \left[1 + \dfrac{1}{2^2} + \dfrac{1}{3^2} + \cdots\right] - 2\left[\dfrac{1}{2^2} + \dfrac{1}{4^2} + \dfrac{1}{6^2} + \cdots\right]$

$$= \dfrac{\pi^2}{6} - \dfrac{2}{2^2}\left[1 + \dfrac{1}{2^2} + \dfrac{1}{3^2} + \cdots\right] = \dfrac{\pi^2}{6} - \dfrac{1}{2}\dfrac{\pi^2}{6} = \dfrac{\pi^2}{12}$$

EXERCISE SET 11.8

1. $\rho = \lim\limits_{k \to +\infty} \dfrac{k+1}{k+2}|x| = |x|$, the series converges if $|x| < 1$ and diverges if $|x| > 1$. If $x = -1$,

$\displaystyle\sum_{k=0}^{\infty} \dfrac{(-1)^k}{k+1}$ converges by the alternating series test; if $x = 1$, $\displaystyle\sum_{k=0}^{\infty} \dfrac{1}{k+1}$ diverges. The radius of

convergence is 1, the interval of convergence is $[-1, 1)$.

3. $\rho = \lim\limits_{k \to +\infty} \dfrac{|x|}{k+1} = 0$, the radius of convergence is $+\infty$, the interval is $(-\infty, +\infty)$.

5. $\rho = \lim\limits_{k \to +\infty} \dfrac{5k^2|x|}{(k+1)^2} = 5|x|$, converges if $|x| < 1/5$ and diverges if $|x| > 1/5$. If $x = -1/5$,

$\displaystyle\sum_{k=1}^{\infty} \dfrac{(-1)^k}{k^2}$ converges; if $x = 1/5$, $\displaystyle\sum_{k=1}^{\infty} 1/k^2$ converges. Radius of convergence is $1/5$, interval of

convergence is $[-1/5, 1/5]$.

7. $\rho = \lim\limits_{k \to +\infty} \dfrac{k|x|}{k+2} = |x|$, converges if $|x| < 1$, diverges if $|x| > 1$. If $x = -1$, $\displaystyle\sum_{k=1}^{\infty} \dfrac{(-1)^k}{k(k+1)}$

converges; if $x = 1$, $\displaystyle\sum_{k=1}^{\infty} \dfrac{1}{k(k+1)}$ converges. Radius of convergence is 1, interval of convergence

is $[-1, 1]$.

9. $\rho = \lim\limits_{k \to +\infty} \dfrac{\sqrt{k}}{\sqrt{k+1}}|x| = |x|$, converges if $|x| < 1$, diverges if $|x| > 1$. If $x = -1$, $\displaystyle\sum_{k=1}^{\infty} \dfrac{-1}{\sqrt{k}}$

diverges; if $x = 1$, $\displaystyle\sum_{k=1}^{\infty} \dfrac{(-1)^{k-1}}{\sqrt{k}}$ converges. Radius of convergence is 1, interval of convergence

is $(-1, 1]$.

11. $\rho = \lim\limits_{k \to +\infty} \dfrac{|x|^2}{(2k+3)(2k+2)} = 0$, radius of convergence is $+\infty$, interval of convergence

is $(-\infty, +\infty)$.

13. $\rho = \lim\limits_{k\to+\infty} \dfrac{3|x|}{k+1} = 0$, radius of convergence is $+\infty$, interval of convergence is $(-\infty, +\infty)$.

15. $\rho = \lim\limits_{k\to+\infty} \dfrac{1+k^2}{1+(k+1)^2}|x| = |x|$, converges if $|x| < 1$, diverges if $|x| > 1$. If $x = -1$, $\sum\limits_{k=0}^{\infty} \dfrac{(-1)^k}{1+k^2}$

converges; if $x = 1$, $\sum\limits_{k=0}^{\infty} \dfrac{1}{1+k^2}$ converges. Radius of convergence is 1, interval of convergence is $[-1, 1]$.

17. $\rho = \lim\limits_{k\to+\infty} \dfrac{k|x+1|}{k+1} = |x+1|$, converges if $|x+1| < 1$, diverges if $|x+1| > 1$. If $x = -2$, $\sum\limits_{k=1}^{\infty} \dfrac{-1}{k}$

diverges; if $x = 0$, $\sum\limits_{k=1}^{\infty} \dfrac{(-1)^{k+1}}{k}$ converges. Radius of convergence is 1, interval of convergence is $(-2, 0]$.

19. $\rho = \lim\limits_{k\to+\infty} (3/4)|x+5| = \dfrac{3}{4}|x+5|$, convergence if $|x+5| < 4/3$, diverges if $|x+5| > 4/3$. If $x = -19/3$, $\sum\limits_{k=0}^{\infty} (-1)^k$ diverges; if $x = -11/3$, $\sum\limits_{k=0}^{\infty} 1$ diverges. Radius of convergence is $4/3$, interval of convergence is $(-19/3, -11/3)$.

21. $\rho = \lim\limits_{k\to+\infty} \dfrac{k^2+4}{(k+1)^2+4}|x+1|^2 = |x+1|^2$, converges if $|x+1| < 1$, diverges if $|x+1| > 1$. If $x = -2$, $\sum\limits_{k=1}^{\infty} \dfrac{(-1)^{3k+1}}{k^2+4}$ converges; if $x = 0$, $\sum\limits_{k=1}^{\infty} \dfrac{(-1)^k}{k^2+4}$ converges. Radius of convergence is 1, interval of convergence is $[-2, 0]$.

23. $\rho = \lim\limits_{k\to+\infty} \dfrac{\pi|x-1|^2}{(2k+3)(2k+2)} = 0$, radius of convergence $+\infty$, interval of convergence $(-\infty, +\infty)$.

25. $x + \dfrac{1}{2}x^2 + \dfrac{3}{14}x^3 + \dfrac{3}{35}x^4 + \cdots$; $\rho = \lim\limits_{k\to+\infty} \dfrac{k+1}{3k+1}|x| = \dfrac{1}{3}|x|$, converges if $\dfrac{1}{3}|x| < 1$, $|x| < 3$ so $R = 3$.

27. $x + \dfrac{3}{2}x^2 + \dfrac{5}{8}x^3 + \dfrac{7}{48}x^4 + \cdots$; $\rho = \lim\limits_{k\to+\infty} \dfrac{2k+1}{(2k)(2k-1)}|x| = 0$ so $R = +\infty$.

29. By the ratio test for absolute convergence, $\rho = \lim\limits_{k\to+\infty} \dfrac{|x-a|}{b} = \dfrac{|x-a|}{b}$; converges if $|x-a| < b$, diverges if $|x-a| > b$. If $x = a - b$, $\sum\limits_{k=0}^{\infty} (-1)^k$ diverges; if $x = a + b$, $\sum\limits_{k=0}^{\infty} 1$ diverges. The interval of convergence is $(a-b, a+b)$.

31. By the ratio test for absolute convergence,

$$\rho = \lim_{k \to +\infty} \frac{(k+1+p)!k!(k+q)!}{(k+p)!(k+1)!(k+1+q)!}|x| = \lim_{k \to +\infty} \frac{k+1+p}{(k+1)(k+1+q)}|x| = 0,$$

radius of convergence is $+\infty$.

33. By assumption $\displaystyle\sum_{k=0}^{\infty} c_k x^k$ converges if $|x| < R$ so $\displaystyle\sum_{k=0}^{\infty} c_k x^{2k} = \sum_{k=0}^{\infty} c_k (x^2)^k$ converges if $|x^2| < R$,

$|x| < \sqrt{R}$. Thus $\displaystyle\sum_{k=0}^{\infty} c_k x^{2k}$ has radius of convergence $\sqrt{R}$.

EXERCISE SET 11.9

1. $1 - 2x + 2x^2 - \dfrac{4}{3}x^3 + \dfrac{2}{3}x^4$

3. $2x - \dfrac{4}{3}x^3$

5. $x + \dfrac{1}{3}x^3$

7. $x + x^2 + \dfrac{1}{2}x^3 + \dfrac{1}{6}x^4$

9. $1 + \dfrac{1}{2}x^2 + \dfrac{5}{24}x^4$

11. $\ln 3 + \dfrac{2}{3}x - \dfrac{2}{9}x^2 + \dfrac{8}{81}x^3 - \dfrac{4}{81}x^4$

13. $e + e(x-1) + \dfrac{e}{2}(x-1)^2 + \dfrac{e}{6}(x-1)^3$

15. $2 + \dfrac{1}{4}(x-4) - \dfrac{1}{64}(x-4)^2 + \dfrac{1}{512}(x-4)^3$

17. $\dfrac{\sqrt{2}}{2} - \dfrac{\sqrt{2}}{2}(x - \pi/4) - \dfrac{\sqrt{2}}{4}(x - \pi/4)^2 + \dfrac{\sqrt{2}}{12}(x - \pi/4)^3$

19. $-\dfrac{\sqrt{3}}{2} + \dfrac{\pi}{2}(x + 1/3) + \dfrac{\sqrt{3}\pi^2}{4}(x + 1/3)^2 - \dfrac{\pi^3}{12}(x + 1/3)^3$

21. $\dfrac{\pi}{4} + \dfrac{1}{2}(x-1) - \dfrac{1}{4}(x-1)^2 + \dfrac{1}{12}(x-1)^3$

23. $f^{(k)}(x) = (-1)^k e^{-x},\ f^{(k)}(0) = (-1)^k;\ \displaystyle\sum_{k=0}^{\infty} \frac{(-1)^k}{k!}x^k$

25. $f^{(k)}(x) = \dfrac{(-1)^k k!}{(1+x)^{k+1}},\ f^{(k)}(0) = (-1)^k k!;\ \displaystyle\sum_{k=0}^{\infty}(-1)^k x^k$

27. $f^{(0)}(0) = 0$; for $k \geq 1$, $f^{(k)}(x) = \dfrac{(-1)^{k+1}(k-1)!}{(1+x)^k},\ f^{(k)}(0) = (-1)^{k+1}(k-1)!;\ \displaystyle\sum_{k=1}^{\infty} \frac{(-1)^{k+1}}{k}x^k$

29. $f^{(k)}(0) = 0$ if k is odd, $f^{(k)}(0)$ is alternately $1/2^k$ $and - 1/2^k$ if k is even.

$$\sum_{k=0}^{\infty} \frac{f^{(k)}(0)}{k!} x^k = 1 - \frac{1}{2^2 2!} x^2 + \frac{1}{2^4 4!} x^4 - \cdots = \sum_{k=0}^{\infty} \frac{(-1)^k}{2^{2k}(2k)!} x^{2k}$$

31. $f^{(k)}(0) = 0$ if k is odd, $f^{(k)}(0) = 1$ if k is even;

$$\sum_{k=0}^{\infty} \frac{f^{(k)}(0)}{k!} x^k = 1 + \frac{1}{2!} x^2 + \frac{1}{4!} x^4 + \cdots = \sum_{k=0}^{\infty} \frac{1}{(2k)!} x^{2k}$$

33. $f^{(k)}(x) = \frac{(-1)^k k!}{x^{k+1}}$, $f^{(k)}(-1) = -k!$; $\displaystyle\sum_{k=0}^{\infty} (-1)(x+1)^k$

35. $f^{(0)}(1) = 0$; for $k \geq 1$, $f^{(k)}(x) = \frac{(-1)^{k+1}(k-1)!}{x^k}$, $f^{(k)}(1) = (-1)^{k+1}(k-1)!$;

$$\sum_{k=1}^{\infty} \frac{(-1)^{k+1}}{k}(x-1)^k$$

37. $f^{(k)}(1/2) = 0$ if k is odd, $f^{(k)}(1/2)$ is alternately π^k and $-\pi^k$ if k is even;

$$\sum_{k=0}^{\infty} \frac{f^{(k)}(1/2)}{k!}(x-1/2)^k = 1 - \frac{\pi^2}{2!}(x-1/2)^2 + \frac{\pi^4}{4!}(x-1/2)^4 - \cdots$$

$$= \sum_{k=0}^{\infty} \frac{(-1)^k \pi^{2k}}{(2k)!}(x-1/2)^{2k}$$

39. $f^{(k)}(\ln 4) = 15/8$ for k even, $f^{(k)}(\ln 4) = 17/8$ for k odd, which can be written as

$$f^{(k)}(\ln 4) = \frac{16 - (-1)^k}{8}; \sum_{k=0}^{\infty} \frac{16 - (-1)^k}{8k!}(x - \ln 4)^k$$

EXERCISE SET 11.10

1. $f^{(6)}(x) = 2^6 e^{2x}$, $R_5(x) = \frac{2^6 e^{2c}}{6!} x^6$ **3.** $f^{(5)}(x) = -5!/(x+1)^6$, $R_4(x) = -\frac{1}{(c+1)^6} x^5$

5. $f^{(4)}(x) = (4+x)e^x$, $R_3(x) = \frac{(4+c)e^c}{4!} x^4$

7. $f^{(3)}(x) = -\frac{2(1 - 3x^2)}{(1 + x^2)^3}$, $R_2(x) = -\frac{1 - 3c^2}{3(1 + c^2)^3} x^3$

9. $f^{(4)}(x) = -\dfrac{15}{16}x^{-7/2}$, $R_3(x) = -\dfrac{5}{128c^{7/2}}(x-4)^4$

11. $f^{(5)}(x) = \cos x$, $R_4(x) = \dfrac{\cos c}{5!}(x - \pi/6)^5$

13. $f^{(6)}(x) = 7!/(1+x)^8$, $R_5(x) = \dfrac{7}{(1+c)^8}(x+2)^6$

15. $f^{(n+1)}(x) = (n+1)!/(1-x)^{n+2}$, $R_n(x) = \dfrac{1}{(1-c)^{n+2}}x^{n+1}$

17. $f^{(n+1)}(x) = 2^{n+1}e^{2x}$, $R_n(x) = \dfrac{2^{n+1}e^{2c}}{(n+1)!}x^{n+1}$

19. $f(x) = \sin x$, $f^{(n+1)}(x) = \pm\sin x$ or $\pm\cos x$, $|f^{(n+1)}(x)| \le 1$,

$|R_n(x)| = \dfrac{|f^{(n+1)}(c)|}{(n+1)!}|x|^{n+1} \le \dfrac{|x|^{n+1}}{(n+1)!}$, $\displaystyle\lim_{n\to+\infty}\dfrac{|x|^{n+1}}{(n+1)!} = 0$, by the Squeezing Theorem

$\displaystyle\lim_{n\to+\infty}|R_n(x)| = 0$ so $\displaystyle\lim_{n\to+\infty}R_n(x) = 0$ for all x.

21. $f(x) = e^x$, $f^{(n+1)}(x) = e^x$, $|R_n(x)| = \dfrac{e^c}{(n+1)!}|x-1|^{n+1}$. If $x \ge 1$, $e^c \le e^x$,

$|R_n(x)| \le e^x\dfrac{|x-1|^{n+1}}{(n+1)!}$, $\displaystyle\lim_{n\to+\infty}e^x\dfrac{|x-1|^{n+1}}{(n+1)!} = e^x(0) = 0$ so $\displaystyle\lim_{n\to+\infty}R_n(x) = 0$.

If $x < 1$, $e^c < e$, $|R_n(x)| < e\dfrac{|x-1|^{n+1}}{(n+1)!}$ and again $\displaystyle\lim_{n\to+\infty}R_n(x) = 0$.

23. $f(x) = e^x$, $f^{(n+1)}(x) = e^x$, $|R_n(x)| = \dfrac{e^c}{(n+1)!}|x-a|^{n+1}$. If $x \ge a$, $e^c \le e^x$,

$|R_n(x)| \le e^x\dfrac{|x-a|^{n+1}}{(n+1)!}$, $\displaystyle\lim_{n\to+\infty}e^x\dfrac{|x-a|^{n+1}}{(n+1)!} = e^x(0) = 0$ so $\displaystyle\lim_{n\to+\infty}R_n(x) = 0$.

If $x < a$, $e^c < e^a$, $|R_n(x)| < e^a\dfrac{|x-a|^{n+1}}{(n+1)!}$ and again $\displaystyle\lim_{n\to+\infty}R_n(x) = 0$.

25. $f(x) = \cos x$, $f^{(n+1)}(x) = \pm\sin x$ or $\pm\cos x$, $|f^{(n+1)}(x)| \le 1$,

$|R_n(x)| = \dfrac{|f^{(n+1)}(c)|}{(n+1)!}|x-a|^{n+1} \le \dfrac{|x-a|^{n+1}}{(n+1)!}$, $\displaystyle\lim_{n\to+\infty}\dfrac{|x-a|^{n+1}}{(n+1)!} = 0$ so $\displaystyle\lim_{n\to+\infty}R_n(x) = 0$

for all x.

27. $1 - 2x + 2x^2 - \dfrac{4}{3}x^3 + \cdots;\ (-\infty, +\infty)$

29. $x\left(1 - x + \dfrac{1}{2!}x^2 - \dfrac{1}{3!}x^3 + \cdots\right) = x - x^2 + \dfrac{1}{2!}x^3 - \dfrac{1}{3!}x^4 + \cdots;\ (-\infty, +\infty)$

31. $2x - \dfrac{2^3}{3!}x^3 + \dfrac{2^5}{5!}x^5 - \dfrac{2^7}{7!}x^7 + \cdots;\ (-\infty, +\infty)$

33. $x^2\left(1 - \dfrac{1}{2!}x^2 + \dfrac{1}{4!}x^4 - \dfrac{1}{6!}x^6 + \cdots\right) = x^2 - \dfrac{1}{2!}x^4 + \dfrac{1}{4!}x^6 - \dfrac{1}{6!}x^8 + \cdots;\ (-\infty, +\infty)$

35. $\dfrac{1}{2}\left[1 - \left(1 - \dfrac{2^2}{2!}x^2 + \dfrac{2^4}{4!}x^4 - \dfrac{2^6}{6!}x^6 + \cdots\right)\right] = x^2 - \dfrac{2^3}{4!}x^4 + \dfrac{2^5}{6!}x^6 - \dfrac{2^7}{8!}x^8 + \cdots;\ (-\infty, +\infty)$

37. $-x^2 - \dfrac{1}{2}x^4 - \dfrac{1}{3}x^6 - \dfrac{1}{4}x^8 - \cdots;\ (-1, 1)$

39. $1 + 4x^2 + 16x^4 + 64x^6 + \cdots;\ (-1/2, 1/2)$

41. $x^2\left(1 - 3x + 9x^2 - 27x^3 + \cdots\right) = x^2 - 3x^3 + 9x^4 - 27x^5 + \cdots;\ (-1/3, 1/3)$

43. $x\left(2x + \dfrac{2^3}{3!}x^3 + \dfrac{2^5}{5!}x^5 + \dfrac{2^7}{7!}x^7 + \cdots\right) = 2x^2 + \dfrac{2^3}{3!}x^4 + \dfrac{2^5}{5!}x^6 + \dfrac{2^7}{7!}x^8 + \cdots;\ (-\infty, +\infty)$

45. $1 + \dfrac{3}{2}x - \dfrac{9}{8}x^2 + \dfrac{27}{16}x^3 - \cdots;\ (-1/3, 1/3)$

47. $(1 - 2x)^{-2} = 1 + 4x + 12x^2 + 32x^3 + \cdots;\ (-1/2, 1/2)$

49. $x(1 - x^2)^{-1/2} = x + \dfrac{1}{2}x^3 + \dfrac{3}{8}x^5 + \dfrac{5}{16}x^7 + \cdots;\ (-1, 1)$

51. $\dfrac{1}{x} = \dfrac{1}{1 + (x - 1)} = \dfrac{1}{1 - [-(x - 1)]} = \sum_{k=0}^{\infty}[-(x - 1)]^k = \sum_{k=0}^{\infty}(-1)^k(x - 1)^k$ which is valid for $-1 < -(x - 1) < 1,\ 0 < x < 2$

53. $(1 + x)^m = \dbinom{m}{0} + \sum_{k=1}^{\infty}\dbinom{m}{k}x^k = \sum_{k=0}^{\infty}\dbinom{m}{k}x^k$

55. $\sin\pi = 0$ 　　　　　　　　　　　　**57.** $e^{-\ln 3} = 1/3$

59. (a) $\cos 2x = 1 - \dfrac{(2x)^2}{2!} + \dfrac{(2x)^4}{4!} - \dfrac{(2x)^6}{6!} + \cdots = 1 - 2x^2 + \dfrac{2}{3}x^4 - \dfrac{4}{45}x^6 + \cdots,$

$x^2 \cos 2x = x^2 - 2x^4 + \dfrac{2}{3}x^6 - \dfrac{4}{45}x^8 + \cdots$

(b) $\dfrac{f^{(5)}(0)}{5!} = 0$ so $f^{(5)}(0) = 0$

EXERCISE SET 11.11

1. (a) $3/(n+1)! < 0.5 \times 10^{-5}, \ n = 9$ **(b)** $3/(n+1)! < 0.5 \times 10^{-10}, \ n = 13$

3. $|R_n(1/2)| = \dfrac{e^c}{(n+1)!}(1/2)^{n+1} < \dfrac{2}{(n+1)!}\dfrac{1}{2^{n+1}} = \dfrac{1}{2^n(n+1)!} < 0.5 \times 10^{-4}$ if $n = 5$

so $\sqrt{e} = e^{0.5} \approx 1 + 0.5 + \dfrac{(0.5)^2}{2!} + \dfrac{(0.5)^3}{3!} + \dfrac{(0.5)^4}{4!} + \dfrac{(0.5)^5}{5!}, \ \sqrt{e} \approx 1.6487$

5. $|R_n(\pi/20)| \le \dfrac{(\pi/20)^{n+1}}{(n+1)!} < 0.5 \times 10^{-4}$ if $n = 3$, $\cos(\pi/20) \approx 1 - \dfrac{(\pi/20)^2}{2!} \approx 0.9877$

7. Let $x = 1/9$ in series (16) to get $\ln 1.25 \approx 0.223$

9. $(0.1)^3/3 < 0.5 \times 10^{-3}$ so $\tan^{-1}(0.1) \approx 0.100$ to three decimal place accuracy.

11. $|R_n(0.1)| = \dfrac{|f^{(n+1)}(c)|}{(n+1)!}(0.1)^{n+1}$ where $f^{(n+1)}(c) = \sinh c$ or $\cosh c$ for $0 < c < 0.1$, but

$\sinh c < \cosh c < \cosh 0.1$, and $\cosh 0.1 = \dfrac{1}{2}\left(e^{0.1} + e^{-0.1}\right) < \dfrac{1}{2}(2+1) = 1.5$

so $|R_n(0.1)| < \dfrac{1.5(0.1)^{n+1}}{(n+1)!} \le 0.5 \times 10^{-4}$ if $n = 3$, $\cosh 0.1 \approx 1 + \dfrac{(0.1)^2}{2!} \approx 1.0050$

13. $\sin x = x - \dfrac{x^3}{3!} + (0)x^4 + R_4(x), \ |R_4(x)| \le \dfrac{|x|^5}{5!} < 0.5 \times 10^{-3}$ if $|x|^5 < 0.06$,

$|x| < (0.06)^{1/5} \approx 0.569$

15. $\cos x = 1 - \dfrac{x^2}{2!} + \dfrac{x^4}{4!} + (0)x^5 + R_5(x), \ |R_5(x)| \le \dfrac{|x|^6}{6!} \le \dfrac{(0.2)^6}{6!} < 9 \times 10^{-8}$

17. **(a)** $\ln 2 = 1 - 1/2 + 1/3 - 1/4 + \cdots$, $|\text{error}| < 1/(n+1) \le 0.5 \times 10^{-6}$ if $n + 1 \ge 2 \times 10^6$, $n \ge 1,999,999$; $n = 1,999,999$

(b) Let $f(x) = \ln \dfrac{1+x}{1-x} = \ln(1+x) - \ln(1-x)$, $f^{(n+1)}(c) = n! \left[\dfrac{(-1)^n}{(1+c)^{n+1}} + \dfrac{1}{(1-c)^{n+1}} \right]$,

$$|f^{n+1}(c)| \le n! \left[\frac{1}{(1+c)^{n+1}} + \frac{1}{(1-c)^{n+1}} \right]$$

$$< n! \left[\frac{1}{(1+0)^{n+1}} + \frac{1}{(1-1/3)^{n+1}} \right] \text{ for } 0 < c < 1/3$$

$$= n![1 + 1/(2/3)^{n+1}],$$

$$|R_n(1/3)| < [1 + 1/(2/3)^{n+1}] \frac{(1/3)^{n+1}}{n+1} = (1/3^{n+1} + 1/2^{n+1})/(n+1) \le 0.5 \times 10^{-6}$$

if $n \ge 16$ so 8 <u>terms</u> in series (16) are sufficient to assure six decimal place accuracy (even powers of x have zero coefficients).

19. $(1/2)^9/9! < 0.5 \times 10^{-3}$ and $(1/3)^7/7! < 0.5 \times 10^{-3}$ so

$$\tan^{-1} 1/2 \approx 1/2 - \frac{(1/2)^3}{3} + \frac{(1/2)^5}{5} - \frac{(1/2)^7}{7} \approx 0.463$$

$$\tan^{-1} 1/3 \approx 1/3 - \frac{(1/3)^3}{3} + \frac{(1/3)^5}{5} \approx 0.322, \ \pi \approx 4(0.463 + 0.322) = 3.140$$

EXERCISE SET 11.12

1. **(a)** $\dfrac{d}{dx} \left(1 + x + x^2/2! + x^3/3! + \cdots \right) = 1 + x + x^2/2! + \cdots = e^x$

(b) $\displaystyle\int \left(1 + x + x^2/2! + \cdots \right) dx = (x + x^2/2! + x^3/3! + \cdots) + C_1$

$$= \left(1 + x + x^2/2! + x^3/3! + \cdots \right) + C_1 - 1 = e^x + C$$

3. **(a)** $\dfrac{d}{dx} \left(x + x^3/3! + x^5/5! + \cdots \right) = 1 + x^2/2! + x^4/4! + \cdots = \cosh x$

(b) $\displaystyle\int \left(x + x^3/3! + x^5/5! + \cdots \right) dx = \left(x^2/2! + x^4/4! + x^6/6! + \cdots \right) + C_1$

$$= \left(1 + x^2/2! + x^4/4! + x^6/6! + \cdots \right) + C_1 - 1$$

$$= \cosh x + C$$

5. $1/(1+x)^2 = \dfrac{d}{dx}\left[-\dfrac{1}{1+x}\right] = \dfrac{d}{dx}\left[-\sum_{k=0}^{\infty}(-1)^k x^k\right]$

$$= \dfrac{d}{dx}\left[\sum_{k=0}^{\infty}(-1)^{k+1}x^k\right] = \sum_{k=1}^{\infty}(-1)^{k+1}kx^{k-1}$$

7. $\ln\dfrac{1}{1-x} = -\ln(1-x) = \displaystyle\int \dfrac{1}{1-x}dx - C = \int\left[\sum_{k=0}^{\infty}x^k\right]dx - C$

$$= \sum_{k=0}^{\infty}\dfrac{x^{k+1}}{k+1} - C = \sum_{k=1}^{\infty}\dfrac{x^k}{k} - C, \ln\dfrac{1}{1-0} = 0 \text{ so } C = 0.$$

9. $x = 1/4,\ S = \ln\dfrac{1}{1-1/4} = \ln\dfrac{4}{3}$

11. $f(x) = \dfrac{1}{1-x} = \displaystyle\sum_{k=0}^{\infty}x^k,\ f'(x) = \dfrac{1}{(1-x)^2} = \sum_{k=1}^{\infty}kx^{k-1},$

$$f''(x) = \dfrac{2}{(1-x)^3} = \sum_{k=2}^{\infty}k(k-1)x^{k-2} = 2 + 6x + 12x^2 + 20x^3 + \cdots$$

13. **(a)** $\rho = \displaystyle\lim_{k\to+\infty}\dfrac{|x|^{k+2}}{k+2}\cdot\dfrac{k+1}{|x|^{k+1}} = |x|\lim_{k\to+\infty}\dfrac{k+1}{k+2} = |x|;$ converges if $|x| < 1$.

(b) $f'(x) = \displaystyle\sum_{k=0}^{\infty}(-1)^k x^k,$ converges on $(-1, 1)$

(c) $f'(x) = \dfrac{1}{1+x}$ so $f(x) = \ln(1+x) + C$ for x in $(-1, 1)$, let $x = 0$ to find that $C = 0$, thus $f(x) = \ln(1+x)$.

15. $\displaystyle\int_0^1 \cos\sqrt{x}\,dx = \int_0^1 \left(1 - x/2! + x^2/4! - x^3/6! + \cdots\right)dx$

$$= x - \dfrac{1}{2\cdot 2!}x^2 + \dfrac{1}{3\cdot 4!}x^3 - \dfrac{1}{4\cdot 6!}x^4 + \cdots\Bigg]_0^1 = 1 - \dfrac{1}{2\cdot 2!} + \dfrac{1}{3\cdot 4!} - \dfrac{1}{4\cdot 6!} + \cdots,$$

but $\dfrac{1}{4\cdot 6!} < 0.5\times 10^{-3}$ so $\displaystyle\int_0^1 \cos\sqrt{x}\,dx \approx 1 - \dfrac{1}{2\cdot 2!} + \dfrac{1}{3\cdot 4!} \approx 0.764$

17. $\displaystyle\int_0^{1/2}\dfrac{dx}{1+x^4} = \int_0^{1/2}\left(1 - x^4 + x^8 - \cdots\right)dx$

$$= x - \dfrac{1}{5}x^5 + \dfrac{1}{9}x^9 - \cdots\Bigg]_0^{1/2} = 1/2 - \dfrac{(1/2)^5}{5} + \dfrac{(1/2)^9}{9} - \cdots,$$

but $\dfrac{(1/2)^9}{9} < 0.5\times 10^{-3}$ so $\displaystyle\int_0^{1/2}\dfrac{dx}{1+x^4} \approx 1/2 - \dfrac{(1/2)^5}{5} \approx 0.494$

19. $\int_0^{0.1} e^{-x^3} dx = \int_0^{0.1} \left(1 - x^3 + x^6/2! - \cdots\right) dx$

$$= x - \frac{1}{4}x^4 + \frac{1}{7\cdot 2!}x^7 - \cdots \Big]_0^{0.1} = 0.1 - \frac{1}{4}(0.1)^4 + \frac{1}{7\cdot 2!}(0.1)^7 - \cdots,$$

but $\dfrac{1}{4}(0.1)^4 < 0.5 \times 10^{-3}$ so $\displaystyle\int_0^{0.1} e^{-x^3} dx \approx 0.100$

21. $\int_0^{1/2} (1 + x^2)^{-1/4} dx = \int_0^{1/2} \left(1 - \frac{1}{4}x^2 + \frac{5}{32}x^4 - \frac{15}{128}x^6 + \cdots\right) dx$

$$= x - \frac{1}{12}x^3 + \frac{1}{32}x^5 - \frac{15}{896}x^7 + \cdots \Big]_0^{1/2}$$

$$= 1/2 - \frac{1}{12}(1/2)^3 + \frac{1}{32}(1/2)^5 - \frac{15}{896}(1/2)^7 + \cdots,$$

but $\dfrac{15}{896}(1/2)^7 < 0.5 \times 10^{-3}$ so $\displaystyle\int_0^{1/2} (1+x^2)^{-1/4} dx \approx 1/2 - \frac{1}{12}(1/2)^3 + \frac{1}{32}(1/2)^5 \approx 0.491$

23. $e^{-x^2} \cos x = \left(1 - x^2 + \dfrac{x^4}{2!} - \dfrac{x^6}{3!} + \cdots\right)\left(1 - \dfrac{x^2}{2!} + \dfrac{x^4}{4!} - \dfrac{x^6}{6!} + \cdots\right)$

$$= 1 - \frac{3}{2}x^2 + \frac{25}{24}x^4 - \frac{331}{720}x^6 + \cdots$$

25. $\dfrac{\sin x}{e^x} = e^{-x}\sin x = \left(1 - x + \dfrac{x^2}{2!} - \dfrac{x^3}{3!} + \dfrac{x^4}{4!} - \cdots\right)\left(x - \dfrac{x^3}{3!} + \dfrac{x^5}{5!} - \cdots\right)$

$$= x - x^2 + \frac{1}{3}x^3 - \frac{1}{30}x^5 + \cdots$$

27. $x\ln\left(1 - x^2\right) = x\left(-x^2 - \dfrac{1}{2}x^4 - \dfrac{1}{3}x^6 - \dfrac{1}{4}x^8 - \cdots\right) = -x^3 - \dfrac{1}{2}x^5 - \dfrac{1}{3}x^7 - \dfrac{1}{4}x^9 - \cdots$

29. $x^2 e^{4x}\sqrt{1+x} = x^2\left(1 + 4x + 8x^2 + \dfrac{32}{3}x^3 + \cdots\right)\left(1 + \dfrac{1}{2}x - \dfrac{1}{8}x^2 + \dfrac{1}{16}x^3 + \cdots\right)$

$$= x^2\left(1 + \frac{9}{2}x + \frac{79}{8}x^2 + \frac{683}{48}x^3 + \cdots\right) = x^2 + \frac{9}{2}x^3 + \frac{79}{8}x^4 + \frac{683}{48}x^5 + \cdots$$

31. (a) $\dfrac{1 - \cos x}{\sin x} = \dfrac{1 - \left(1 - x^2/2! + x^4/4! - x^6/6! + \cdots\right)}{x - x^3/3! + x^5/5! - \cdots}$

$$= \frac{x^2/2! - x^4/4! + x^6/6! - \cdots}{x - x^3/3! + x^5/5! - \cdots} = \frac{x/2! - x^3/4! + x^5/6! - \cdots}{1 - x^2/3! + x^4/5! - \cdots}, x \neq 0$$

$$\lim_{x \to 0} \frac{1 - \cos x}{\sin x} = \frac{0}{1} = 0$$

(b) $\ln\sqrt{1+x} - \sin 2x = \dfrac{1}{2}\ln(1+x) - \sin 2x$

$$= \frac{1}{2}\left(x - \frac{1}{2}x^2 + \frac{1}{3}x^3 - \cdots\right) - \left(2x - \frac{4}{3}x^3 + \frac{4}{15}x^5 - \cdots\right)$$

$$= -\frac{3}{2}x - \frac{1}{4}x^2 + \frac{3}{2}x^3 + \cdots,$$

$$\lim_{x \to 0} \frac{\ln\sqrt{1+x} - \sin 2x}{x} = \lim_{x \to 0}\left(-\frac{3}{2} - \frac{1}{4}x + \frac{3}{2}x^2 + \cdots\right) = -3/2$$

33. (a) $\sinh^{-1} x = \displaystyle\int (1+x^2)^{-1/2}\, dx - C = \int \left(1 - \frac{1}{2}x^2 + \frac{3}{8}x^4 - \frac{5}{16}x^6 + \cdots\right) dx - C$

$$= \left(x - \frac{1}{6}x^3 + \frac{3}{40}x^5 - \frac{5}{112}x^7 + \cdots\right) - C,$$

$\sinh^{-1} 0 = 0$ so $C = 0$

(b) $(1+x^2)^{-1/2} = 1 + \displaystyle\sum_{k=1}^{\infty} \frac{(-1/2)(-3/2)(-5/2)\cdots(-1/2-k+1)}{k!}(x^2)^k$

$$= 1 + \sum_{k=1}^{\infty}(-1)^k \frac{1\cdot 3\cdot 5\cdots(2k-1)}{2^k k!}x^{2k},$$

$$\sinh^{-1} x = x + \sum_{k=1}^{\infty}(-1)^k \frac{1\cdot 3\cdot 5\cdots(2k-1)}{2^k k!(2k+1)}x^{2k+1}$$

(c) $R = 1$

TECHNOLOGY EXERCISES 11

1. (a) $n = 500$

(b) $\pi \approx 4s_{500} = 3.139593;\ |\text{error}| < 4 \times 10^{-3}$

(c) $|\text{error}| \approx 2 \times 10^{-3}$

3. (a) Increase the number of digits to obtain $\pi \approx 3.14159265358979388$.

(b) $|\text{error}| \approx 6.4 \times 10^{-16}$

7. (a) $\displaystyle\int_{n}^{+\infty} \frac{1}{x^{3.7}}\, dx < 10^{-5}$ if $n > 49.3$; let $n = 50$.

(b) $s_{50} = 1.106279.$

9. $B_0 = 1$, $B_1 = -1/2$, $B_2 = 1/6$, $B_3 = 0$, $B_4 = -1/30$, $B_5 = 0$, $B_6 = 1/42$

11. (a) 0.001615   (b) 0.000333

CHAPTER 12
Topics In Analytic Geometry

EXERCISE SET 12.2

1.

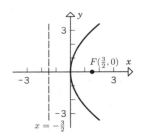

3.

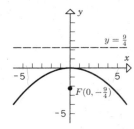

5. $y^2 = (12/5)x$

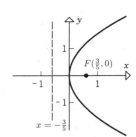

7.

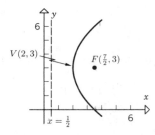

9.

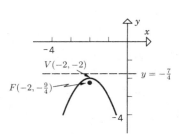

11. $(x-2)^2 = -2(y-5/2)$

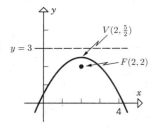

13. $(y-2)^2 = x + 2$

15. $(y-1)^2 = x + 1$

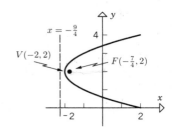

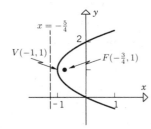

17. $y^2 = 4px$, $p = 3$, $y^2 = 12x$

19. $y^2 = -4px$, $p = 7$, $y^2 = -28x$

21. $y^2 = ax$, $2^2 = a(2)$, $a = 2$, $y^2 = 2x$

23. $x^2 = -4py$, $p = 3$, $x^2 = -12y$

25. $y^2 = a(x - h)$, $4 = a(3 - h)$ and $9 = a(2 - h)$, solve simultaneously to get $h = 19/5$, $a = -5$ so $y^2 = -5(x - 19/5)$

27. The vertex is half way between the focus and directrix so the vertex is at $(3/2, 0)$ and $p = 3/2$, $y^2 = 6(x - 3/2)$.

29. The vertex is 3 units above the directrix so $p = 3$, $(x - 1)^2 = 12(y - 1)$.

31. $(x - 5)^2 = a(y + 3)$, $(9 - 5)^2 = a(5 + 3)$ so $a = 2$, $(x - 5)^2 = 2(y + 3)$

33. **(a)** $y = Ax^2 + Bx + C$; use $(0, 3)$, $(2, 0)$ and $(3, 2)$ to get the system of equations $C = 3$, $9A + 3B + C = 2$, and $4A + 2B + C = 0$ which when solved yields $A = 7/6$, $B = -23/6$, $C = 3$ so $y = \dfrac{7}{6}x^2 - \dfrac{23}{6}x + 3$.

(b) $x = Ay^2 + By + C$; $9A + 3B + C = 0$, $4A + 2B + C = 3$, and $C = 2$. Solve to get $A = -7/6$, $B = 17/6$, $C = 2$ so $x = -\dfrac{7}{6}y^2 + \dfrac{17}{6}y + 2$

35. Complete the square to get $\left(x + \dfrac{B}{2A}\right)^2 = \dfrac{1}{A}\left(y - C + \dfrac{B^2}{4A}\right)$ so the vertex is at $\left(-\dfrac{B}{2A}, \dfrac{4AC - B^2}{4A}\right)$, the focus is at $\left(-\dfrac{B}{2A}, \dfrac{4AC - B^2 + 1}{4A}\right)$, and the directrix is $y = \dfrac{4AC - B^2 - 1}{4A}$.

37. $y = ax^2 + b$, $(20, 0)$ and $(10, 12)$
are on the curve so $400a + b = 0$
and $100a + b = 12$. Solve for b
to get $b = 16$ = height of arch.

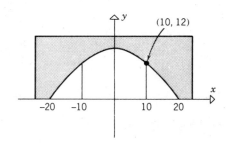

39. Let (x_0, y_0) be a point on the parabola $y^2 = 4px$, then
$$PF = \sqrt{(x_0 - p)^2 + y_0^2} = \sqrt{x_0^2 - 2px_0 + p^2 + 4px_0} = \sqrt{(x_0 + p)^2}$$
so $PF = x_0 + p$ where $x_0 \geq 0$ and PF is a minimum when $x_0 = 0$ (the vertex).

41. Use an xy-coordinate system so that $y^2 = 4px$ is an equation of the parabola, then $(1, 1/2)$ is a point on the curve so $(1/2)^2 = 4p(1)$, $p = 1/16$. The light source should be placed at the focus which is $1/16$ ft. from the vertex.

43. Similar to proof in text.

EXERCISE SET 12.3

1. $c^2 = 16 - 9 = 7$, $c = \sqrt{7}$

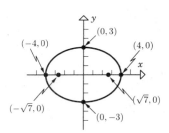

3. $\dfrac{x^2}{1} + \dfrac{y^2}{9} = 1$

$c^2 = 9 - 1 = 8, c = \sqrt{8}$

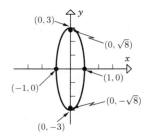

5. $\dfrac{x^2}{2} + \dfrac{y^2}{2/3} = 1$

$c^2 = 2 - 2/3 = 4/3, c = 2/\sqrt{3}$

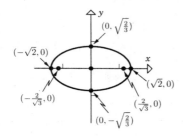

7. $\dfrac{(x-1)^2}{16} + \dfrac{(y-3)^2}{9} = 1$

$c^2 = 16 - 9 = 7, c = \sqrt{7}$

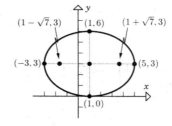

9. $\dfrac{(x+2)^2}{4} + \dfrac{(y+1)^2}{3} = 1$

$c^2 = 4 - 3 = 1, c = 1$

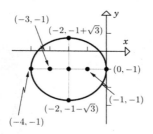

11. $\dfrac{(x+1)^2}{9} + \dfrac{(y-1)^2}{1} = 1$

$c^2 = 9 - 1 = 8, c = \sqrt{8}$

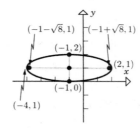

13. $\dfrac{(x+1)^2}{4} + \dfrac{(y-5)^2}{16} = 1$

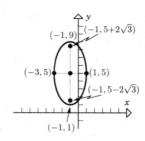

15. $x^2/9 + y^2/4 = 1$

17. $a = 26/2 = 13$, $c = 5$, $b^2 = a^2 - c^2 = 169 - 25 = 144$; $x^2/169 + y^2/144 = 1$

19. $c = 1$, $a^2 = b^2 + c^2 = 2 + 1 = 3$; $x^2/3 + y^2/2 = 1$

21. $b^2 = 16 - 12 = 4$; $x^2/16 + y^2/4 = 1$ and $x^2/4 + y^2/16 = 1$

23. $a = 6$, $(2, 3)$ satisfies $x^2/36 + y^2/b^2 = 1$ so $4/36 + 9/b^2 = 1$, $b^2 = 81/8$; $x^2/36 + y^2/(81/8) = 1$

25. The center is midway between the foci so it is at $(1, 3)$ thus $c = 1$, $b = 1$, $a^2 = 1 + 1 = 2$; $(x - 1)^2 + (y - 3)^2/2 = 1$

27. $(4, 1)$ and $(4, 5)$ are the foci so the center is at $(4, 3)$ thus $c = 2$, $a = 12/2 = 6$, $b^2 = 36 - 4 = 32$; $(x - 4)^2/32 + (y - 3)^2/36 = 1$

29. $x^2/16 + y^2/4 = 1$; $a = 4$, $b = 2$, $c = \sqrt{16 - 4} = 2\sqrt{3}$; $e = c/a = \sqrt{3}/2$.

31. $c = 4$; $e = c/a = 4/a = 2/3$, $a = 6$; $b^2 = a^2 - c^2 = 20$; $x^2/36 + y^2/20 = 1$.

33. $\sqrt{x^2 + y^2} = (4/5)|x - 2|$, $x^2 + y^2 = (16/25)(x^2 - 4x + 4)$, $9x^2 + 25y^2 + 64x - 64 = 0$.

35. Substitute $x = 8 - 2y$ into $x^2 + 4y^2 = 40$ to get $y^2 - 4y + 3 = 0$ which yields $y = 1, 3$. Substitute these into $x = 8 - 2y$ to get $x = 6, 2$ so the points of intersection are $(2, 3)$ and $(6, 1)$.

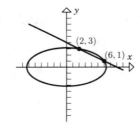

37. Substitute $x^2 = 20 - y^2$ into $x^2 + 9y^2 = 36$ to get $y^2 = 2$, $y = \pm\sqrt{2}$. Use $x^2 = 20 - y^2$ to get $x = \pm 3\sqrt{2}$ when $y = \sqrt{2}$ or $-\sqrt{2}$. The points of intersection are $(3\sqrt{2}, \sqrt{2})$, $(3\sqrt{2}, -\sqrt{2})$, $(-3\sqrt{2}, \sqrt{2})$, and $(-3\sqrt{2}, -\sqrt{2})$.

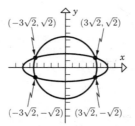

39. Use implicit differentiation on $x^2 + 4y^2 = 8$ to get $\left.\dfrac{dy}{dx}\right|_{(x_0, y_0)} = -\dfrac{x_0}{4y_0}$ where (x_0, y_0) is the point of tangency, but $-x_0/(4y_0) = -1/2$ because the slope of the line is $-1/2$ so $x_0 = 2y_0$. (x_0, y_0) is on the ellipse so $x_0^2 + 4y_0^2 = 8$ which when solved with $x_0 = 2y_0$ yields the points of tangency $(2, 1)$ and $(-2, -1)$. Substitute these into the equation of the line to get $k = \pm 4$.

41. $y = (b/a)\sqrt{a^2 - x^2}$ is the upper half of the ellipse,

$$A = 4\int_0^a \frac{b}{a}\sqrt{a^2 - x^2}dx = \frac{4b}{a}\int_0^a \sqrt{a^2 - x^2}dx = \frac{4b}{a}\left(\frac{1}{4}\pi a^2\right) = \pi ab$$

43. $y = \frac{b}{a}\sqrt{a^2 - x^2}$, $\dfrac{dy}{dx} = -\dfrac{b}{a}\dfrac{x}{\sqrt{a^2 - x^2}}$, $1 + \left(\dfrac{dy}{dx}\right)^2 = \dfrac{a^4 - c^2x^2}{a^2(a^2 - x^2)}$,

$$S = 2(2\pi)\int_0^a y\sqrt{1 + (dy/dx)^2}\,dx = 4\pi\int_0^a \frac{b}{a}\sqrt{a^2 - x^2}\frac{\sqrt{a^4 - c^2x^2}}{a\sqrt{a^2 - x^2}}dx$$

$$= \frac{4\pi b}{a^2}\int_0^a \sqrt{a^4 - c^2x^2}\,dx = \frac{4\pi b}{a^2 c}\int_0^{ac} \sqrt{a^4 - u^2}\,du \quad (u = cx)$$

$$= \frac{4\pi b}{a^2 c}\left[\frac{u}{2}\sqrt{a^4 - u^2} + \frac{a^4}{2}\sin^{-1}\frac{u}{a^2}\right]_0^{ac} = \frac{4\pi b}{a^2 c}\left[\frac{1}{2}ac\sqrt{a^4 - a^2c^2} + \frac{1}{2}a^4\sin^{-1}\frac{c}{a}\right]$$

$$= 2\pi ab\left[\frac{b}{a} + \frac{a}{c}\sin^{-1}\frac{c}{a}\right].$$

45. Use $\dfrac{x^2}{9} + \dfrac{y^2}{4} = 1$, $x = \dfrac{3}{2}\sqrt{4 - y^2}$,

$$V = \int_{-2}^{-2+h} (2)(3/2)\sqrt{4 - y^2}(18)dy = 54\int_{-2}^{-2+h} \sqrt{4 - y^2}\,dy$$

$$= 54\left[\frac{y}{2}\sqrt{4 - y^2} + 2\sin^{-1}\frac{y}{2}\right]_{-2}^{-2+h} = 27\left[4\sin^{-1}\frac{h - 2}{2} + (h - 2)\sqrt{4h - h^2} + 2\pi\right]$$

47. The vertex in the first quadrant is at the point where $y = x > 0$ so $x^2/a^2 + x^2/b^2 = 1$, $x^2 = a^2b^2/(a^2 + b^2)$, $x = ab/\sqrt{a^2 + b^2}$. $A = (2x)^2 = 4x^2 = 4a^2b^2/(a^2 + b^2)$.

49. Open the compass to the length of half the major axis, place the point of the compass at an end of the minor axis and draw arcs that cross the major axis to both sides of the center of the ellipse. Place the tacks where the arcs intersect the major axis.

51. $L = 2a = \sqrt{D^2 + p^2 D^2} = D\sqrt{1 + p^2}$ (see figure),

so $a = \dfrac{1}{2}D\sqrt{1 + p^2}$, but $b = \dfrac{1}{2}D$,

$T = c = \sqrt{a^2 - b^2}$

$$= \sqrt{\frac{1}{4}D^2(1 + p^2) - \frac{1}{4}D^2} = \frac{1}{2}pD.$$

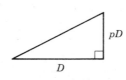

53. Let R be the radius of a circle C that is tangent to both C_1 and C_2. The distances between the center of C and the centers of C_1 and C_2 are, respectively, $r_1 + R$ and $r_2 - R$ (see accompanying diagram). Their sum is $(r_1 + R) + (r_2 - R) = r_1 + r_2$ which is a constant so the centers lie on an ellipse with foci at the centers of C_1 and C_2. The length of the major axis is

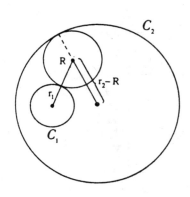

$2a = r_1 + r_2$; the center of the ellipse is at the midpoint of the line segment that joins the centers of C_1 and C_2.

55. Similar to derivation in text.

EXERCISE SET 12.4

1. $c^2 = a^2 + b^2 = 16 + 4 = 20, c = 2\sqrt{5}$

3. $y^2/4 - x^2/9 = 1$
$c^2 = 4 + 9 = 13, c = \sqrt{13}$

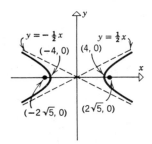

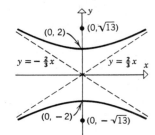

5. $x^2/1 - y^2/8 = 1$
$c^2 = 1 + 8 = 9, c = 3$

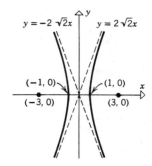

7. $c^2 = 1 + 1 = 2, c = \sqrt{2}$

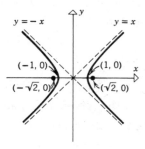

9. $c^2 = 9 + 4 = 13, c = \sqrt{13}$

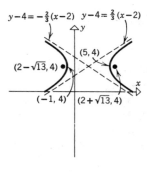

11. $(y+3)^2/36 - (x+2)^2/4 = 1$
$c^2 = 36 + 4 = 40, c = 2\sqrt{10}$

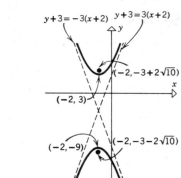

13. $(x+1)^2/4 - (y-1)^2/1 = 1$
$c^2 = 4 + 1 = 5, c = \sqrt{5}$

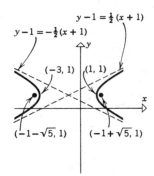

15. $(x-1)^2/4 - (y+3)^2/64 = 1$
$c^2 = 4 + 64 = 68, c = 2\sqrt{17}$

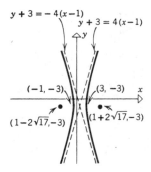

17. $a = 2, c = 3, b^2 = 9 - 4 = 5; x^2/4 - y^2/5 = 1$

19. $a = 1, b/a = 2, b = 2; x^2 - y^2/4 = 1$

21. vertices along x-axis: $b/a = 3/2$ so $a = 8/3; x^2/(64/9) - y^2/16 = 1$
vertices along y-axis: $a/b = 3/2$ so $a = 6; y^2/36 - x^2/16 = 1$

23. The form of the equation is $y^2/a^2 - x^2/b^2$ because $(5,9)$ is above the asymptote $y = x, a/b = 1$ and $81/a^2 - 25/b^2 = 1$, solve to get $a^2 = b^2 = 56; y^2/56 - x^2/56 = 1$

25. $a = 2$ so $x^2/4 - y^2/b^2 = 1$, $(4,2)$ is on the curve so $4 - 4/b^2 = 1, b^2 = 4/3; x^2/4 - y^2/(4/3) = 1$

27. the center is at $(2,-3), a = 2, c = 3, b^2 = 9 - 4 = 5; (x-2)^2/4 - (y+3)^2/5 = 1$

29. the center is at $(6,4), a = 4, c = 5, b^2 = 25 - 16 = 9; (x-6)^2/16 - (y-4)^2/9 = 1$

31. From the definition of a hyperbola, $\left| \sqrt{(x-1)^2 + (y-1)^2} - \sqrt{x^2 + y^2} \right| = 1$,

$\sqrt{(x-1)^2 + (y-1)^2} - \sqrt{x^2 + y^2} = \pm 1$, transpose the second radical to the right hand side of the equation and square and simplify to get $\pm 2\sqrt{x^2 + y^2} = -2x - 2y + 1$, square and simplify again to get $8xy - 4x - 4y + 1 = 0$.

33. $x^2/9 - y^2/16 = 1; a = 3, b = 4, c = \sqrt{9+16} = 5; e = c/a = 5/3$.

35. $c = 5; e = c/a = 5/a = 5/3, a = 3; b^2 = c^2 - a^2 = 16; x^2/9 - y^2/16 = 1$.

37. $\sqrt{x^2 + y^2} = 3|x - 1|$, $x^2 + y^2 = 9(x^2 - 2x + 1)$, $8x^2 - y^2 - 18x + 9 = 0$.

39. Substitute $x = 2y + 20$ into
$x^2 - 4y^2 = 36$ to get $y = -91/20$,
so $x = 2(-91/20) + 20 = 109/10$.
The curves intersect at $(109/10, -91/20)$.

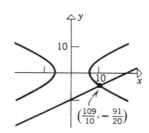

41. Eliminate x to get $y^2 = 1$, $y = \pm 1$.
Use either equation to find that
$x = \pm 2$ if $y = 1$ or if $y = -1$.
The curves intersect at $(2, 1)$,
$(2, -1)$, $(-2, 1)$, and $(-2, -1)$.

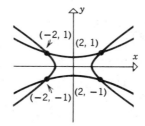

43. Let (x_0, y_0) be such a point. The foci are at $(-\sqrt{5}, 0)$ and $(\sqrt{5}, 0)$, the lines are perpendicular if the product of their slopes is -1 so $\dfrac{y_0}{x_0 + \sqrt{5}} \cdot \dfrac{y_0}{x_0 - \sqrt{5}} = -1, y_0^2 = 5 - x_0^2$ and $4x_0^2 - y_0^2 = 4$. Solve to get $x_0 = \pm 3/\sqrt{5}, y_0 = \pm 4/\sqrt{5}$. The coordinates are $(\pm 3/\sqrt{5}, 4/\sqrt{5}), (\pm 3/\sqrt{5}, -4/\sqrt{5})$.

45. Let (x_0, y_0) be one of the points then $dy/dx\big|_{(x_0, y_0)} = 4x_0/y_0$, the tangent line is
$y = (4x_0/y_0)x + 4$, but (x_0, y_0) is on both the line and the curve which leads to
$4x_0^2 - y_0^2 + 4y_0 = 0$ and $4x_0^2 - y_0^2 = 36$, solve to get $x_0 = \pm 3\sqrt{13}/2$, $y_0 = -9$.

47. c approaches a as e approaches 1 so $b = \sqrt{c^2 - a^2}$ approaches 0; the hyperbola flattens and approaches the focal axis, excluding the segment between the vertices.

c approaches $+\infty$ as e approaches $+\infty$ so b approaches $+\infty$; the hyperbola approaches the lines that are perpendicular to the focal axis at the vertices.

49. Let d_1 and d_2 be the distances of the first and second observers, respectively, from the point where the gun was fired, and let v be the speed of sound. Then $t =$ (time for sound to reach the second observer) $-$ (time for sound to reach the first observer) $= d_2/v - d_1/v$ so $d_2 - d_1 = vt$. For constant v and t the difference of distances, d_2 and d_1 is constant so the gun was fired somewhere on a branch of a hyperbola whose foci are where the observers are.

51. Similar to the derivation in the text.

53. **(a)** Use $x^2/a^2 + y^2/b^2 = 1$ and $x^2/A^2 - y^2/B^2$ as the equations of the ellipse and hyperbola. If (x_0, y_0) is a point of intersection then $b^2 x_0^2 + a^2 y_0^2 = a^2 b^2$ and $B^2 x_0^2 - A^2 y_0^2 = A^2 B^2$, solve to get

$$x_0^2 = \frac{a^2 A^2 (b^2 + B^2)}{a^2 B^2 + A^2 b^2} \qquad \text{and} \qquad y_0^2 = \frac{b^2 B^2 (a^2 - A^2)}{a^2 B^2 + A^2 b^2} \tag{1}$$

From Exercises 40 (Section 12.3) and 44 the slopes of the tangent lines to the ellipse and hyperbola at (x_0, y_0) are, respectively, $-\dfrac{b^2}{a^2} \dfrac{x_0}{y_0}$ and $\dfrac{B^2}{A^2} \dfrac{x_0}{y_0}$ so their product is $-\dfrac{b^2 B^2 x_0^2}{a^2 A^2 y_0^2}$ which, using (1), gives $-\dfrac{b^2 + B^2}{a^2 - A^2}$. The ellipse and hyperbola have the same foci so $c^2 = a^2 - b^2 = A^2 + B^2$, $a^2 - A^2 = b^2 + B^2$, $(b^2 + B^2)/(a^2 - A^2) = 1$, thus the product of the slopes of the tangent lines is -1 and the lines are perpendicular.

(b) From the figure,
$2(\alpha + \beta) = 180°$ so $\alpha + \beta = 90°$.

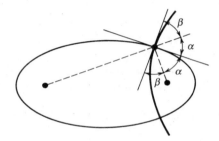

EXERCISE SET 12.5

1. **(a)** $\sin\theta = \sqrt{3}/2$, $\cos\theta = 1/2$
$x' = (-2)(1/2) + (6)(\sqrt{3}/2) = -1 + 3\sqrt{3}$, $y' = -(-2)(\sqrt{3}/2) + 6(1/2) = 3 + \sqrt{3}$

(b) $x = \dfrac{1}{2}x' - \dfrac{\sqrt{3}}{2}y' = \dfrac{1}{2}(x' - \sqrt{3}y')$, $y = \dfrac{\sqrt{3}}{2}x' + \dfrac{1}{2}y' = \dfrac{1}{2}(\sqrt{3}x' + y')$

$\sqrt{3}\left[\dfrac{1}{2}(x' - \sqrt{3}y')\right]\left[\dfrac{1}{2}(\sqrt{3}x' + y')\right] + \left[\dfrac{1}{2}(\sqrt{3}x' + y')\right]^2 = 6$

$\dfrac{\sqrt{3}}{4}(\sqrt{3}x'^2 - 2x'y' - \sqrt{3}y'^2) + \dfrac{1}{4}(3x'^2 + 2\sqrt{3}x'y' + y'^2) = 6$

$\dfrac{3}{2}x'^2 - \dfrac{1}{2}y'^2 = 6$, $\dfrac{x'^2}{4} - \dfrac{y'^2}{12} = 1$

1. (c)

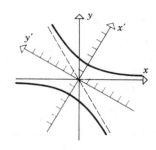

3. $\cot 2\theta = (0-0)/1 = 0$, $2\theta = 90°$, $\theta = 45°$
$x = (\sqrt{2}/2)(x' - y')$, $y = (\sqrt{2}/2)(x' + y')$
$y'^2/18 - x'^2/18 = 1$, hyperbola

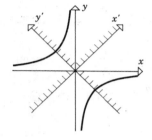

5. $\cot 2\theta = [1 - (-2)]/4 = 3/4$
$\cos 2\theta = 3/5$
$\sin\theta = \sqrt{(1 - 3/5)/2} = 1/\sqrt{5}$
$\cos\theta = \sqrt{(1 + 3/5)/2} = 2/\sqrt{5}$
$x = (1/\sqrt{5})(2x' - y')$
$y = (1/\sqrt{5})(x' + 2y')$
$x'^2/3 - y'^2/2 = 1$, hyperbola

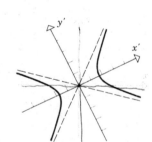

7. $\cot 2\theta = (1 - 3)/(2\sqrt{3}) = -1/\sqrt{3}$,
$2\theta = 120°$, $\theta = 60°$
$x = (1/2)(x' - \sqrt{3}y')$
$y = (1/2)(\sqrt{3}x' + y')$
$y' = x'^2$, parabola

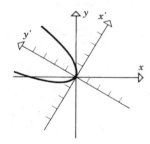

9. $\cot 2\theta = (9-16)/(-24) = 7/24$
$\cos 2\theta = 7/25,$
   $\sin \theta = 3/5, \qquad \cos \theta = 4/5$
   $x = (1/5)(4x' - 3y'),$
   $y = (1/5)(3x' + 4y')$
   $y'^2 = 4(x' - 1),$ parabola

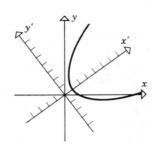

11. $\cot 2\theta = (52 - 73)/(-72) = 7/24$
$\cos 2\theta = 7/25, \qquad \sin \theta = 3/5,$
   $\cos \theta = 4/5$
   $x = (1/5)(4x' - 3y'),$
   $y = (1/5)(3x' + 4y')$
$(x' + 1)^2/4 + y'^2 = 1,$ ellipse

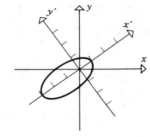

13. Let $x = x'\cos\theta - y'\sin\theta$, $y = x'\sin\theta + y'\cos\theta$ then $x^2 + y^2 = r^2$ becomes
$(\sin^2\theta + \cos^2\theta)x'^2 + (\sin^2\theta + \cos^2\theta)y'^2 = r^2$, $x'^2 + y'^2 = r^2$. Under a rotation transformation the center of the circle stays at the origin of both coordinate systems.

15. $x' = (\sqrt{2}/2)(x + y)$, $y' = (\sqrt{2}/2)(-x + y)$ which when substituted into $3x'^2 + y'^2 = 6$ yields $x^2 + xy + y^2 = 3$.

17. $\sqrt{x} + \sqrt{y} = 1$, $\sqrt{x} = 1 - \sqrt{y}$, $x = 1 - 2\sqrt{y} + y$, $2\sqrt{y} = 1 - x + y$, $4y = 1 + x^2 + y^2 - 2x + 2y - 2xy$, $x^2 - 2xy + y^2 - 2x - 2y + 1 = 0$. $\cot 2\theta = \dfrac{1-1}{2} = 0$, $2\theta = \pi/2$, $\theta = \pi/4$. Let $x = x'/\sqrt{2} - y'/\sqrt{2}$, $y = x'/\sqrt{2} + y'/\sqrt{2}$ to get $2y'^2 - 2\sqrt{2}x' + 1 = 0$, which is a parabola. From $\sqrt{x} + \sqrt{y} = 1$ we see that $0 \le x \le 1$ and $0 \le y \le 1$, so the graph is just a portion of a parabola.

19. Use (9) to express $B' - 4A'C'$ in terms of A, B, C, and θ, then simplify.

21. $\cot 2\theta = (A - C)/B = 0$ if $A = C$ so $2\theta = 90°$, $\theta = 45°$.

23. $B^2 - 4AC = (-1)^2 - 4(1)(1) = -3 < 0$; ellipse, point, or no graph

25. $B^2 - 4AC = (2\sqrt{3})^2 - 4(1)(3) = 0$; parabola, line, pair of parallel lines, or no graph

27. $B^2 - 4AC = (-24)^2 - 4(34)(41) = -5000 < 0$; ellipse, point, or no graph

29. Part (b): from (15), $A'C' < 0$ so A' and C' have opposite signs. By multiplying (14) through by -1, if necessary, assume that $A' < 0$ and $C' > 0$ so $(x' - h)^2/C' - (y' - k)^2/|A'| = K$. If $K \neq 0$ then the graph is a hyperbola (divide both sides by K), if $K = 0$ then we get the pair of intersecting lines $(x' - h)/\sqrt{C'} = \pm(y' - k)/\sqrt{|A'|}$.

Part (c): from (15), $A'C' = 0$ so either $A' = 0$ or $C' = 0$ but not both (this would imply that $A = B = C = 0$ which results in (14) being linear). Suppose $A' \neq 0$ and $C' = 0$ then complete the square to get $(x' - h)^2 = -E'y'/A' + K$. If $E' \neq 0$ the graph is a parabola, if $E' = 0$ and $K = 0$ the graph is the line $x' = h$, if $E' = 0$ and $K > 0$ the graph is the pair of parallel lines $x' = h \pm \sqrt{K}$, if $E' = 0$ and $K < 0$ there is no graph.

TECHNOLOGY EXERCISES 12

1. The distance is given by $D = \sqrt{(x-1)^2 + (y+1)^2}$ where $y = (\sqrt{5}/3)\sqrt{9 - x^2}$ or $y = -(\sqrt{5}/3)\sqrt{9 - x^2}$. $D_{\max} = 4.207900$ and occurs on $y = (\sqrt{5}/3)\sqrt{9 - x^2}$ at $x = -2.849866$; $D_{\min} = 1.056175$ and occurs on $y = -(\sqrt{5}/3)\sqrt{9 - x^2}$ at $x = 1.380655$.

3. The volume is given by $\dfrac{450}{1521}\pi \displaystyle\int_0^{h/2} (1521 + y^2)\, dy = \dfrac{50}{169}\pi \left(\dfrac{1521}{2}h + \dfrac{1}{24}h^3\right) = 50,000$; solve for h to get $h = 59.3$ ft.

5. **(a)** With $y = (b/a)\sqrt{a^2 - x^2}$, the distance traveled in one orbit about the sun is
$$4\int_0^a \sqrt{1 + (dy/dx)^2}\, dx = 9.36129 \times 10^8 \text{ km.}$$
(b) The average speed is $9.36129 \times 10^8/(365 \times 24) = 106,864$ km/hr.

7. **(a)** With $h = 4$ ft we find that the total volume of the tank is 108π ft^3. Solve $V = 27\pi$, $V = 54\pi$, and $V = 81\pi$ for h and convert to inches to obtain 14.3 inches, 24 inches, and 33.7 inches.
(b) $\dfrac{dV}{dt} = \dfrac{dV}{dh}\dfrac{dh}{dt}$ so $\dfrac{dh}{dt} = \dfrac{dV/dt}{dV/dh}$; evaluate dV/dh at the first value of h (in feet) found in part (a), let $dV/dt = 0.5$ and find dh/dt, then convert to inches/sec to get 0.06 inches/sec.

9. **(a)** $B^2 - 4AC = (9)^2 - 4(2)(1) > 0$ so the conic is a hyperbola.
(b) $y = -\dfrac{9}{2}x - \dfrac{1}{2} - \dfrac{1}{2}\sqrt{73x^2 + 42x + 17}$ or $y = -\dfrac{9}{2}x - \dfrac{1}{2} + \dfrac{1}{2}\sqrt{73x^2 + 42x + 17}$

CHAPTER 13
Polar Coordinates And Parametric Equations

EXERCISE SET 13.1

1.

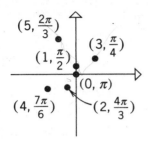

3. (a) $(3\sqrt{3}, 3)$ (b) $(-7/2, 7\sqrt{3}/2)$ (c) $(4\sqrt{2}, 4\sqrt{2})$
    (d) $(5, 0)$ (e) $(-7\sqrt{3}/2, 7/2)$ (f) $(0, 0)$

5. (a) $(5, \pi)$ (b) $(4, 11\pi/6)$ (c) $(2, 3\pi/2)$
    (d) $(8\sqrt{2}, 5\pi/4)$ (e) $(6, 2\pi/3)$ (f) $(\sqrt{2}, \pi/4)$

7. (a) $(-5, 0)$ (b) $(-4, 5\pi/6)$ (c) $(-2, \pi/2)$
    (d) $(-8\sqrt{2}, \pi/4)$ (e) $(-6, 5\pi/3)$ (f) $(-\sqrt{2}, 5\pi/4)$

9. $x^2 + y^2 = 4$; circle   **11.** $y = 4$; horizontal line

13. $r^2 = 3r\cos\theta$, $x^2 + y^2 = 3x$, $(x - 3/2)^2 + y^2 = 9/4$; circle

15. $r^2(2\sin\theta\cos\theta) = 8$, $2(r\sin\theta)(r\cos\theta) = 8$, $xy = 4$; hyperbola

17. $r + r\sin\theta = 2$, $r = 2 - y$, $r^2 = (2 - y)^2$, $x^2 + y^2 = 4 - 4y + y^2$, $x^2 + 4y = 4$; parabola

19. $3r\cos\theta + 2r\sin\theta = 6$, $3x + 2y = 6$; line   **21.** $r\cos\theta = 7$

23. $r = 3$ **25.** $r^2 - 6r\sin\theta = 0$, $r = 6\sin\theta$

27. $r^2\cos^2\theta = 9r\sin\theta$, $r = 9\dfrac{\sin\theta}{\cos^2\theta} = 9\tan\theta\sec\theta$

29. $4(r\cos\theta)(r\sin\theta) = 9$, $4r^2\sin\theta\cos\theta = 9$, $r^2\sin 2\theta = 9/2$

31. $r^4 = 2r^2\sin\theta\cos\theta$, $r^2 = \sin 2\theta$

33. $(r-1)(r-2) = 0$, $r = 1$ or $r = 2$; the graph consists of the concentric circles $r = 1$ and $r = 2$.

35.

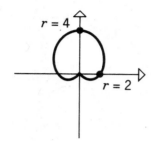

37.

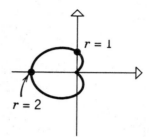

39. $r^2 = ar\sin\theta + br\cos\theta$, $x^2 + y^2 = ay + bx$ which is a circle.

41. (r, θ) and $(-r, \theta + \pi)$ are polar coordinates of the same point so if $r < 0$ then $-r > 0$ and $x = (-r)\cos(\theta + \pi) = (-r)(-\cos\theta) = r\cos\theta$, $y = (-r)\sin(\theta + \pi) = (-r)(-\sin\theta) = r\sin\theta$.

EXERCISE SET 13.2

1.

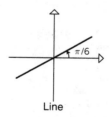

Line

3.

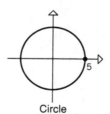

Circle

5.

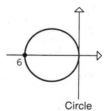

Circle

7.

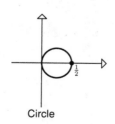

Circle

9.

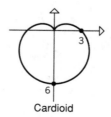

Cardioid

11.

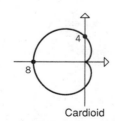

Cardioid

13.

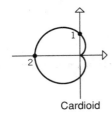

Cardioid

15.

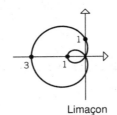

Limaçon

17.

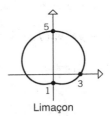

Limaçon

19.

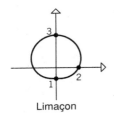

Limaçon

21.

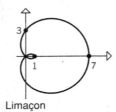

Limaçon

23.

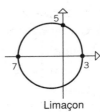

Limaçon

25.

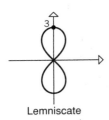

Lemniscate

27.

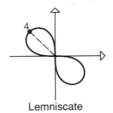

Lemniscate

29.

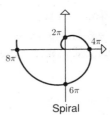

Spiral

31.

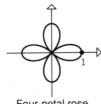

Four-petal rose

33.

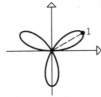

Three-petal rose

35.

Eight-petal rose

37. $r^2 = 4\cos\theta + 4r\sin\theta$,
$x^2 + y^2 = 4x + 4y$,
$(x-2)^2 + (y-2)^2 = 8$

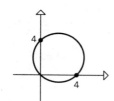

39.

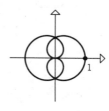

41. Note that $r \to \pm\infty$ as θ approches
odd multiples of $\pi/2$;
$x = r\cos\theta = 4\tan\theta\cos\theta = 4\sin\theta$,
$y = r\sin\theta = 4\tan\theta\sin\theta$
so $x \to \pm 4$ and $y \to \pm\infty$ as
θ approaches odd multiples of $\pi/2$.

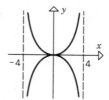

43. Note that $r \to \pm\infty$ as θ approaches
odd multiples of $\pi/2$;
$x = r\cos\theta = (2\sin\theta\tan\theta)\cos\theta = 2\sin^2\theta$,
$y = r\sin\theta = (2\sin\theta\tan\theta)\sin\theta = 2\sin^2\theta\tan\theta$
so $x \to 2$ and $y \to \pm\infty$ as
θ approaches odd multiples of $\pi/2$.

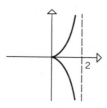

45. $\displaystyle\lim_{\theta \to 0^+} y = \lim_{\theta \to 0^+} \frac{\sin\theta}{\sqrt{\theta}} = 0$

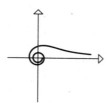

47. $y = r\sin\theta = (1 + \cos\theta)\sin\theta = \sin\theta + \sin\theta\cos\theta$,
$dy/d\theta = \cos\theta - \sin^2\theta + \cos^2\theta = 2\cos^2\theta + \cos\theta - 1 = (2\cos\theta - 1)(\cos\theta + 1)$

$dy/d\theta = 0$ if $\cos\theta = 1/2$ or if $\cos\theta = -1$; $\theta = \pi/3$ or π.

If $\theta = 0, \pi/3, \pi$, then $y = 0, 3\sqrt{3}/4, 0$ so the maximum value of y is $3\sqrt{3}/4$.

49. $y = r\sin\theta = \cos 2\theta \sin\theta$, $dy/d\theta = \cos 2\theta \cos\theta - 2\sin 2\theta \sin\theta$, use the identities

$\cos 2\theta = 1 - 2\sin^2\theta$ and $\sin 2\theta = 2\sin\theta\cos\theta$ to get $dy/d\theta = \cos\theta(1 - 6\sin^2\theta)$, $dy/d\theta = 0$

for $0 \le \theta \le \pi/4$ if $\sin^2\theta = 1/6$, $\sin\theta = 1/\sqrt{6}$ where y attains its maximum value of

$y = \cos 2\theta \sin\theta = (1 - 2\sin^2\theta)\sin\theta = [1 - 2(1/6)](1/\sqrt{6}) = \sqrt{6}/9$ so the width of the

petal is $2y = 2\sqrt{6}/9$.

51. $x = r\cos\theta = (1 + \cos\theta)\cos\theta = \cos\theta + \cos^2\theta$,

$dx/d\theta = -\sin\theta - 2\sin\theta\cos\theta = -\sin\theta(1 + 2\cos\theta)$,

$dx/d\theta = 0$ if $\sin\theta = 0$ or if $\cos\theta = -1/2$; $\theta = 0, 2\pi/3$, or π. If $\theta = 0, 2\pi/3, \pi$, then

$x = 2, -1/4, 0$ so the minimum value of x is $-1/4$.

53. $r = \dfrac{3/2}{1 - \cos\theta}, e = 1$

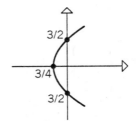

55. $r = \dfrac{3/2}{1 + \frac{1}{2}\sin\theta}, e = 1/2$

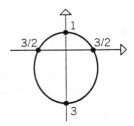

57. $r = \dfrac{2}{1 + \frac{3}{2}\cos\theta}, e = 3/2$

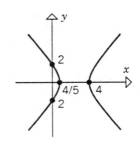

59. The graph of $r = f(\theta + \alpha)$ is the graph of $r = f(\theta)$ rotated α radians clockwise about the pole if $\alpha > 0$, $|\alpha|$ radians counterclockwise if $\alpha < 0$.

EXERCISE SET 13.3

1. $A = \int_{\pi/6}^{\pi/3} \frac{1}{2}\theta^2 d\theta = 7\pi^3/1296$

3. $A = 2\int_0^{\pi} \frac{1}{2}(2 + 2\cos\theta)^2 d\theta = 6\pi$

5. $A = 2\int_{\pi/6}^{\pi/2} \frac{1}{2}[25\sin^2\theta - (2 + \sin\theta)^2]d\theta = 8\pi/3 + \sqrt{3}$

7. $A = 2\int_0^{\pi/3} \frac{1}{2}[(2 + 2\cos\theta)^2 - 9]d\theta = 9\sqrt{3}/2 - \pi$

9. $A = 2\int_0^{\pi/2} \frac{1}{2}\sin 2\theta \, d\theta = 1$

11. $A = 6\int_0^{\pi/6} \frac{1}{2}(16\cos^2 3\theta)d\theta = 4\pi$

13. $A = 2\left[\int_0^{\pi/3} \frac{1}{2}(1 + \cos\theta)^2 d\theta + \int_{\pi/3}^{\pi/2} \frac{1}{2}(9\cos^2\theta)d\theta\right] = 5\pi/4$

15. $A = 2\int_0^{\cos^{-1}(3/5)} \frac{1}{2}(100 - 36\sec^2\theta)d\theta = 100\cos^{-1}(3/5) - 48$

17. $A = \int_0^{\pi/2} \frac{1}{2}a^2 \sec^4(\theta/2)d\theta = 4a^2/3$

19. $A = \int_0^{\pi} \frac{9}{2}e^{-4\theta}d\theta = \frac{9}{8}(1 - e^{-4\pi})$

21. $A = \int_{1/9}^{4} \frac{1}{2}\frac{1}{\theta}d\theta = \ln 6$

23. $A = 4\int_0^{\pi/6} \frac{1}{2}(4\cos 2\theta - 2)d\theta = 2\sqrt{3} - 2\pi/3$

25. (a) $x = r\cos\theta$ and $y = r\sin\theta$ so $r^3\cos^3\theta - r^2\sin\theta\cos\theta + r^3\sin^3\theta = 0$,
$r^2[r(\cos^3\theta + \sin^3\theta) - \sin\theta\cos\theta] = 0$, $r = \sin\theta\cos\theta/(\cos^3\theta + \sin^3\theta)$.

(b) Divide numerator and denominator of the expression in part (a) by $\cos^3\theta$ to get
$r = \sec\theta\tan\theta/(1 + \tan^3\theta)$. The loop is traced out for $0 \le \theta \le \pi/2$ so

$$A = \frac{1}{2}\int_0^{\pi/2} r^2 d\theta = \frac{1}{2}\int_0^{\pi/2} \frac{\sec^2\theta\tan^2\theta}{(1 + \tan^3\theta)^2}d\theta, \text{ let } u = 1 + \tan^3\theta \text{ to get}$$

$$A = \frac{1}{6}\int_1^{+\infty} \frac{1}{u^2}du = -\frac{1}{6}\frac{1}{u}\Big]_1^{+\infty} = \frac{1}{6}.$$

EXERCISE SET 13.4

1. (a)

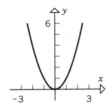

(b) $y = x^2$

3. $\cos^2 t + \sin^2 t = 1$;
$x^2 + y^2 = 1$

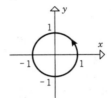

5. $t = (x+4)/3$;
$y = 2x + 10$

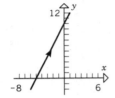

7. $\cos t = x/2, \ \sin t = y/5$;
$x^2/4 + y^2/25 = 1$

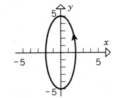

9. $\cos t = (x-3)/2, \sin t = (y-2)/4$;
$(x-3)^2/4 + (y-2)^2/16 = 1$

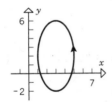

11. $\sin 2\pi t = x/4, \ \cos 2\pi t = y/4$;
$x^2/16 + y^2/16 = 1$

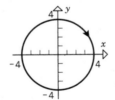

13. $\cos 2t = 1 - 2\sin^2 t$;
$x = 1 - 2y^2, -1 \le y \le 1$

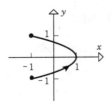

15. $y = \ln t^2 = \ln x, x \ge 1$

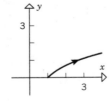

17. $x = 3y^2 - 1$,
$-1 < x \le 2, 0 < y \le 1$

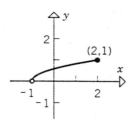

19. $x/2 + y/3 = 1$,
$0 \le x \le 2, 0 \le y \le 3$

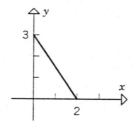

21. $y = 1 - x^2$,
$-1 \le x \le 1, 0 \le y \le 1$

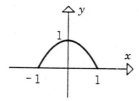

23. $x + (y - 1)^2 = 1$,
$0 \le x \le 1, 0 \le y \le 2$

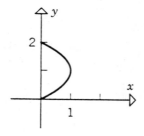

25. $x^2 + y^2 = 1$,
$\cos 1 \le x < 1,\ 0 < y \le \sin 1$

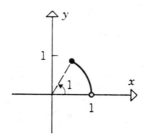

27. $(x_0, \sqrt{x_0})$ is on the curve for $x_0 \ge 0$. $y' = 1/(2\sqrt{x})$; the tangent line at $(x_0, \sqrt{x_0})$ is
$y - \sqrt{x_0} = (x - x_0)/(2\sqrt{x_0})$, which crosses the x-axis at $x = -x_0 = t$ so $x_0 = -t$ hence
$x = -t$, $y = \sqrt{-t}$ for $t \le 0$.

29. $dy/dx = \dfrac{\cos t}{-\sin t} = -\cot t$, $dy/dx|_{t=3\pi/4} = 1$

31. $dy/dx = \dfrac{2}{1/(2\sqrt{t})} = 4\sqrt{t}$, $dy/dx|_{t=9} = 12$

33. $dy/dx = \dfrac{6\pi \cos 2\pi s}{-8\pi \sin 2\pi s} = -\dfrac{3}{4}\cot 2\pi s$, $dy/dx|_{s=-1/4} = 0$

35. $dy/dx = \dfrac{t^2}{t} = t$, $d^2y/dx^2 = \dfrac{1}{t}$, $d^2y/dx^2|_{t=2} = 1/2$

37. $dy/dx = \dfrac{2}{1/(2\sqrt{t})} = 4\sqrt{t}$, $d^2y/dx^2 = \dfrac{2/\sqrt{t}}{1/(2\sqrt{t})} = 4$, $d^2y/dx^2|_{t=1} = 4$

39. $(dx/dt)^2 + (dy/dt)^2 = (4)^2 + (3)^2 = 25$, $L = \displaystyle\int_0^2 5dt = 10$

41. $(dx/dt)^2 + (dy/dt)^2 = (t^2)^2 + (t)^2 = t^2(t^2+1)$, $L = \displaystyle\int_0^1 t(t^2+1)^{1/2}dt = (2\sqrt{2}-1)/3$

43. $(dx/dt)^2 + (dy/dt)^2 = (-2\sin 2t)^2 + (2\cos 2t)^2 = 4$, $L = \displaystyle\int_0^{\pi/2} 2\,dt = \pi$

45. $(dx/dt)^2 + (dy/dt)^2 = [2(1+t)]^2 + [3(1+t)^2]^2 = (1+t)^2[4 + 9(1+t)^2]$,

$L = \displaystyle\int_0^1 (1+t)[4 + 9(1+t)^2]^{1/2}dt = (80\sqrt{10} - 13\sqrt{13})/27$

47. $(dx/dt)^2 + (dy/dt)^2 = [a(1-\cos t)]^2 + [a\sin t]^2 = 2a^2(1-\cos t) = 4a^2\sin^2(t/2)$,

$L = \displaystyle\int_0^{2\pi} 2a\sin(t/2)dt = 8a$

49. **(a)** $(dx/dt)^2 + (dy/dt)^2 = 4\sin^2 t + \cos^2 t = 4\sin^2 t + (1-\sin^2 t) = 1 + 3\sin^2 t$,

$L = \displaystyle\int_0^{2\pi}\sqrt{1 + 3\sin^2 t}\,dt = 4\displaystyle\int_0^{\pi/2}\sqrt{1 + 3\sin^2 t}\,dt$

(b) 9.69

(c) distance traveled $= \displaystyle\int_{1.5}^{4.8}\sqrt{1 + 3\sin^2 t}\,dt \approx 5.16$ cm

51. $x = (2 + 3\sin\theta)\cos\theta$, $y = (2 + 3\sin\theta)\sin\theta$

53. $dy/dx = \dfrac{-e^{-t}}{e^t} = -e^{-2t}$; for $t = 2$, $dy/dx = -e^{-4}$, $(x,y) = (e^2, e^{-2})$;
$y - e^{-2} = -e^{-4}(x - e^2)$, $y = -e^{-4}x + 2e^{-2}$

55. $dy/dx = \dfrac{2t + 1}{6t^2 - 30t + 24} = \dfrac{2t + 1}{6(t - 1)(t - 4)}$

  (a) $dy/dx = 0$ if $t = -1/2$

  (b) $dx/dy = \dfrac{6(t - 1)(t - 4)}{2t + 1} = 0$ if $t = 1, 4$

57. If $x = 3$ then $t^2 - 3t + 5 = 3$, $t^2 - 3t + 2 = 0$, $(t - 1)(t - 2) = 0$, $t = 1$ or 2. If $t = 1$ or 2 then $y = 1$ so $(3,1)$ is reached when $t = 1$ or 2. $dy/dx = (3t^2 + 2t - 10)/(2t - 3)$. For $t = 1$, $dy/dx = 5$, the tangent line is $y - 1 = 5(x - 3)$, $y = 5x - 14$. For $t = 2$, $dy/dx = 6$, the tangent line is $y - 1 = 6(x - 3)$, $y = 6x - 17$.

59. Assuming that $a \neq 0$ and $b \neq 0$, eliminate the parameter to get $(x - h)^2/a^2 + (y - k)^2/b^2 = 1$. If $|a| = |b|$ the curve is a circle with center (h, k) and radius $|a|$; if $|a| \neq |b|$ the curve is an ellipse with center (h, k) and major axis parallel to the x-axis when $|a| > |b|$ or major axis parallel to the y-axis when $|a| < |b|$.

61. (a)

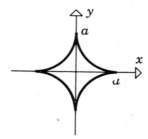

  (b) Use $b = a/4$ in the equations of Exercise 60 to get $x = \dfrac{3}{4}a\cos\phi + \dfrac{1}{4}a\cos 3\phi$, $y = \dfrac{3}{4}a\sin\phi - \dfrac{1}{4}a\sin 3\phi$; but trigonometric identities yield $\cos 3\phi = 4\cos^3\phi - 3\cos\phi$, $\sin 3\phi = 3\sin\phi - 4\sin^3\phi$ so $x = a\cos^3\phi$, $y = a\sin^3\phi$.

  (c) From the result in part (b), $\cos\phi = (x/a)^{1/3}$, $\sin\phi = (y/a)^{1/3}$ so $(x/a)^{2/3} + (y/a)^{2/3} = 1$, $x^{2/3} + y^{2/3} = a^{2/3}$

63. $x' = e^t(\cos t - \sin t)$, $y' = e^t(\cos t + \sin t)$, $(x')^2 + (y')^2 = 2e^{2t}$

$S = 2\pi \displaystyle\int_0^{\pi/2} (e^t \sin t)\sqrt{2e^{2t}}\,dt = 2\sqrt{2}\pi \int_0^{\pi/2} e^{2t}\sin t\,dt$

$= 2\sqrt{2}\pi \left[\dfrac{1}{5}e^{2t}(2\sin t - \cos t)\right]_0^{\pi/2} = \dfrac{2\sqrt{2}}{5}\pi(2e^\pi + 1)$

65. $x' = 1$, $y' = 4t$, $(x')^2 + (y')^2 = 1 + 16t^2$, $S = 2\pi \int_0^1 t\sqrt{1 + 16t^2}\, dt = \dfrac{\pi}{24}(17\sqrt{17} - 1)$

67. $\dfrac{dx}{d\phi} = a(1 - \cos\phi)$, $\dfrac{dy}{d\phi} = a\sin\phi$, $\left(\dfrac{dx}{d\phi}\right)^2 + \left(\dfrac{dy}{d\phi}\right)^2 = 2a^2(1 - \cos\phi)$

$S = 2\pi \int_0^{2\pi} a(1 - \cos\phi)\sqrt{2a^2(1 - \cos\phi)}\, d\phi = 2\sqrt{2}\pi a^2 \int_0^{2\pi} (1 - \cos\phi)^{3/2} d\phi$,

but $1 - \cos\phi = 2\sin^2\dfrac{\phi}{2}$ so $(1 - \cos\phi)^{3/2} = 2\sqrt{2}\sin^3\dfrac{\phi}{2}$ for $0 \le \phi \le \pi$ and, taking advantage

of the symmetry of the cycloid, $S = 16\pi a^2 \int_0^\pi \sin^3\dfrac{\phi}{2}d\phi = 64\pi a^2/3$

EXERCISE SET 13.5

1. $\theta = \pi/3$; $dr/d\theta = -\sqrt{3}$, $r = 1$, $m = 1/\sqrt{3}$

3. $\theta = 2$; $dr/d\theta = -1/4$, $r = 1/2$, $m = \dfrac{\tan 2 - 2}{2\tan 2 + 1}$

5. $\theta = 3\pi/4$; $dr/d\theta = -3\sqrt{2}/2$, $r = \sqrt{2}/2$, $m = -2$

7. $r^2 + (dr/d\theta)^2 = (e^{3\theta})^2 + (3e^{3\theta})^2 = 10e^{6\theta}$, $L = \int_0^2 \sqrt{10}e^{3\theta}d\theta = \sqrt{10}(e^6 - 1)/3$

9. $r^2 + (dr/d\theta)^2 = (2a\cos\theta)^2 + (-2a\sin\theta)^2 = 4a^2$, $L = \int_0^\pi 2a\,d\theta = 2\pi a$

11. $r^2 + (dr/d\theta)^2 = (a\theta^2)^2 + (2a\theta)^2 = a^2\theta^2(\theta^2 + 4)$,

$L = \int_0^\pi a\theta(\theta^2 + 4)^{1/2}d\theta = \dfrac{a}{3}[(\pi^2 + 4)^{3/2} - 8]$

13. $r^2 + (dr/d\theta)^2 = [a(1 - \cos\theta)]^2 + [a\sin\theta]^2 = 4a^2\sin^2(\theta/2)$, $L = 2\int_0^\pi 2a\sin(\theta/2)d\theta = 8a$

15. **(a)** $r^2 + (dr/d\theta)^2 = (\cos n\theta)^2 + (-n\sin n\theta)^2 = \cos^2 n\theta + n^2\sin^2 n\theta$
$= (1 - \sin^2 n\theta) + n^2\sin^2 n\theta = 1 + (n^2 - 1)\sin^2 n\theta$,

$L = 2\int_0^{\pi/(2n)} \sqrt{1 + (n^2 - 1)\sin^2 n\theta}\, d\theta$

(b) $L = 2\int_0^{\pi/4} \sqrt{1 + 3\sin^2 2\theta}\, d\theta \approx 2.42$

17. (a) $\dfrac{dr}{dt} = 2$ and $\dfrac{d\theta}{dt} = 0.5$ so $\dfrac{dr}{d\theta} = \dfrac{dr/dt}{d\theta/dt} = \dfrac{2}{0.5} = 4$, $r = 4\theta + C$, $r = 10$ when $\theta = 0$ so $10 = C, r = 4\theta + 10$.

  (b) $r^2 + (dr/d\theta)^2 = (4\theta + 10)^2 + 16$, during the first 5 seconds the rod rotates through an angle of $(0.5)(5) = 2.5$ radians so $L = \displaystyle\int_0^{2.5} \sqrt{(4\theta + 10)^2 + 16}\,d\theta$, let $u = 4\theta + 10$ to get

$$L = \frac{1}{4}\int_{10}^{20} \sqrt{u^2 + 16}\,du = \frac{1}{4}\left[\frac{u}{2}\sqrt{u^2+16} + 8\ln|u + \sqrt{u^2+16}|\right]_{10}^{20}$$

$$= \frac{1}{4}\left[10\sqrt{416} - 5\sqrt{116} + 8\ln\frac{20+\sqrt{416}}{10+\sqrt{116}}\right] \approx 38.9 \text{ mm}$$

19. $dx/d\theta = 4\sin^2\theta - \sin\theta - 2$, $dy/d\theta = \cos\theta(1 - 4\sin\theta)$. $dy/d\theta = 0$ when $\cos\theta = 0$ or $\sin\theta = 1/4$ so $\theta = \pi/2$, $3\pi/2$, $\sin^{-1}(1/4)$, or $\pi - \sin^{-1}(1/4)$; $dx/d\theta \neq 0$ at these points so there is a horizontal tangent at each one.

TECHNOLOGY EXERCISES 13

1. $\alpha = 8\pi$

3. $\displaystyle\int_0^{\pi/3} \sqrt{r^2 + (dr/d\theta)^2}\,d\theta = 4.454964$

5. The length of the orbit is $L = \displaystyle\int_0^{2\pi} \sqrt{r^2 + (dr/d\theta)^2}\,d\theta$, thus the average speed is $L/[(1.88)(365)(24)] = 86{,}792$ km/hr.

9. The distance between the particles is $D = \sqrt{(t - 1 - \cos t)^2 + (e^{-t} - \sin t)^2}$; solve $dD/dt = 0$ for t to get $t = 1.168585$ sec, so the minimum distance is about 0.65 m.

11. (b) $\displaystyle\int_0^4 \sqrt{(dx/dt)^2 + (dy/dt)^2}\,dt = 49.64$ ft

  (c) The angle is $\theta = \tan^{-1}[(dy/dt)/(dx/dt)]$ where dy/dt and dx/dt are evaluated at $t = 4$, thus $\theta = 69°$.

CHAPTER 14
Second-Order Differential Equations

EXERCISE SET 14.1

1. (a) $y = e^{2x}, y' = 2e^{2x}, y'' = 4e^{2x}; y'' - y' - 2y = 0$
   $y = e^{-x}, y' = -e^{-x}, y'' = e^{-x}; y'' - y' - 2y = 0.$

   (b) $y = c_1 e^{2x} + c_2 e^{-x}, y' = 2c_1 e^{2x} - c_2 e^{-x}, y'' = 4c_1 e^{2x} + c_2 e^{-x}; y'' - y' - 2y = 0$

3. $m^2 + 3m - 4 = 0, (m-1)(m+4) = 0; m = 1, -4$ so $y = c_1 e^x + c_2 e^{-4x}.$

5. $m^2 - 2m + 1 = 0, (m-1)^2 = 0; m = 1,$ so $y = c_1 e^x + c_2 x e^x.$

7. $m^2 + 5 = 0, m = \pm\sqrt{5}\, i$ so $y = c_1 \cos \sqrt{5}\, x + c_2 \sin \sqrt{5}\, x.$

9. $m^2 - m = 0, m(m-1) = 0; m = 0, 1$ so $y = c_1 + c_2 e^x.$

11. $m^2 + 4m + 4 = 0, (m+2)^2 = 0; m = -2$ so $y = c_1 e^{-2t} + c_2 t e^{-2t}.$

13. $m^2 - 4m + 13 = 0, m = 2 \pm 3i$ so $y = e^{2x}(c_1 \cos 3x + c_2 \sin 3x).$

15. $8m^2 - 2m - 1 = 0, (4m+1)(2m-1) = 0; m = -1/4, 1/2$ so $y = c_1 e^{-x/4} + c_2 e^{x/2}.$

17. $m^2 + 2m - 3 = 0, (m+3)(m-1) = 0; m = -3, 1$ so $y = c_1 e^{-3x} + c_2 e^x$ and $y' = -3c_1 e^{-3x} + c_2 e^x.$
    Solve the system $c_1 + c_2 = 1, -3c_1 + c_2 = 5$ to get $c_1 = -1, c_2 = 2$ so $y = -e^{-3x} + 2e^x.$

19. $m^2 - 6m + 9 = 0, (m-3)^2 = 0; m = 3$ so $y = (c_1 + c_2 x)e^{3x}$ and $y' = (3c_1 + c_2 + 3c_2 x)e^{3x}.$
    Solve the system $c_1 = 2, 3c_1 + c_2 = 1$ to get $c_1 = 2, c_2 = -5$ so $y = (2 - 5x)e^{3x}.$

21. $m^2 + 4m + 5 = 0, m = -2 \pm i$ so $y = e^{-2x}(c_1 \cos x + c_2 \sin x),$
    $y' = e^{-2x}[(c_2 - 2c_1)\cos x - (c_1 + 2c_2)\sin x].$ Solve the system $c_1 = -3, c_2 - 2c_1 = 0$
    to get $c_1 = -3, c_2 = -6$ so $y = -e^{-2x}(3\cos x + 6\sin x).$

23. (a) $m = 5, -2$ so $(m-5)(m+2) = 0, m^2 - 3m - 10 = 0; y'' - 3y' - 10y = 0.$
    (b) $m = 4, 4$ so $(m-4)^2 = 0, m^2 - 8m + 16 = 0; y'' - 8y' + 16y = 0.$
    (c) $m = -1 \pm 4i$ so $(m+1-4i)(m+1+4i) = 0, m^2 + 2m + 17 = 0; y'' + 2y' + 17y = 0.$

25. $m^2 + km + k = 0$, $m = \left(-k \pm \sqrt{k^2 - 4k}\right)/2$

(a) $k^2 - 4k > 0$, $k(k-4) > 0$; $k < 0$ or $k > 4$

(b) $k^2 - 4k = 0$; $k = 0, 4$ (c) $k^2 - 4k < 0$, $k(k-4) < 0$; $0 < k < 4$

27. (a) $\dfrac{d^2y}{dz^2} + 2\dfrac{dy}{dz} + 2y = 0$, $m^2 + 2m + 2 = 0$; $m = -1 \pm i$ so

$y = e^{-z}(c_1 \cos z + c_2 \sin z) = \dfrac{1}{x}[c_1 \cos(\ln x) + c_2 \sin(\ln x)]$.

(b) $\dfrac{d^2y}{dz^2} - 2\dfrac{dy}{dz} - 2y = 0$, $m^2 - 2m - 2 = 0$; $m = 1 \pm \sqrt{3}$ so

$y = c_1 e^{(1+\sqrt{3})z} + c_2 e^{(1-\sqrt{3})z} = c_1 x^{1+\sqrt{3}} + c_2 x^{1-\sqrt{3}}$

29. (a) $W(x) = \begin{vmatrix} e^{m_1 x} & e^{m_2 x} \\ m_1 e^{m_1 x} & m_2 e^{m_2 x} \end{vmatrix} = m_2 e^{(m_1+m_2)x} - m_1 e^{(m_1+m_2)x}$

$= (m_2 - m_1)e^{(m_1+m_2)x} \neq 0$ if $m_1 \neq m_2$.

(b) $W(x) = \begin{vmatrix} e^{mx} & xe^{mx} \\ me^{mx} & (mx+1)e^{mx} \end{vmatrix} = e^{2mx} \neq 0$.

31. (a) The general solution is $c_1 e^{\mu x} + c_2 e^{mx}$; let $c_1 = 1/(\mu - m)$, $c_2 = -1/(\mu - m)$.

(b) $\displaystyle\lim_{\mu \to m} \dfrac{e^{\mu x} - e^{mx}}{\mu - m} = \lim_{\mu \to m} xe^{\mu x} = xe^{mx}$.

EXERCISE SET 14.2

1. $m^2 + 6m + 5 = 0$, $(m+1)(m+5) = 0$; $m = -1, -5$ so $y_c = c_1 e^{-x} + c_2 e^{-5x}$. Let
$y_p = Ae^{3x}$, then $y_p' = 3Ae^{3x}$, $y_p'' = 9Ae^{3x}$, $(9A + 18A + 5A)e^{3x} = 32Ae^{3x} = 2e^{3x}$,
$A = 1/16$; $y = c_1 e^{-x} + c_2 e^{-5x} + \frac{1}{16}e^{3x}$.

3. $m^2 - 9m + 20 = 0$, $(m-4)(m-5) = 0$; $m = 4, 5$ so $y_c = c_1 e^{4x} + c_2 e^{5x}$. Let $y_p = Axe^{5x}$,
then $y_p' = (5Ax + A)e^{5x}$, $y_p'' = (25Ax + 10A)e^{5x}$,
$(25Ax + 10A - 45Ax - 9A + 20Ax)e^{5x} = Ae^{5x} = -3e^{5x}$, $A = -3$; $y = c_1 e^{4x} + c_2 e^{5x} - 3xe^{5x}$.

5. $m^2 + 2m + 1 = 0$, $(m+1)^2 = 0$; $m = -1$ so $y_c = (c_1 + c_2 x)e^{-x}$. Let $y_p = Ax^2 e^{-x}$,
then $y_p' = (-Ax^2 + 2Ax)e^{-x}$, $y_p'' = (Ax^2 - 4Ax + 2A)e^{-x}$,
$(Ax^2 - 4Ax + 2A - 2Ax^2 + 4Ax + Ax^2)e^{-x} = 2Ae^{-x} = e^{-x}$, $A = 1/2$; $y = (c_1 + c_2 x)e^{-x} + \frac{1}{2}x^2 e^{-x}$.

7. $m^2 + m - 12 = 0$, $(m-3)(m+4) = 0$; $m = 3, -4$ so $y_c = c_1 e^{3x} + c_2 e^{-4x}$. Let
$y_p = A_0 + A_1 x + A_2 x^2$, then $y_p' = A_1 + 2A_2 x$, $y_p'' = 2A_2$,
$2A_2 + A_1 + 2A_2 x - 12A_0 - 12A_1 x - 12A_2 x^2$
$\quad = (-12A_0 + A_1 + 2A_2) + 2(-6A_1 + A_2)x - 12A_2 x^2 = 4x^2$;
solve the system $-12A_0 + A_1 + 2A_2 = 0$, $-6A_1 + A_2 = 0$, $-12A_2 = 4$ to get
$A_0 = -13/216$, $A_1 = -1/18$, $A_2 = -1/3$ so $y = c_1 e^{3x} + c_2 e^{-4x} - \frac{13}{216} - \frac{1}{18}x - \frac{1}{3}x^2$.

9. $m^2 - 6m = 0$; $m = 0, 6$ so $y_c = c_1 + c_2 e^{6x}$. Let $y_p = A_0 x + A_1 x^2$, then $y_p' = A_0 + 2A_1 x$, $y_p'' = 2A_1$,
$2A_1 - 6A_0 - 12A_1 x = (-6A_0 + 2A_1) - 12A_1 x = x - 1$; solve the system $-6A_0 + 2A_1 = -1$,
$-12A_1 = 1$ to get $A_0 = 5/36$, $A_1 = -1/12$, so $y = c_1 + c_2 e^{6x} + \frac{5}{36}x - \frac{1}{12}x^2$.

11. $m^2 = 0$, $m = 0$ so $y_c = c_1 + c_2 x$. Let $y_p = A_0 x^2 + A_1 x^3 + A_2 x^4 + A_3 x^5$,
then $y_p' = 2A_0 x + 3A_1 x^2 + 4A_2 x^3 + 5A_3 x^4$, $y_p'' = 2A_0 + 6A_1 x + 12A_2 x^2 + 20A_3 x^3$;
$A_0 = -1/2$, $A_1 = 0$, $A_2 = 0$, $A_3 = 1/20$ so $y = c_1 + c_2 x - x^2/2 + x^5/20$.

13. $m^2 - m - 2 = 0$, $(m+1)(m-2) = 0$; $m = -1, 2$ so $y_c = c_1 e^{-x} + c_2 e^{2x}$. Let
$y_p = A_1 \cos x + A_2 \sin x$, then $y_p' = -A_1 \sin x + A_2 \cos x$, $y_p'' = -A_1 \cos x - A_2 \sin x$,
$-A_1 \cos x - A_2 \sin x + A_1 \sin x - A_2 \cos x - 2A_1 \cos x - 2A_2 \sin x$
$\quad = (-3A_1 - A_2) \cos x + (A_1 - 3A_2) \sin x = 10 \cos x$; solve the system $-3A_1 - A_2 = 10$,
$A_1 - 3A_2 = 0$ to get $A_1 = -3$, $A_2 = -1$ so $y = c_1 e^{-x} + c_2 e^{2x} - 3 \cos x - \sin x$.

15. $m^2 - 4 = 0$; $m = \pm 2$ so $y_c = c_1 e^{-2x} + c_2 e^{2x}$. Let $y_p = A_1 \cos 2x + A_2 \sin 2x$,
then $y_p' = -2A_1 \sin 2x + 2A_2 \cos 2x$, $y_p'' = -4A_1 \cos 2x - 4A_2 \sin 2x$,
$-8A_1 \cos 2x - 8A_2 \sin 2x = 2 \sin 2x + 3 \cos 2x$, $A_1 = -3/8$, $A_2 = -1/4$;
$y = c_1 e^{-2x} + c_2 e^{2x} - \frac{3}{8} \cos 2x - \frac{1}{4} \sin 2x$.

17. $m^2 + 1 = 0$; $m = \pm i$ so $y_c = c_1 \cos x + c_2 \sin x$. Let $y_p = A_1 x \cos x + A_2 x \sin x$,
then $y_p' = (A_1 + A_2 x) \cos x + (A_2 - A_1 x) \sin x$, $y_p'' = (2A_2 - A_1 x) \cos x - (2A_1 + A_2 x) \sin x$,
$2A_2 \cos x - 2A_1 \sin x = \sin x$, $A_1 = -1/2$, $A_2 = 0$; $y = c_1 \cos x + c_2 \sin x - \frac{1}{2}x \cos x$.

19. $m^2 - 3m + 2 = 0$, $(m-1)(m-2) = 0$; $m = 1, 2$ so $y_c = c_1 e^x + c_2 e^{2x}$. Let $y_p = A_0 + A_1 x$,
then $y_p' = A_1$, $y_p'' = 0$; solve the system $2A_0 - 3A_1 = 0$, $2A_1 = 1$ to get $A_0 = 3/4$, $A_1 = 1/2$;
$y = c_1 e^x + c_2 e^{2x} + 3/4 + x/2$.

21. $m^2 + 4m + 9 = 0$; $m = -2 \pm \sqrt{5}\, i$ so $y_c = e^{-2x}(c_1 \cos \sqrt{5}\, x + c_2 \sin \sqrt{5}\, x)$.
Let $y_p = A_0 + A_1 x + A_2 x^2$, then $y_p' = A_1 + 2A_2 x$, $y_p'' = 2A_2$; solve the system
$9A_0 + 4A_1 + 2A_2 = 0$, $9A_1 + 8A_2 = 3$, $9A_2 = 1$, to get $A_0 = -94/729$, $A_1 = 19/81$, $A_2 = 1/9$
so $y = e^{-2x}(c_1 \cos \sqrt{5}\, x + c_2 \sin \sqrt{5}\, x) - 94/729 + 19x/81 + x^2/9$.

23. $m^2 + 4 = 0$; $m = \pm 2i$ so $y_c = c_1 \cos 2x + c_2 \sin 2x$; $\sin x \cos x = \frac{1}{2} \sin 2x$,

let $y_p = A_1 x \cos 2x + A_2 x \sin 2x$, then $y_p' = (A_1 + 2A_2 x) \cos 2x + (A_2 - 2A_1 x) \sin 2x$,

$y_p'' = (4A_2 - 4A_1 x) \cos 2x - (4A_1 + 4A_2 x) \sin 2x$, $4A_2 \cos 2x - 4A_1 \sin 2x = \frac{1}{2} \sin 2x$,

$A_1 = -1/8$, $A_2 = 0$; $y = c_1 \cos 2x + c_2 \sin 2x - \frac{1}{8} x \cos 2x$.

25. **(a)** Let $y = y_1 + y_2$, then $y' = y_1' + y_2'$, $y'' = y_1'' + y_2''$ so

$y'' + p(x)y' + q(x)y = [y_1'' + p(x)y_1' + q(x)y_1] + [y_2'' + p(x)y_2' + q(x)y_2] = r_1(x) + r_2(x)$.

(b) $m^2 + 3m - 4 = 0$, $(m-1)(m+4) = 0$; $m = 1, -4$ so $y_c = c_1 e^x + c_2 e^{-4x}$. For $y'' + 3y' - 4y = x$
let $y_1 = A_0 + A_1 x$, then $y_1' = A_1$, $y_1'' = 0$; $-4A_0 + 3A_1 = 0$ and $-4A_1 = 1$ so $A_0 = -3/16$,
$A_1 = -1/4$; $y_1 = -3/16 - x/4$, for $y'' + 3y' - 4y = e^x$ let $y_2 = Axe^x$, then $y_2' = (Ax + A)e^x$,
$y_2'' = (Ax + 2A)e^x$, $5A = 1$, $A = 1/5$; $y_2 = \frac{1}{5} xe^x$ thus $y_1 + y_2 = -\frac{3}{16} - \frac{1}{4} x + \frac{1}{5} xe^x$ is a
particular solution.

(c) If $y_i(x)$ is a solution of $y'' + p(x)y' + q(x)y = r_i(x)$ for $i = 1, 2, \cdots, n$ then
$y_1(x) + y_2(x) + \cdots + y_n(x)$ is a solution of
$y'' + p(x)y' + q(x)y = r_1(x) + r_2(x) + \cdots + r_n(x)$.

27. $m^2 - 1 = 0$; $m = \pm 1$ so $y_c = c_1 e^{-x} + c_2 e^x$. Let $r_1(x) = 1$ and $r_2(x) = e^x$, then $y_1 = A_0$,
$y_1' = y_2'' = 0$, $A_0 = -1$; $y_2 = Axe^x$, $y_2' = (Ax + A)e^x$, $y_2'' = (Ax + 2A)e^x$, $2A = 1$, $A = 1/2$ so
$y = c_1 e^{-x} + c_2 e^x - 1 + \frac{1}{2} xe^x$.

29. $m^2 + 4 = 0$; $m = \pm 2i$ so $y_c = c_1 \cos 2x + c_2 \sin 2x$. Let $r_1(x) = 1 + x$ and $r_2(x) = \sin x$, then
$y_1 = A_0 + A_1 x$, $y_1' = A_1$, $y_1'' = 0$, $4A_0 = 1$, and $4A_1 = 1$ so $A_0 = A_1 = 1/4$;
$y_2 = B_1 \cos x + B_2 \sin x$, $y_2' = -B_1 \sin x + B_2 \cos x$, $y_2'' = -B_1 \cos x - B_2 \sin x$,
$3B_1 = 0$ and $3B_2 = 1$ so $B_1 = 0$, $B_2 = 1/3$; $y = c_1 \cos 2x + c_2 \sin 2x + \frac{1}{4} + \frac{1}{4} x + \frac{1}{3} \sin x$.

31. $m^2 - 2m + 1 = 0$, $(m-1)^2 = 0$; $m = 1$ so $y_c = (c_1 + c_2 x)e^x$. Let
$r_1(x) = \frac{1}{2} e^x$ and $r_2(x) = -\frac{1}{2} e^{-x}$, then $y_1 = Ax^2 e^x$, $y_1' = (Ax^2 + 2Ax)e^x$
$y_1'' = (Ax^2 + 4Ax + 2A)e^x$, $2A = 1/2$, $A = 1/4$; $y_2 = Be^{-x}$, $y_2' = -Be^{-x}$,
$y_2'' = Be^{-x}$, $4B = 1/2$, $B = 1/8$; $y = (c_1 + c_2 x)e^x + \frac{1}{4} x^2 e^x + \frac{1}{8} e^{-x}$.

33. $m^2 + 1 = 0$; $m = \pm i$ so $y_c = c_1 \cos x + c_2 \sin x$. Let $r_1(x) = 6$ and
$r_2(x) = 6 \cos 2x$, then $y_1 = A_0$, $y_1' = y_1'' = 0$, $A_0 = 6$; $y_2 = A_1 \cos 2x + A_2 \sin 2x$,
$y_2' = -2A_1 \sin 2x + 2A_2 \cos 2x$, $y_2'' = -4A_1 \cos 2x - 4A_2 \sin 2x$, $-3A_1 = 6$ and $-3A_2 = 0$
so $A_1 = -2$, $A_2 = 0$; $y = c_1 \cos x + c_2 \sin x + 6 - 2 \cos 2x$.

35. (a) $m^2 + \mu^2 = 0$; $m = \pm\mu i$ so $y_c = c_1 \cos \mu x + c_2 \sin \mu x$. Let $y_p = A_1 \cos bx + A_2 \sin bx$, then $y_p' = -bA_1 \sin bx + bA_2 \cos bx$, $y_p'' = -b^2 A_1 \cos bx - b^2 A_2 \sin bx$, $A_1 = 0$, $A_2 = a/(\mu^2 - b^2)$; $y = c_1 \cos \mu x + c_2 \sin \mu x + \dfrac{a}{\mu^2 - b^2} \sin bx$.

(b) $y = c_1 \cos \mu x + c_2 \sin \mu x + \displaystyle\sum_{k=1}^{n} \dfrac{a_k}{\mu^2 - k^2 \pi^2} \sin k\pi x$.

37. $m^2 - m = 0$; $m = 0, 1$ so $y_c = c_1 + c_2 e^x$. Let $y_p = A_0 x + A_1 x^2$, then $y_p' = A_0 + 2A_1 x$, $y_p'' = 2A_1$; $A_0 = 0$, $A_1 = 2$, $y = c_1 + c_2 e^x + 2x^2$. If $y'(x_0) = y''(x_0) = 0$, then $4 - 4x_0 = 0$, $x_0 = 1$ so $y'(1) = 0$ and $y''(1) = 0$; $y' = c_2 e^x + 4x$, $y'' = c_2 e^x + 4$, $y'(1) = c_2 e + 4 = y''(1) = 0$ if $c_2 = -4/e$ so $y = c_1 - 4e^{x-1} + 2x^2$, or simply $y = c - 4e^{x-1} + 2x^2$ where c is an arbitrary constant.

EXERCISE SET 14.3

1. $m^2 + 1 = 0$; $m = \pm i$ so $y_c = c_1 \cos x + c_2 \sin x$. $u' \cos x + v' \sin x = 0$ and $-u' \sin x + v' \cos x = x^2$; $u' = -x^2 \sin x$, $v' = x^2 \cos x$ so $u = x^2 \cos x - 2x \sin x - 2 \cos x$, $v = x^2 \sin x + 2x \cos x - 2 \sin x$, $y_p = u \cos x + v \sin x = x^2 - 2$, $y = c_1 \cos x + c_2 \sin x + x^2 - 2$.

3. $m^2 + m - 2 = 0$, $(m-1)(m+2) = 0$; $m = 1, -2$ so $y_c = c_1 e^x + c_2 e^{-2x}$. $u' e^x + v' e^{-2x} = 0$ and $u' e^x - 2v' e^{-2x} = 2e^x$; $u' = \frac{2}{3}$, $v' = -\frac{2}{3} e^{3x}$ so $u = \frac{2}{3} x$, $v = -\frac{2}{9} e^{3x}$, $y_p = u e^x + v e^{-2x} = \frac{2}{3} x e^x - \frac{2}{9} e^x$. But $-\frac{2}{9} e^x$ satisfies the complementary equation so $y = c_1 e^x + c_2 e^{-2x} + \frac{2}{3} x e^x$.

5. $m^2 + 4 = 0$; $m = \pm 2i$ so $y_c = c_1 \cos 2x + c_2 \sin 2x$. $u' \cos 2x + v' \sin 2x = 0$ and $-2u' \sin 2x + v' \cos 2x = \sin 2x$; $u' = -\frac{1}{2} \sin^2 2x$, $v' = \frac{1}{2} \sin 2x \cos 2x$ so $u = -\frac{1}{4} x + \frac{1}{16} \sin 4x = -\frac{1}{4} x + \frac{1}{8} \sin 2x \cos 2x$, $v = -\frac{1}{8} \cos^2 2x$, $y_p = u \cos 2x + v \sin 2x = -\frac{1}{4} x \cos 2x$, $y = c_1 \cos 2x + c_2 \sin 2x - \frac{1}{4} x \cos 2x$.

7. $m^2 + 1 = 0$; $m = \pm i$ so $y_c = c_1 \cos x + c_2 \sin x$. $u' \cos x + v' \sin x = 0$ and $-u' \sin x + v' \cos x = \tan x$; $u' = -\tan x \sin x = \cos x - \sec x$, $v' = \tan x \cos x = \sin x$ so $u = \sin x - \ln|\sec x + \tan x|$, $v = -\cos x$, $y_p = u \cos x + v \sin x = -\cos x \ln|\sec x + \tan x|$, $y = c_1 \cos x + c_2 \sin x - \cos x \ln|\sec x + \tan x|$.

9. $m^2 - 2m + 1 = 0$, $(m-1)^2 = 0$; $m = 1$ so $y_c = c_1 e^x + c_2 x e^x$. $u'e^x + v'xe^x = 0$ and
$u'e^x + v'(x+1)e^x = e^x/x$; $u' = -1$, $v' = 1/x$ so $u = -x$, $v = \ln|x|$,
$y_p = ue^x + vxe^x = -xe^x + xe^x \ln|x|$. But $-xe^x$ satisfies the complementary equation so
$y = c_1 e^x + c_2 xe^x + xe^x \ln|x|$.

11. $m^2 + 1 = 0$; $m = \pm i$ so $y_c = c_1 \cos x + c_2 \sin x$. $u' \cos x + v' \sin x = 0$ and
$-u' \sin x + v' \cos x = 3 \sin^2 x$; $u' = -3 \sin^3 x$, $v' = 3 \sin^2 x \cos x$ so $u = 3 \cos x - \cos^3 x$,
$v = \sin^3 x$, $y_p = u \cos x + v \sin x = 3 \cos^2 x - \cos^4 x + \sin^4 x = 3 \cos^2 x - (\cos^4 x - \sin^4 x)$
$\qquad = \frac{3}{2}(1 + \cos 2x) - (\cos^2 x - \sin^2 x)(\cos^2 x + \sin^2 x) = \frac{3}{2} + \frac{1}{2} \cos 2x$,
$y = c_1 \cos x + c_2 \sin x + \frac{3}{2} + \frac{1}{2} \cos 2x$.

13. $m^2 + 1 = 0$; $m = \pm i$ so $y_c = c_1 \cos x + c_2 \sin x$. $u' \cos x + v' \sin x = 0$ and
$-u' \sin x + v' \cos x = \csc x$; $u' = -1$, $v' = \cos x \csc x = \cot x$ so $u = -x$,
$v = \ln|\sin x|$, $y_p = u \cos x + v \sin x = -x \cos x + \sin x \ln|\sin x|$,
$y = c_1 \cos x + c_2 \sin x - x \cos x + \sin x \ln|\sin x|$.

15. $m^2 + 1 = 0$; $m = \pm i$ so $y_c = c_1 \cos x + c_2 \sin x$. $u' \cos x + v' \sin x = 0$ and
$-u' \sin x + v' \cos x = \sec x \tan x$; $u' = -\tan^2 x$, $v' = \tan x$ so $u = x - \tan x$,
$v = -\ln|\cos x|$, $y_p = u \cos x + v \sin x = x \cos x - \sin x - \sin x \ln|\cos x|$. But $-\sin x$ satisfies
the complementary equation so $y = c_1 \cos x + c_2 \sin x + x \cos x - \sin x \ln|\cos x|$.

17. $m^2 + 2m + 1 = 0$, $(m+1)^2 = 0$; $m = -1$ so $y_c = c_1 e^{-x} + c_2 x e^{-x}$. $u'e^{-x} + v'xe^{-x} = 0$
and $-u'e^{-x} + v'(1-x)e^{-x} = e^{-x}/x^2$; $u' = -1/x$, $v' = 1/x^2$ so $u = -\ln|x|$, $v = -1/x$,
$y_p = ue^{-x} + vxe^{-x} = -e^{-x} \ln|x| - e^{-x}$. But $-e^{-x}$ satisfies the complementary equation so
$y = c_1 e^{-x} + c_2 xe^{-x} - e^{-x} \ln|x|$.

19. $m^2 + 4m + 4 = 0$, $(m+2)^2 = 0$; $m = -2$ so $y_c = c_1 e^{-2x} + c_2 x e^{-2x}$. $u'e^{-2x} + v'xe^{-2x} = 0$
and $-2u'e^{-2x} + v'(1-2x)e^{-2x} = xe^{-x}$; $u' = -x^2 e^x$, $v' = xe^x$ so $u = (-x^2 + 2x - 2)e^x$,
$v = (x-1)e^x$, $y_p = ue^{-2x} + vxe^{-2x} = (x-2)e^{-x}$, $y = c_1 e^{-2x} + c_2 xe^{-2x} + (x-2)e^{-x}$.

21. $m^2 + 1 = 0$; $m = \pm i$ so $y_c = c_1 \cos x + c_2 \sin x$. $u' \cos x + v' \sin x = 0$ and
$-u' \sin x + v' \cos x = \sec^2 x$; $u' = -\sec x \tan x$, $v' = \sec x$ so $u = -\sec x$,
$v = \ln|\sec x + \tan x|$, $y_p = u \cos x + v \sin x = -1 + \sin x \ln|\sec x + \tan x|$,
$y = c_1 \cos x + c_2 \sin x - 1 + \sin x \ln|\sec x + \tan x|$.

23. $m^2 - 2m + 1 = 0$, $(m-1)^2 = 0$; $m = 1$ so $y_c = c_1 e^x + c_2 x e^x$. $u'e^x + v'xe^x = 0$ and
$u'e^x + v'(x+1)e^x = e^x/x^2$; $u' = -1/x$, $v' = 1/x^2$ so $u = -\ln|x|$, $v = -1/x$,
$y_p = ue^x + vxe^x = -e^x \ln|x| - e^x$. But $-e^x$ satisfies the complementary equation so
$y = c_1 e^x + c_2 xe^x - e^x \ln|x|$.

25. $m^2 - 1 = 0$; $m = \pm 1$ so $y_c = c_1 e^x + c_2 e^{-x}$. $u'e^x + v'e^{-x} = 0$ and $u'e^x - v'e^{-x} = e^x \cos x$;
$u' = \frac{1}{2}\cos x$, $v' = -\frac{1}{2}e^{2x}\cos x$ so $u = \frac{1}{2}\sin x$, $v = -\frac{1}{10}e^{2x}(2\cos x + \sin x)$,
$y_p = ue^x + ve^{-x} = \frac{1}{2}e^x \sin x - \frac{1}{10}e^x(2\cos x + \sin x) = \frac{1}{5}e^x(2\sin x - \cos x)$,
$y = c_1 e^x + c_2 e^{-x} + \frac{1}{5}e^x(2\sin x - \cos x)$.

27. $m^2 + 2m + 1 = 0$, $(m+1)^2 = 0$; $m = -1$ so $y_c = c_1 e^{-x} + c_2 x e^{-x}$. $u'e^{-x} + v'xe^{-x} = 0$
and $-u'e^{-x} + v'(1-x)e^{-x} = e^{-x}\ln|x|$; $u' = -x\ln|x|$, $v' = \ln|x|$ so $u = -\frac{1}{2}x^2\ln|x| + \frac{1}{4}x^2$,
$v = x\ln|x| - x$,
$y_p = ue^{-x} + vxe^{-x} = -\frac{1}{2}x^2 e^{-x}\ln|x| + \frac{1}{4}x^2 e^{-x} + x^2 e^{-x}\ln|x| - x^2 e^{-x} = \frac{1}{2}x^2 e^{-x}\ln|x| - \frac{3}{4}x^2 e^{-x}$,
$y = c_1 e^{-x} + c_2 x e^{-x} + \frac{1}{2}x^2 e^{-x}\ln|x| - \frac{3}{4}x^2 e^{-x}$.

29. $m^2 + 1 = 0$; $m = \pm i$ so $y_c = c_1 \cos x + c_2 \sin x$. $u'\cos x + v'\sin x = 0$, $-u'\sin x + v'\cos x = r(x)$;
$u' = -r(x)\sin x$, $v' = r(x)\cos x$ so $u = -\displaystyle\int r(x)\sin x\, dx$,
$v = \displaystyle\int r(x)\cos x\, dx$, $y = c_1 \cos x + c_2 \sin x - \left[\displaystyle\int r(x)\sin x\, dx\right]\cos x + \left[\displaystyle\int r(x)\cos x\, dx\right]\sin x$.

EXERCISE SET 14.4

1. **(a)** $k/M = 0.25/1 = 0.25$; $y'' + 0.25y = 0$, $y(0) = 0.3$, $y'(0) = 0$.
   (b) $m^2 + 0.25 = 0$; $m = \pm 0.5i$ so $y = c_1 \cos 0.5t + c_2 \sin 0.5t$,
   $y' = -0.5c_1 \sin 0.5t + 0.5c_2 \cos 0.5t$, $c_1 = 0.3$ and $c_2 = 0$, $y = 0.3\cos 0.5t$.

3. **(a)** $k/M = g/\ell = 9.8/0.05 = 196$; $y'' + 196y = 0$, $y(0) = -0.12$, $y'(0) = 0$.
   (b) $m = \pm 14i$ so $y = c_1 \cos 14t + c_2 \sin 14t$, $y' = -14c_1 \sin 14t + 14c_2 \cos 14t$, $c_1 = -0.12$ and
   $c_2 = 0$, $y = -0.12\cos 14t$.

5. **(a)** $k/M = 64$, $y_0 = 0.2$; $y = 0.2\cos 8t$.
   (b) $|y_0| = 0.2$ m   **(c)** $T = \pi/4$ sec   **(d)** $f = 4/\pi$ Hz

7. **(a)** $k/M = g/\ell = 32/(1/12) = 384$, $y_0 = -3/12 = -1/4$; $y = -\dfrac{1}{4}\cos 8\sqrt{6}\,t$.
   (b) $|y_0| = 1/4$ ft   **(c)** $T = \pi/(4\sqrt{6})$ sec   **(d)** $f = 4\sqrt{6}/\pi$ Hz

9. **(a)** $M = w/g = 32/32 = 1$, $k/M = 8$; $y'' + 4y' + 8y = 0$, $y(0) = -3$, $y'(0) = 0$.
   (b) $m^2 + 4m + 8 = 0$; $m = -2 \pm 2i$ so $y = e^{-2t}(c_1 \cos 2t + c_2 \sin 2t)$,
   $y' = 2e^{-2t}[(c_2 - c_1)\cos 2t - (c_1 + c_2)\sin 2t]$, $c_1 = c_2 = -3$, $y = -3e^{-2t}(\cos 2t + \sin 2t)$.

9. **(d)** $\alpha = \beta = 2$, $\omega = \tan^{-1}(\alpha/\beta) = \tan^{-1} 1 = \pi/4$; $y = -3\sqrt{2}\, e^{-2t} \cos(2t - \pi/4)$.

   (e) $T = \pi$ sec   **(f)** $f = 1/\pi$ Hz

11. **(a)** $\alpha = 1/5$, $\beta = \sqrt{2}/5$; $y = \dfrac{5}{2}\sqrt{6}e^{-t/5} \cos[\sqrt{2}t/5 - \tan^{-1}(1/\sqrt{2})]$.

    (b) $T = 10\pi/\sqrt{2}$ sec   **(c)** $\sqrt{2}/(10\pi)$ Hz

13. **(a)** $y'' + 64y = 0$, $y(0) = 0$, $y'(0) = -2$. $y = c_1 \cos 8t + c_2 \sin 8t$, $y' = -8c_1 \sin 8t + 8c_2 \cos 8t$,

    $c_1 = 0$ and $c_2 = -1/4$; $y = -\dfrac{1}{4} \sin 8t$.

    (b) $|y_0| = 1/4$ ft   **(c)** $T = \pi/4$ sec   **(d)** $f = 4/\pi$ Hz

15. $M = w/g = 1/8$, $k/M = 50$; $y'' + 2y' + 50y = 0$, $y(0) = 1/3$, $y'(0) = -5$.

    $m^2 + 2m + 50 = 0$; $m = -1 \pm 7i$ so $y = e^{-t}(c_1 \cos 7t + c_2 \sin 7t)$,

    $y' = e^{-t}[(7c_2 - c_1) \cos 7t - (7c_1 + c_2) \sin 7t]$, $c_1 = 1/3$ and $c_2 = -2/3$,

    $y = \dfrac{1}{3} e^{-t}(\cos 7t - 2 \sin 7t)$.

17. $T = 2\pi\sqrt{M/k} = 2\pi\sqrt{w/(kg)}$, $T^2 = 4\pi^2 w/(32k)$, $\pi^2 w = 8T^2 k$; for w and $w + 4$, $\pi^2 w = 72k$ and $\pi^2(w + 4) = 200k$. Solve this system of equations to get

    (a) $k = \pi^2/32$ lb/ft   **(b)** $w = 9/4$ lb

19. Let ℓ be the depth to which the cylinder is submerged in the water at equilibrium, then $\rho \pi r^2 \ell = \delta \pi r^2 h$ so $\ell = \delta h/\rho$, $M/k = \ell/g = \delta h/(\rho g)$, $T = 2\pi\sqrt{\delta h/(\rho g)}$ sec.

21. **(a)** $y = y_0 \cos(\sqrt{k/M}\, t)$, $y' = -y_0 \sqrt{k/M} \sin(\sqrt{k/M}\, t)$; when $\sin(\sqrt{k/M}t)$ is 1 or -1, $\cos(\sqrt{k/M}\, t)$ is 0 so $y = 0$ when $|y'| = |y_0|\sqrt{k/M} = 2\pi|y_0|/T$.

    (b) $y'' = -y_0(k/M) \cos(\sqrt{k/M}\, t)$; $|y''|$ is maximum when $\cos(\sqrt{k/M}\, t)$ is 1 or -1, and hence $|y|$ is maximum. The maximum value of $|y''|$ is $|y_0|k/M = 4\pi^2|y_0|/T^2$.

23. Let $\omega = \tan^{-1}(\alpha/\beta)$, then $\cos \omega = \beta/\sqrt{\alpha^2 + \beta^2}$ and $\sin \omega = \alpha/\sqrt{\alpha^2 + \beta^2}$,

    $\beta \cos \beta t + \alpha \sin \beta t = \sqrt{\alpha^2 + \beta^2}(\cos \beta t \cos \omega + \sin \beta t \sin \omega) = \sqrt{\alpha^2 + \beta^2} \cos(\beta t - \omega)$

    so $y(t) = \dfrac{y_0 \sqrt{\alpha^2 + \beta^2}}{\beta} e^{-\alpha t} \cos(\beta t - \omega)$.

25. **(a)** $m = -\dfrac{c}{2M} = -\alpha$ so $y = (c_1 + c_2 t)e^{-\alpha t}$, $y' = (c_2 - \alpha c_1 - \alpha c_2 t)e^{-\alpha t}$, $c_1 = y_0$ and $c_2 = \alpha y_0$;

    $y = y_0(1 + \alpha t)e^{-\alpha t}$.

(b) $\displaystyle\lim_{t\to+\infty} y_0(1+\alpha t)e^{-\alpha t} = \lim_{t\to+\infty} \frac{y_0(1+\alpha t)}{e^{\alpha t}} = \lim_{t\to+\infty} \frac{y_0}{e^{\alpha t}} = 0.$

(c) $\alpha > 0$ so $y_0(1+\alpha t)e^{-\alpha t} \neq 0$ for $t > 0$.

TECHNOLOGY EXERCISES 14

1. (a) $y = e^{-6t/5} + \dfrac{16}{5}\, te^{-6t/5}$

   (b) $y'(t) = 0$ when $t = t_1 = 25/48 \approx 0.520833$, $y(t_1) = 1.427364$ cm

   (c) $y = \dfrac{16}{5}\, e^{-6t/5}(t + 5/16) = 0$ only if $t = -5/16$, so $y \neq 0$ for $t \geq 0$.

3. (a) $y = e^{-t/2}\cos(\sqrt{19}\,t/2) - (6/\sqrt{19})e^{-t/2}\sin(\sqrt{19}\,t/2)$

   (b) $y'(t) = 0$ for the first time when $t = t_1 = 0.905533$, $y(t_1) = -1.054466$ cm so the maximum distance below the equilibrium position is 1.054466 cm.

   (c) $y(t) = 0$ for the first time when $t = t_2 = 0.288274$, $y'(t_2) = -3.210357$ cm/sec.

   (d) The acceleration is $y''(t)$ so from the differential equation $y'' = -y' - 5y$. But $y = 0$ when the object passes through the equilibrium position, thus $y'' = -y' = 3.210357$ cm/sec^2.

5. (a) The natural frequency of the system is $3/\pi$ Hz, that of the external force is $5/(2\pi)$ Hz.

   (b) $y = \dfrac{1}{11}\sin(5t) - \dfrac{5}{66}\sin(6t)$

   (c) $y = \dfrac{4}{23}\sin(5.5t) - \dfrac{11}{69}\sin(6t)$, $y = \dfrac{25}{59}\sin(5.8t) - \dfrac{145}{354}\sin(6t)$

7. (a) $y = (4 + 2v_0)e^{-3t/2} - (3 + 2v_0)e^{-2t}$

APPENDIX B
Trigonometry Review

EXERCISE SET B.I

1. (a) $5\pi/12$ (b) $13\pi/6$ (c) $\pi/9$ (d) $23\pi/30$

3. (a) $12°$ (b) $(270/\pi)°$ (c) $288°$ (d) $540°$

5.

	$\sin\theta$	$\cos\theta$	$\tan\theta$	$\csc\theta$	$\sec\theta$	$\cot\theta$
(a)	$\sqrt{21}/5$	$2/5$	$\sqrt{21}/2$	$5/\sqrt{21}$	$5/2$	$2/\sqrt{21}$
(b)	$3/4$	$\sqrt{7}/4$	$3/\sqrt{7}$	$4/3$	$4/\sqrt{7}$	$\sqrt{7}/3$
(c)	$3/\sqrt{10}$	$1/\sqrt{10}$	3	$\sqrt{10}/3$	$\sqrt{10}$	$1/3$

7. $\sin\theta = 3/\sqrt{10}$, $\cos\theta = 1/\sqrt{10}$ **9.** $\tan\theta = \sqrt{21}/2$, $\csc\theta = 5/\sqrt{21}$

11. Let x be the length of the side adjacent to θ, then $\cos\theta = x/6 = 0.3$, $x = 1.8$.

13.

	θ	$\sin\theta$	$\cos\theta$	$\tan\theta$	$\csc\theta$	$\sec\theta$	$\cot\theta$
(a)	$225°$	$-1/\sqrt{2}$	$-1/\sqrt{2}$	1	$-\sqrt{2}$	$-\sqrt{2}$	1
(b)	$-210°$	$1/2$	$-\sqrt{3}/2$	$-1/\sqrt{3}$	2	$-2/\sqrt{3}$	$-\sqrt{3}$
(c)	$5\pi/3$	$-\sqrt{3}/2$	$1/2$	$-\sqrt{3}$	$-2/\sqrt{3}$	2	$-1/\sqrt{3}$
(d)	$-3\pi/2$	1	0	—	1	—	0

15.

	$\sin\theta$	$\cos\theta$	$\tan\theta$	$\csc\theta$	$\sec\theta$	$\cot\theta$
(a)	$4/5$	$3/5$	$4/3$	$5/4$	$5/3$	$3/4$
(b)	$-4/5$	$3/5$	$-4/3$	$-5/4$	$5/3$	$-3/4$
(c)	$1/2$	$-\sqrt{3}/2$	$-1/\sqrt{3}$	2	$-2\sqrt{3}$	$-\sqrt{3}$
(d)	$-1/2$	$\sqrt{3}/2$	$-1/\sqrt{3}$	-2	$2/\sqrt{3}$	$-\sqrt{3}$
(e)	$1/\sqrt{2}$	$1/\sqrt{2}$	1	$\sqrt{2}$	$\sqrt{2}$	1
(f)	$1/\sqrt{2}$	$-1/\sqrt{2}$	-1	$\sqrt{2}$	$-\sqrt{2}$	-1

17. **(a)** $x = 3\sin 25° \approx 1.2679$ **(b)** $x = 3/\tan(2\pi/9) \approx 3.5753$

19.

	$\sin\theta$	$\cos\theta$	$\tan\theta$	$\csc\theta$	$\sec\theta$	$\cot\theta$
(a)	$a/3$	$\sqrt{9-a^2}/3$	$a/\sqrt{9-a^2}$	$3/a$	$3/\sqrt{9-a^2}$	$\sqrt{9-a^2}/a$
(b)	$a/\sqrt{a^2+25}$	$5/\sqrt{a^2+25}$	$a/5$	$\sqrt{a^2+25}/a$	$\sqrt{a^2+25}/5$	$5/a$
(c)	$\sqrt{a^2-1}/a$	$1/a$	$\sqrt{a^2-1}$	$a/\sqrt{a^2-1}$	a	$1/\sqrt{a^2-1}$

21. **(a)** $\theta = 3\pi/4 \pm n\pi,\ n = 0, 1, 2, \ldots$
    (b) $\theta = \pi/3 \pm 2n\pi$ and $\theta = 5\pi/3 \pm 2n\pi,\ n = 0, 1, 2, \ldots$

23. **(a)** $\theta = \pi/6 \pm n\pi,\ n = 0, 1, 2, \ldots$
    (b) $\theta = 4\pi/3 \pm 2n\pi$ and $\theta = 5\pi/3 \pm 2n\pi,\ n = 0, 1, 2, \ldots$

25. **(a)** $\theta = 3\pi/4 \pm n\pi,\ n = 0, 1, 2, \ldots$ **(b)** $\theta = \pi/6 \pm n\pi,\ n = 0, 1, 2, \ldots$

27. **(a)** $\theta = \pi/3 \pm 2n\pi$ and $\theta = 2\pi/3 \pm 2n\pi,\ n = 0, 1, 2, \ldots$
    (b) $\theta = \pi/6 \pm 2n\pi$ and $\theta = 11\pi/6 \pm 2n\pi,\ n = 0, 1, 2, \ldots$

29. $\sin\theta = 2/5,\ \cos\theta = -\sqrt{21}/5,\ \tan\theta = -2/\sqrt{21},\ \csc\theta = 5/2,\ \sec\theta = -5/\sqrt{21},\ \cot\theta = -\sqrt{21}/2$

31. **(a)** $\theta = \pm n\pi,\ n = 0, 1, 2, \ldots$ **(b)** $\theta = \pi/2 \pm n\pi,\ n = 0, 1, 2, \ldots$
    (c) $\theta = \pm n\pi,\ n = 0, 1, 2, \ldots$ **(d)** $\theta = \pm n\pi,\ n = 0, 1, 2, \ldots$
    (e) $\theta = \pi/2 \pm n\pi,\ n = 0, 1, 2, \ldots$ **(f)** $\theta = \pm n\pi,\ n = 0, 1, 2, \ldots$

33. **(a)** $s = r\theta = 4(\pi/6) = 2\pi/3$ cm **(b)** $s = r\theta = 4(5\pi/6) = 10\pi/3$ cm

35. $\theta = s/r = 2/5$

37. **(a)** $2\pi r = R(2\pi - \theta),\ r = \dfrac{2\pi - \theta}{2\pi}R$

    (b) $h = \sqrt{R^2 - r^2} = \sqrt{R^2 - (2\pi - \theta)^2 R^2/(4\pi^2)} = \dfrac{\sqrt{4\pi\theta - \theta^2}}{2\pi}R$

39. Let h be the altitude as shown
in the figure, then $h = 3\sin 60° = 3\sqrt{3}/2$
so $A = \dfrac{1}{2}(3\sqrt{3}/2)(7) = 21\sqrt{3}/4$.

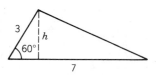

41. Let x be the distance above the ground, then $x = 10\sin 67° \approx 9.2$ ft.

43. From the figure, $h = x - y$
but $x = d \tan \beta$, $y = d \tan \alpha$
so $h = d(\tan \beta - \tan \alpha)$.

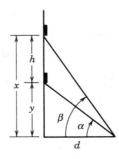

45. **(a)** $\sin 2\theta = 2 \sin \theta \cos \theta = 2(\sqrt{5}/3)(2/3) = 4\sqrt{5}/9$
(b) $\cos 2\theta = 2 \cos^2 \theta - 1 = 2(2/3)^2 - 1 = -1/9$

47. $\sin 3\theta = \sin(2\theta + \theta) = \sin 2\theta \cos \theta + \cos 2\theta \sin \theta = (2 \sin \theta \cos \theta) \cos \theta + (\cos^2 \theta - \sin^2 \theta) \sin \theta$
$= 2 \sin \theta \cos^2 \theta + \sin \theta \cos^2 \theta - \sin^3 \theta = 3 \sin \theta \cos^2 \theta - \sin^3 \theta;$

similarly, $\cos 3\theta = \cos^3 \theta - 3 \sin^2 \theta \cos \theta$

49. $\dfrac{\cos \theta \tan \theta + \sin \theta}{\tan \theta} = \dfrac{\cos \theta (\sin \theta / \cos \theta) + \sin \theta}{\sin \theta / \cos \theta} = 2 \cos \theta$

51. $\tan \theta + \cot \theta = \dfrac{\sin \theta}{\cos \theta} + \dfrac{\cos \theta}{\sin \theta} = \dfrac{\sin^2 \theta + \cos^2 \theta}{\sin \theta \cos \theta} = \dfrac{1}{\sin \theta \cos \theta} = \dfrac{2}{2 \sin \theta \cos \theta} = \dfrac{2}{\sin 2\theta} = 2 \csc 2\theta$

53. $\dfrac{\sin \theta + \cos 2\theta - 1}{\cos \theta - \sin 2\theta} = \dfrac{\sin \theta + (1 - 2 \sin^2 \theta) - 1}{\cos \theta - 2 \sin \theta \cos \theta} = \dfrac{\sin \theta (1 - 2 \sin \theta)}{\cos \theta (1 - 2 \sin \theta)} = \tan \theta$

55. Using (47), $2 \cos 2\theta \sin \theta = 2(1/2)[\sin(-\theta) + \sin 3\theta] = \sin 3\theta - \sin \theta$

57. $\tan(\theta/2) = \dfrac{\sin(\theta/2)}{\cos(\theta/2)} = \dfrac{2 \sin(\theta/2) \cos(\theta/2)}{2 \cos^2(\theta/2)} = \dfrac{\sin \theta}{1 + \cos \theta}$

59. From the figure, area $= \dfrac{1}{2} hc$ but $h = b \sin A$

so area $= \dfrac{1}{2} bc \sin A$. The formulas

area $= \dfrac{1}{2} ac \sin B$ and area $= \dfrac{1}{2} ab \sin C$

follow by drawing altitudes from

vertices B and C, respectively.

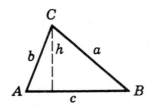

61. (a) $\sin(\pi/2 + \theta) = \sin(\pi/2)\cos\theta + \cos(\pi/2)\sin\theta = (1)\cos\theta + (0)\sin\theta = \cos\theta$
    (b) $\cos(\pi/2 + \theta) = \cos(\pi/2)\cos\theta - \sin(\pi/2)\sin\theta = (0)\cos\theta - (1)\sin\theta = -\sin\theta$
    (c) $\sin(3\pi/2 - \theta) = \sin(3\pi/2)\cos\theta - \cos(3\pi/2)\sin\theta = (-1)\cos\theta - (0)\sin\theta = -\cos\theta$
    (d) $\cos(3\pi/2 + \theta) = \cos(3\pi/2)\cos\theta - \sin(3\pi/2)\sin\theta = (0)\cos\theta - (-1)\sin\theta = \sin\theta$

63. (a) Add (34) and (36) to get $\sin(\alpha - \beta) + \sin(\alpha + \beta) = 2\sin\alpha\cos\beta$ so
    $\sin\alpha\cos\beta = (1/2)[\sin(\alpha - \beta) + \sin(\alpha + \beta)]$.
    (b) Subtract (35) from (37).   (c) Add (35) and (37).

65. $\sin\alpha + \sin(-\beta) = 2\sin\dfrac{\alpha - \beta}{2}\cos\dfrac{\alpha + \beta}{2}$, but $\sin(-\beta) = -\sin\beta$ so

$\sin\alpha - \sin\beta = 2\cos\dfrac{\alpha + \beta}{2}\sin\dfrac{\alpha - \beta}{2}$.

67. Consider the triangle having a, b, and d as sides. The angle formed by sides a and b is $\pi - \theta$
so from the law of cosines, $d^2 = a^2 + b^2 - 2ab\cos(\pi - \theta) = a^2 + b^2 + 2ab\cos\theta$,
$d = \sqrt{a^2 + b^2 + 2ab\cos\theta}$.

EXERCISE SET B.II

1. (a) $2\pi/5$ (b) 6π (c) 8π (d) $\pi/7$
     (e) 2 (f) 10 (g) 1 (h) $2k\pi$

3. (a) 5 (b) $1/3$ (c) $1/2$ (d) 1

5.

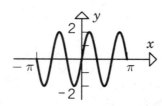

7.

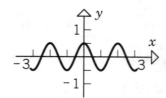

9.

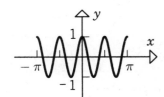

11.

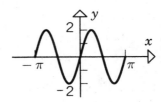

13. (a) odd   (b) even   (c) odd   (d) even
     (e) even   (f) even   (g) odd   (h) even

15.

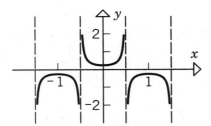

17.

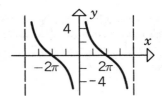

19.

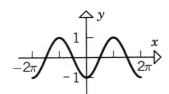

21.

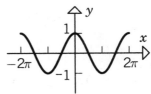

23.

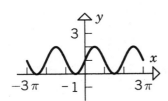

25.

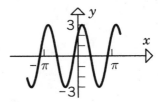

27.

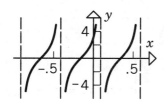

29.

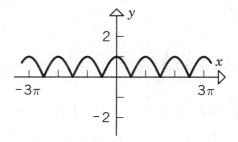

31.

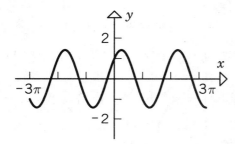

APPENDIX C

Proofs

EXERCISE SET C.I

1. If $x > 2$ then $|(x+1) - 3| = |x - 2| = x - 2 < \epsilon$ when $x < 2 + \epsilon$; take $\delta = \epsilon$.

3. If $x > 4$ then $|\sqrt{x-4} - 0| = \sqrt{x-4} < \epsilon$ when $x - 4 < \epsilon^2$, $x < 4 + \epsilon^2$; take $\delta = \epsilon^2$.

5. If $x > 2$ then $|f(x) - 2| = |x - 2| = x - 2 < \epsilon$ when $x < 2 + \epsilon$; take $\delta = \epsilon$.

7. If $x > 0$ then $|1/x^2 - 0| = 1/x^2 < \epsilon$ when $x^2 > 1/\epsilon$, $x > 1/\sqrt{\epsilon}$; take $N = 1/\sqrt{\epsilon}$.

9. If $x < -2$ then $|1/(x+2) - 0| = 1/|x+2| = -1/(x+2) < \epsilon$ when $x + 2 < -1/\epsilon$, $x < -2 - 1/\epsilon$; take $N = -2 - 1/\epsilon$.

11. If $x > -1$ then $|x/(x+1) - 1| = |-1/(x+1)| = 1/(x+1) < \epsilon$ when $x + 1 > 1/\epsilon$, $x > -1 + 1/\epsilon$; take $N = 1/\epsilon$.

13. If $x < -5/2$ then $\left| \dfrac{4x-1}{2x+5} - 2 \right| = \left| \dfrac{-11}{2x+5} \right| = \dfrac{11}{|2x+5|} = -\dfrac{11}{2x+5} < \epsilon$ when $2x + 5 < -\dfrac{11}{\epsilon}$, $x < -\dfrac{5}{2} - \dfrac{11}{2\epsilon}$; take $N = -\dfrac{5}{2} - \dfrac{11}{2\epsilon}$.

15. If $x \neq 3$ and $N > 0$ then $1/(x-3)^2 > N$ when $(x-3)^2 < 1/N$, $|x - 3| < 1/\sqrt{N}$; take $\delta = 1/\sqrt{N}$.

17. If $x \neq 0$ and $N > 0$ then $1/|x| > N$ when $|x| < 1/N$; take $\delta = 1/N$.

19. If $x \neq 0$ and $N < 0$ then $-1/x^4 < N$ when $x^4 < -1/N = 1/|N|$, $x < 1/\sqrt[4]{|N|}$; take $\delta = 1/\sqrt[4]{|N|}$.

21. (2): Let f be defined on some open interval extending to the right from a. We will write $\lim\limits_{x \to a^+} f(x) = +\infty \ (-\infty)$ if given any positive (negative) number N we can find a number $\delta > 0$ such that $f(x)$ satisfies $f(x) > N \ (f(x) < N)$ whenever x satisfies $a < x < a + \delta$.

    (3): Similar to (2) with "right" replaced by "left", a^+ replaced by a^-, and $a < x < a + \delta$ replaced by $a - \delta < x < a$.

23. **(a)** If $x > 1$ and $N < 0$ then $1/(1-x) < N$ when $1 - x > 1/N = -1/|N|$, $x < 1 + 1/|N|$; take $\delta = 1/|N|$.

     (b) If $x < 1$ and $N > 0$ then $1/(1-x) > N$ when $1 - x < 1/N$, $x > 1 - 1/N$; take $\delta = 1/N$.

25. **(a)** If $M > 0$ then $x + 1 > M$ when $x > M - 1$; take $N = M$

     (b) If $M < 0$ then $x + 1 < M$ when $x < M - 1$; take $N = M - 1$.

EXERCISE SET C.II

1. $|k - k| = |0| = 0 < \epsilon$ when $x > N$ for any $N > 0$.

3. $|[f(x) + g(x)] - [L_1 + L_2]| = |[f(x) - L_1] + [g(x) - L_2]| \le |f(x) - L_1| + |g(x) - L_2|$.

Given $\epsilon > 0$, there exist negative numbers N_1 and N_2 such that $|f(x) - L_1| < \epsilon/2$ and $|g(x) - L_2| < \epsilon/2$ whenever $x < N_1$ and $x < N_2$, respectively. Let $N = \min(N_1, N_2)$, if $x < N$ then $|f(x) - L_1| + |g(x) - L_2| < \epsilon/2 + \epsilon/2 = \epsilon$ so $|[f(x) + g(x)] - [L_1 + L_2]| < \epsilon$.

5. From part (a) of Theorem 7, $\lim_{x \to a}(-1) = -1$ so from part (c) of Theorem 7,

$$\lim_{x \to a}[(-1)g(x)] = \lim_{x \to a}(-1) \lim_{x \to a} g(x) = (-1)L_2 = -L_2 \text{ thus, using part (b) of Theorem 7,}$$

$$\lim_{x \to a}[f(x) - g(x)] = \lim_{x \to a}[f(x) + (-g(x))] = \lim_{x \to a} f(x) + \lim_{x \to a}[-g(x)] = L_1 + (-L_2) = L_1 - L_2$$

7. **(a)** Given $N < 0$, there exist numbers $\delta_1 > 0$ and $\delta_2 > 0$ such that $f(x) < N/2$ and $g(x) > -N/2$ whenever $0 < |x - a| < \delta_1$ and $0 < |x - a| < \delta_2$, respectively. Let $\delta = \min(\delta_1, \delta_2)$ then $f(x) < N/2$ and $-g(x) < N/2$ whenever $0 < |x - a| < \delta$ so $f(x) - g(x) < N/2 + N/2 = N$.

     (b) No, for example $\lim_{x \to 0}[(-1/x^2) + (1/x^4)] = \lim_{x \to 0}[(1 - x^2)/x^4] = +\infty$

9. If $\lim_{x \to a} f(x) = L$ then given $\epsilon > 0$, there exists a $\delta > 0$ such that $|f(x) - L| < \epsilon$ whenever $0 < |x - a| < \delta$. But $|f(x) - L| = |[f(x) - L] - 0| < \epsilon$ whenever $0 < |x - a| < \delta$ so $\lim_{x \to a}[f(x) - L] = 0$. If $\lim_{x \to a}[f(x) - L] = 0$ then given $\epsilon > 0$, there exists a $\delta > 0$ such that $|[f(x) - L] - 0| < \epsilon$ whenever $0 < |x - a| < \delta$, but $|[f(x) - L] - 0| = |f(x) - L| < \epsilon$ whenever $0 < |x - a| < \delta$ so $\lim_{x \to a} f(x) = L$.

11. **(a)** In the proof of part (a), interchange the words "increasing" and "decreasing", and reverse the order of the inequality symbols.

     (b) If $f'(x)$ has the same sign on the intervals (a, x_0) and (x_0, b), then f is increasing on (a, b) or decreasing on (a, b) so f does not have a relative extremum at x_0.

13. **(a)**

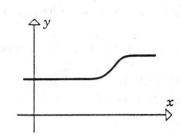

(b)

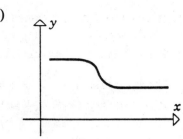

15. Let $h(x) = f(x) - g(x)$, then $h'(x) = f'(x) - g'(x) < 0$ thus h is decreasing on (a, b), that is $h(x_2) < h(x_1)$ if $x_2 > x_1$ so $f(x_2) - g(x_2) < f(x_1) - g(x_1)$, $f(x_2) - f(x_1) < g(x_2) - g(x_1)$.

17. **(a)** From Exercise 16(a) it follows that $f'(x)$ is increasing on some open interval I containing x_0 so f is concave up on I.

(b) Similar to the proof in part (a).

APPENDIX D

Cramer's Rule

EXERCISE SET D

1. $\begin{vmatrix} 3 & -4 \\ 2 & 1 \end{vmatrix} = 11$, $\begin{vmatrix} -5 & -4 \\ 4 & 1 \end{vmatrix} = 11$, $\begin{vmatrix} 3 & -5 \\ 2 & 4 \end{vmatrix} = 22$; $x = 1$, $y = 2$

3. $\begin{vmatrix} 2 & -5 \\ 4 & 6 \end{vmatrix} = 32$, $\begin{vmatrix} -2 & -5 \\ 1 & 6 \end{vmatrix} = -7$, $\begin{vmatrix} 2 & -2 \\ 4 & 1 \end{vmatrix} = 10$; $x_1 = -7/32$, $x_2 = 5/16$

5. $\begin{vmatrix} 1 & 2 & 1 \\ 2 & 1 & -1 \\ 1 & -1 & 1 \end{vmatrix} = -9$, $\begin{vmatrix} 3 & 2 & 1 \\ 0 & 1 & -1 \\ 6 & -1 & 1 \end{vmatrix} = -18$, $\begin{vmatrix} 1 & 3 & 1 \\ 2 & 0 & -1 \\ 1 & 6 & 1 \end{vmatrix} = 9$, $\begin{vmatrix} 1 & 2 & 3 \\ 2 & 1 & 0 \\ 1 & -1 & 6 \end{vmatrix} = -27$;

   $x = 2$, $y = -1$, $z = 3$

7. $\begin{vmatrix} 1 & 1 & -2 \\ 2 & -1 & 1 \\ 1 & -2 & -4 \end{vmatrix} = 21$, $\begin{vmatrix} 1 & 1 & -2 \\ 2 & -1 & 1 \\ -4 & -2 & -4 \end{vmatrix} = 26$, $\begin{vmatrix} 1 & 1 & -2 \\ 2 & 2 & 1 \\ 1 & -4 & -4 \end{vmatrix} = 25$, $\begin{vmatrix} 1 & 1 & 1 \\ 2 & -1 & 2 \\ 1 & -2 & -4 \end{vmatrix} = 15$;

   $x_1 = 26/21$, $x_2 = 25/21$, $x_3 = 5/7$

9. $\begin{vmatrix} \cos\theta & -\sin\theta \\ \sin\theta & \cos\theta \end{vmatrix} = 1$, $\begin{vmatrix} x & -\sin\theta \\ y & \cos\theta \end{vmatrix} = x\cos\theta + y\sin\theta$, $\begin{vmatrix} \cos\theta & x \\ \sin\theta & y \end{vmatrix} = y\cos\theta - x\sin\theta$;

   $x' = x\cos\theta + y\sin\theta$, $y' = -x\sin\theta + y\cos\theta$

APPENDIX E

Complex Numbers

EXERCISE SET E.I

1.

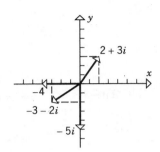

3. **(a)** $x = -2; -y = 3$ so $y = -3$
(b) $x + y = 3$ and $x - y = 1$; solve to get $x = 2, y = 1$

5. **(a)** $z = (3 + 2i) - (1 - i) = 2 + 3i$
(b) $z = -1 - 2i$
(c) $z = (-2 + 7i) + 2i = -2 + 9i$

7. **(a)**

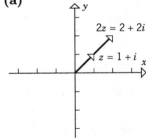

(b)

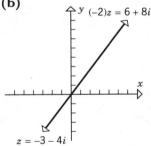

(c)

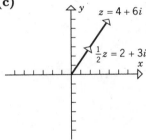

9. **(a)** $z_1 z_2 = (3i)(1 - i) = 3 + 3i, z_1^2 = (3i)^2 = -9, z_2^2 = (1 - i)^2 = -2i$
(b) $z_1 z_2 = (4 + 6i)(2 - 3i) = 26, z_1^2 = (4 + 6i)^2 = -20 + 48i, z_2^2 = (2 - 3i)^2 = -5 - 12i$
(c) $z_1 z_2 = \dfrac{1}{3}(2 + 4i)\dfrac{1}{2}(1 - 5i) = \dfrac{11}{3} - i, z_1^2 = \left[\dfrac{1}{3}(2 + 4i)\right]^2 = -\dfrac{4}{3} + \dfrac{16}{9}i,$

$z_2^2 = \left[\dfrac{1}{2}(1 - 5i)\right]^2 = -6 - \dfrac{5}{2}i$

11. $(1+2i)(4-6i)^2 = (1+2i)(-20-48i) = 76-88i$

13. $(1-3i)^3 = (1-3i)(1-3i)^2 = (1-3i)(-8-6i) = -26+18i$

15. $\left[(2+i)\left(\dfrac{1}{2}+\dfrac{3}{4}i\right)\right]^2 = \left(\dfrac{1}{4}+2i\right)^2 = -\dfrac{63}{16}+i$

17. $(1+i+i^2+i^3)^{100} = 0^{100} = 0$

19. **(a)** $2-7i$ **(b)** $-3+5i$ **(c)** $-5i$ **(d)** i **(e)** -9 **(f)** 0

21. **(a)** $z\bar{z} = (2-4i)(2+4i) = 20 = |z|^2$ **(b)** $z\bar{z} = (-3+5i)(-3-5i) = 34 = |z|^2$

   (c) $z\bar{z} = (\sqrt{2}-\sqrt{2}i)(\sqrt{2}+\sqrt{2}i) = 4 = |z|^2$

23. **(a)** $\dfrac{1}{i} = -i$ **(b)** $\dfrac{1}{1-5i} = \dfrac{1}{26}+\dfrac{5}{26}i$ **(c)** $-\dfrac{7}{i} = 7i$

25. $\dfrac{i}{1+i} = \dfrac{1}{2}+\dfrac{1}{2}i$ **27.** $\dfrac{1}{(3+4i)^2} = \dfrac{1}{-7+24i} = -\dfrac{7}{625}-\dfrac{24}{625}i$

29. $\dfrac{\sqrt{3}+i}{(1-i)(\sqrt{3}-i)} = \dfrac{\sqrt{3}+i}{\sqrt{3}-1-(\sqrt{3}+1)i} = \dfrac{1}{4}(1-\sqrt{3})+\dfrac{1}{4}(1+\sqrt{3})i$

31. $\dfrac{i}{(1-i)(1-2i)(1+2i)} = \dfrac{i}{5(1-i)} = -\dfrac{1}{10}+\dfrac{1}{10}i$

33. **(a)** $z = \dfrac{2-i}{i} = -1-2i$

   (b) $\bar{z} = \dfrac{i}{4-3i} = -\dfrac{3}{25}+\dfrac{4}{25}i,\ z = -\dfrac{3}{25}-\dfrac{4}{25}i$

35. **(a)** $|z| = 2$ if and only if the distance from z to the origin is 2; all such z lie on the circle of radius 2 with center at the origin.

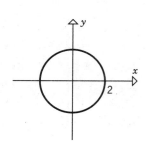

(b) $|z - (1 + i)| = 1$ if and only if the distance between z and $1 + i$ is 1; all such z lie on the circle of radius 1 with center at $1 + i$.

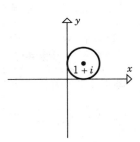

(c) $|z - i| = |z + i|$ if and only if z is equidistant from i and $-i$; all such z lie on the y-axis.

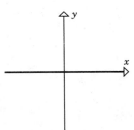

(d) With $z = x + yi$,
$\text{Im}(\bar{z} + i) = \text{Im}(x - iy + i) = 1 - y$, so
$1 - y = 3$, $y = -2$.

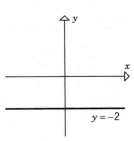

37. Let $z = a + bi$

(a) $\text{Im}(iz) = \text{Im}(-b + ai) = a = \text{Re}(z)$

(b) $\text{Re}(iz) = \text{Re}(-b + ai) = -b = -\text{Im}(z)$

(c) $\text{Re}(\overline{iz}) = \text{Re}\left(\overline{-b + ai}\right) = \text{Re}(-b - ai) = -b = -\text{Im}(z)$

(d) $\text{Im}(\overline{iz}) = \text{Im}\left(\overline{-b + ai}\right) = \text{Im}(-b - ai) = -a = -\text{Re}(z)$

(e) $\text{Re}(i\bar{z}) = \text{Re}(b + ai) = b = \text{Im}(z)$

(f) $\text{Im}(i\bar{z}) = \text{Im}(b + ai) = a = \text{Re}(z)$

39. (a) $i^1 = i, i^2 = -1, i^3 = -i$, and $i^4 = 1$. Suppose that $n > 4$. If n is divisible by 4, then $n/4 = p$ where $p = 1, 2, 3, \ldots$ so $n = 4p$ and $i^n = i^{4p} = (i^4)^p = 1^p = 1$. If n is not divisible by 4, then $n/4 = p$ with a remainder of q where $p = 1, 2, 3, \ldots$ and $q = 1, 2$, or 3 so $n = 4p + q$ and $i^n = i^{4p+q} = (i^{4p})(i^q) = i^q$; but i^q can only take on the values $i, -1$ and $-i$ for $q = 1, 2$, or 3. Thus the only possible values for i^n are $1, -1, i$, and $-i$.

(b) $i^{2509} = i^{4(627)+1} = i^1 = i$

41. Let $z = a + bi$

(a) $\dfrac{1}{2}(z + \bar{z}) = \dfrac{1}{2}(a + bi + a - bi) = a = \text{Re}(z)$

(b) $\dfrac{1}{2i}(z - \bar{z}) = \dfrac{1}{2i}(a + bi - a + bi) = b = \text{Im}(z)$

43. $\dfrac{z_1}{z_2} = \dfrac{x_1 + iy_1}{x_2 + iy_2} = \dfrac{x_1 x_2 + y_1 y_2}{x_2^2 + y_2^2} + \dfrac{x_2 y_1 - x_1 y_2}{x_2^2 + y_2^2} i$

(a) $\text{Re}\left(\dfrac{z_1}{z_2}\right) = \dfrac{x_1 x_2 + y_1 y_2}{x_2^2 + y_2^2}$ $\qquad\qquad$ **(b)** $\text{Im}\left(\dfrac{z_1}{z_2}\right) = \dfrac{x_2 y_1 - x_1 y_2}{x_2^2 + y_2^2}$

45. Let $z = a + bi$, then $|z| = \sqrt{a^2 + b^2}$ and $|\bar{z}| = |a - bi| = \sqrt{a^2 + b^2}$ so $|z| = |\bar{z}|$.

47. (a) $\overline{z^2} = \overline{zz} = \bar{z}\bar{z} = (\bar{z})^2$

(b) $\overline{z^3} = \overline{zz^2} = \bar{z}\overline{z^2} = \bar{z}(\bar{z})^2$ from part (a), so $\overline{z^3} = (\bar{z})^3$; similarly $\overline{z^4} = \overline{zz^3} = \bar{z}\overline{z^3} = \bar{z}(\bar{z})^3 = (\bar{z})^4$ and so on.

(c) Suppose that $n = -m$ where $m > 0$, then if $z \neq 0$ $\overline{z^n} = \overline{z^{-m}} = \overline{(1/z^m)} = \bar{1}/\overline{z^m} = 1/(\bar{z})^m$, so the result is true if n is a negative integer provided $z \neq 0$.

49. Let $z = a + bi$, then $|z| = \sqrt{a^2 + b^2}$, $|\text{Re}(z)| = |a|$, and $|\text{Im}(z)| = |b|$ so $|\text{Re}(z)| \leq |z|$ and $|\text{Im}(z)| \leq |z|$.

51. If $z_1 z_2 = 0$, then $|z_1 z_2| = |z_1||z_2| = 0$ so $|z_1|$ or $|z_2| = 0$, thus $z_1 = 0$ or $z_2 = 0$.

EXERCISES SET E.II

1. (a) 0 $\qquad$ **(b)** $\pi/2$ $\qquad$ **(c)** $-\pi/2$ $\qquad$ **(d)** $\pi/4$ $\qquad$ **(e)** $2\pi/3$ $\qquad$ **(f)** $-\pi/4$

3. **(a)** $2[\cos(\pi/2) + i\sin(\pi/2)]$ **(b)** $4[\cos\pi + i\sin\pi]$

    (c) $5\sqrt{2}[\cos(\pi/4) + i\sin(\pi/4)]$ **(d)** $12[\cos(2\pi/3) + i\sin(2\pi/3)]$

    (e) $3\sqrt{2}[\cos(-3\pi/4) + i\sin(-3\pi/4)]$ **(f)** $4[\cos(-\pi/6) + i\sin(-\pi/6)]$

5. $z_1 = \cos(\pi/2) + i\sin(\pi/2)$, $z_2 = 2[\cos(-\pi/3) + i\sin(-\pi/3)]$, $z_3 = 2[\cos(\pi/6) + i\sin(\pi/6)]$,

    $|z_1 z_2/z_3| = |z_1||z_2|/|z_3| = 1$, $\arg(z_1 z_2/z_3) = (\pi/2) + (-\pi/3) - (\pi/6) = 0$,

    $z_1 z_2/z_3 = \cos(0) + i\sin(0) = 1$

7. **(a)** $-i = \cos(-\pi/2) + i\sin(-\pi/2)$,

    $(-i)^{1/2} = \cos(-\pi/4 + k\pi) + i\sin(-\pi/4 + k\pi), k = 0, 1$;

    the roots are

    $1/\sqrt{2} - 1/\sqrt{2}\,i$,

    $-1/\sqrt{2} + 1/\sqrt{2}\,i$

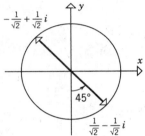

    (b) $1 + \sqrt{3}\,i = 2[\cos(\pi/3) + i\sin(\pi/3)]$,

    $(1 + \sqrt{3}\,i)^{1/2} = \sqrt{2}[\cos(\pi/6 + k\pi) + i\sin(\pi/6 + k\pi)], k = 0, 1$;

    the roots are

    $\sqrt{6}/2 + \sqrt{2}/2i$,

    $-\sqrt{6}/2 - \sqrt{2}/2i$

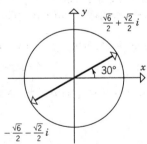

7.   **(c)** $-27 = 27\left[\cos\pi + i\sin\pi\right]$,

$(-27)^{1/3} = \sqrt[3]{27}\left[\cos(\pi/3 + 2k\pi/3) + i\sin(\pi/3 + 2k\pi/3)\right], k = 0, 1, 2;$

the roots are

$3/2 + 3\sqrt{3}/2i, -3,$
$3/2 - 3\sqrt{3}/2i$

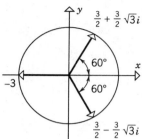

(d) $i = \cos(\pi/2) + i\sin(\pi/2)$,

$(i)^{1/3} = \cos(\pi/6 + 2k\pi/3) + i\sin(\pi/6 + 2k\pi/3), k = 0, 1, 2;$

the roots are

$\sqrt{3}/2 + 1/2i,$
$-\sqrt{3}/2 + 1/2i, -i$

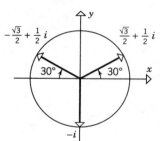

(e) $-1 = \cos\pi + i\sin\pi$,

$(-1)^{1/4} = \cos(\pi/4 + k\pi/2) + i\sin(\pi/4 + k\pi/2), k = 0, 1, 2, 3;$

the roots are

$1/\sqrt{2} + 1/\sqrt{2}\,i,$
$-1/\sqrt{2} + 1/\sqrt{2}i,$
$-1/\sqrt{2} - 1/\sqrt{2}i,$
$1/\sqrt{2} - 1/\sqrt{2}\,i$

(f) $-8 + 8\sqrt{3}\,i = 16\,[\cos(2\pi/3) + i\sin(2\pi/3)]$,
$(-8 + 8\sqrt{3}\,i)^{1/4} = 2\,[\cos(\pi/6 + k\pi/2) + i\sin(\pi/6 + k\pi/2)], k = 0, 1, 2, 3$;
the roots are
$\sqrt{3} + i, -1 + \sqrt{3}\,i,$
$-\sqrt{3} - i, 1 - \sqrt{3}\,i$

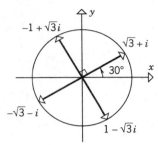

9.

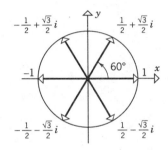

11. **(a)** $z^4 = 16, z = 16^{1/4} = 2\,[\cos(k\pi/2) + i\sin(k\pi/2)], k = 0, 1, 2, 3$; the solutions are $\pm 2, \pm 2i$
(b) $z^4 = -64, z = (-64)^{1/4} = 2\sqrt{2}\,[\cos(\pi/4 + k\pi/2) + i\sin(\pi/4 + k\pi/2)], k = 0, 1, 2, 3$; the solutions are $\pm(2 + 2i), \pm(2 - 2i)$

13. $z = r(\cos\theta + i\sin\theta), i = \cos(\pi/2) + i\sin(\pi/2)$ so from (5)
$z/i = r[\cos(\theta - \pi/2) + i\sin(\theta - \pi/2)]$; division by i rotates z clockwise by $90°$

15. **(a)** $3e^{i\pi} = -3; \text{Re}(z) = -3, \text{Im}(z) = 0$
(b) $3e^{-i\pi} = -3; \text{Re}(z) = -3, \text{Im}(z) = 0$
(c) $z = \sqrt{2}\,e^{-\pi i/2} = -\sqrt{2}\,i; \text{Re}(z) = 0, \text{Im}(z) = -\sqrt{2}$
(d) $z = -3e^{2\pi i} = -3; \text{Re}(z) = -3, \text{Im}(z) = 0$

17. If $n = 0$, then $z^0 = 1 = r^0[\cos(0\theta) + i\sin(0\theta)]$; if $n < 0$, then from (5) and (6) with
$z = r(\cos\theta + i\sin\theta), z^n = \dfrac{1}{z^{-n}} = \dfrac{\cos(0) + i\sin(0)}{r^{-n}[\cos(-n\theta) + i\sin(-n\theta)]} = r^n(\cos n\theta + i\sin n\theta)$

NOTES

NOTES

NOTES

NOTES

NOTES

NOTES

NOTES

NOTES

NOTES

NOTES

NOTES

NOTES

NOTES